Richard M. Meyer

Essential Mathematics for Applied Fields

Springer-Verlag
New York Heidelberg Berlin

Universitext

Richard M. Meyer

Essential Mathematics for Applied Fields

Springer-Verlag
New York Heidelberg Berlin

Richard M. Meyer
Niagara University
College of Arts and Sciences
Niagara University, New York 14109
USA

AMS Subject Classifications: 00A06, 00A69

Library of Congress Cataloging in Publication Data

Meyer, Richard M 1939-
Essential mathematics for applied fields.

(Universitext)
Includes bibliographies and index.
1. Mathematics—1961- I. Title.
QA37.2.M49 510 79-20650

9 8 7 6 5 4 3 2 1

ISBN 0-387-90450-6 Springer-Verlag New York Heidelberg Berlin
ISBN 3-540-90450-6 Springer-Verlag Berlin Heidelberg New York

Preface

1. Purpose

The purpose of this work is to provide, in one volume, a wide spectrum of essential (non-measure theoretic) Mathematics for use by workers in the variety of applied fields. To obtain the background developed here in one volume would require studying a prohibitive number of separate Mathematics courses (assuming they were available). Before, much of the material now covered was (a) unavailable, (b) too widely scattered, or (c) too advanced as presented, to be of use to those who need it. Here, however, we present a sound basis requiring only Calculus through Differential Equations. It provides the needed flexibility to cope, in a rigorous manner, with the every-day, non-standard and new situations that present themselves. There is no substitute for this.

2. Arrangement

The volume consists of twenty Sections, falling into several natural units:

Basic Real Analysis

1. Sets, Sequences, Series, and Functions
2. Doubly Infinite Sequences and Series
3. Sequences and Series of Functions
4. Real Power Series
5. Behavior of a Function Near a Point: Various Types of Limits
6. Orders of Magnitude: the O, o, $\sim$ Notation
7. Some Abelian and Tauberian Theorems

Riemann-Stieltjes Integration

8. 1-Dimensional Cumulative Distribution Functions and Bounded Variation Functions
9. 1-Dimensional Riemann-Stieltjes Integral
10. n-Dimensional Cumulative Distribution Functions and Bounded Variation Functions
11. n-Dimensional Riemann-Stieltjes Integral

The Finite Calculus

12. Finite Differences and Difference Equations

Basic Complex Analysis

13. Complex Variables

Applied Linear Algebra

14. Matrices and Determinants
15. Vectors and Vector Spaces
16. Linear Equations and Generalized Inverse
17. Characteristic Roots and Related Topics

Miscellaneous

18. Convex Sets and Convex Functions
19. Max-Min Problems
20. Some Basic Inequalities

3. Development

Each Section develops its topic rigorously, based upon material previously established; that is, it is self-contained. Throughout the body of the text, at appropriate locations, are found solved Examples and Exercises requiring solution; both are critical parts of the development. Complete Hints or Answers are provided for the Exercises following each Section, as are References to Additional and Related Material.

A serious attempt has been made to include essential Mathematics, and to allow the References to provide entry into the vast literature that exists. Accordingly, it should be noted that Section 13 develops Complex Variables only through Elementary Contour Integration, but enough theory is developed to pursue more advanced topics.

4. Use

The volume can be used as a basis of a one or two semester course covering some or all of the topics. It can be used to supplement existing courses by serving as a remedial reference when deficiencies are noticed. It can also be used in an independent (or guided) study plan. It is also a source for review, or entry into more advanced and/or related literature.

5. Acknowledgements

During the ten years within which this work matured into its present form, many individuals at the State University of New York at Buffalo, and elsewhere, have helped. It was appreciated. Carefully preserved lecture notes taken at the University at Chapel Hill provided useful Exercises and Examples at various points in the text.

Rosanna Bello of Buffalo did the final typescript during a long six-month period of cooperation. Inger (Tulle) Abbott of Williamsville was the artist who rendered the diagrams and drafting details. Professor Steven L. Siegel of Niagara University aided in final proofreading, suggestions, and indexing. Their cooperation, and Helen Meyer's understanding, speeded completion of the project under otherwise adverse conditions.

Finally, I note the freedom and cooperation extended to me by Springer-Verlag through their representatives, Mathematics Editor Kaufmann-Bühler and Editorial Assistant Jane Walsh.

R.M.M.
Williamsville, New York
June 8, 1979

Contents

1. Sets, Sequences, Series, and Functions

Sets, sequences, series, and functions occur in every area of Applied Mathematics. Sets will be designated by capital letters $A, B, A_1, C_2, \ldots$ and so on; individual members of sets will be designated by lower case letters $x, y, a_1, a_2, b, \ldots$ and so on.

If x is (is not) a member of A we write $x \in A$ $(x \notin A)$; then A is termed a subset of B, written $A \subseteq B$, iff $x \in A$ implies $x \in B$. We assume that all sets in a given discussion are subsets of some fixed "Universal Set" U.[1]

The basic operations involving sets are:

(1) Complement A^c: $x \in A^c$ iff $x \notin A$

(2) Union $A \cup B$: $x \in A \cup B$ iff $x \in A$ or $x \in B$ (or both)

(3) Intersection $A \cap B$: $x \in A \cap B$ iff $x \in A$ and $x \in B$

(4) Difference $A \backslash B$: $x \in A \backslash B$ iff $x \in A$ and $x \notin B$

(5) Symmetric Difference $A \triangle B$: $x \in A \triangle B$ iff $x \in A \backslash B$ or $x \in B \backslash A$.

We can picture these and subsequent definitions quite easily by means of so-called Venn Diagrams. In the following diagrams, the square regions represent the Universal set U. By convention, the members (elements) of a set are assumed to be all different (distinct), so even if the sets A and B have an element x in common, the set $A \cup B$ contains the member x represented only once.

---

[1]Frequently the Universal Set is E_n, Euclidean n-space, consisting of all ordered n-tuples of real numbers. For $n=1$ we are dealing with sets of ordinary real numbers.

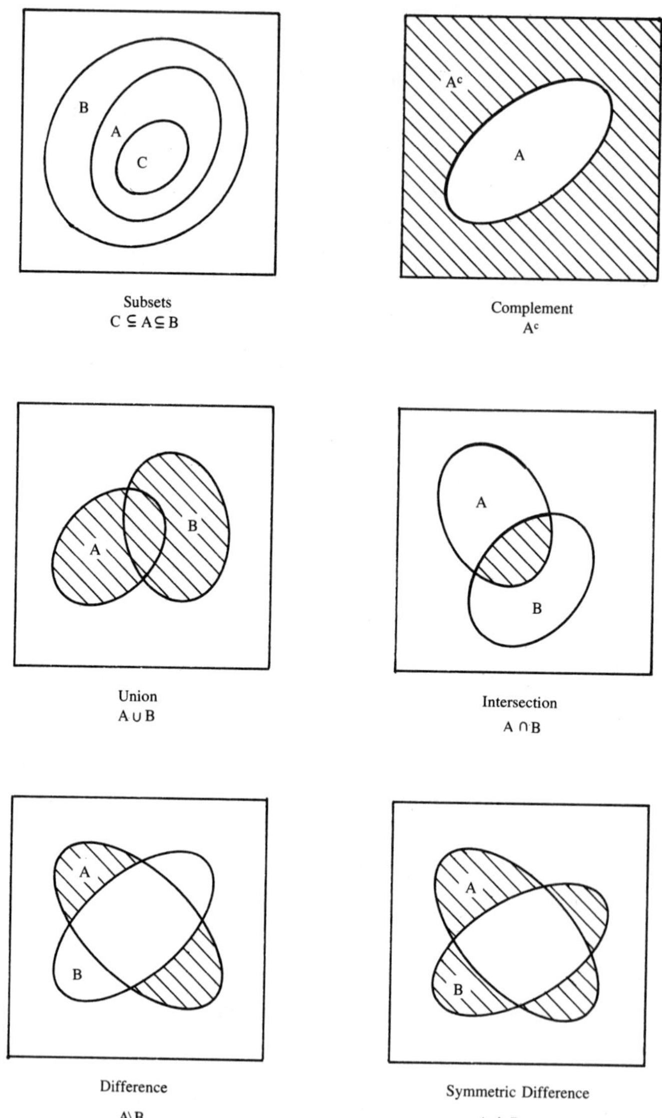
B
A
C
Subsets
$C \subseteq A \subseteq B$
A^c
A
Complement
A^c
A
B
Union
$A \cup B$
A
B
Intersection
$A \cap B$
A
B
Difference
$A \backslash B$
A
B
Symmetric Difference
$A \triangle B$

Figure 1-1

Two sets A and B may or may not have members in common. If A and B have no members in common, they are called disjoint (or mutually exclusive), written $A \cap B = \emptyset$ (the empty set).

Sets differ also in the number of elements they contain. The set A is termed countable iff its members can be placed in a one-to-one correspondence (mating) with a subset (possibly all) of the positive integers. Otherwise the set is called uncountable.

EXAMPLE 1.1 (Finite, countable and uncountable sets) Clearly any set containing a finite number of elements is automatically countable. A non-finite set that is countable is the set of all positive fractions, i.e. numbers of the form p/q where p and q are positive integers. The correspondence is as follows: p/q corresponds with the integer

$$\frac{(p+q-1)(p+q-2)}{2} + p .$$

This correspondence can be arrived at by arranging the fractions systematically in a rectangular array with "p" as the column heading and "q" as the row heading, then counting out to a given fraction by working along diagonals.

EXERCISE 1.1 Construct an example of an uncountable set of real numbers. Verify that it cannot be placed into a one-to-one correspondence with the positive integers. (This is a difficult problem).

A sequence of sets $A_1, A_2, \ldots$ is a collection of sets indexed on a subset (possibly all) of the positive integers. In contrast with a set, the individual terms of a sequence need not all be different. The notions of union, intersection, disjointness etc. can be extended naturally to such countable collections. Accordingly, $A_1, A_2, \ldots$ are termed (pairwise) mutually exclusive iff $A_i \cap A_j = \emptyset$ $(i \neq j,\ i,j = 1,2\ldots)$. Furthermore we write $\cup A_n$ to designate their <u>union</u>, defined to be the set of elements x such that x is a member of at least one set in the collection. Finally, we write $\cap A_n$ to designate their <u>intersection</u> defined as the set of elements x such that x is a member of each set in the collection.

Note that the notions of union and intersection can be extended naturally to collections of sets $\{A_r : r \in R\}$ indexed on an arbitrary (not necessarily countable) indexing set R. For example, if, for every real number r we let A_r be the set of points (x,y) in E_2 satisfying the equation $x+y = r$, then $\bigcup_{R'} A_r$ (where R' is the set of indices r satisfying $-1 < r \leq 1$) has a very simple geometric interpretation.

Let $A_1, A_2, \ldots$ be a sequence of sets. The limit supremum, lim sup A_n, of this sequence of sets is defined to be the set of elements x such that x is contained in infinitely many (but not necessarily all) of the sets in the (infinite) sequence. The limit infimum, lim inf A_n, is defined to be the set of elements x such that x is contained in all but perhaps a finite number of sets in the (infinite) sequence.[1]

[1] 'lim sup' is often abbreviated as $\overline{\lim}$ and 'lim inf' by $\underline{\lim}$.

EXERCISE 1.2 Prove that:

$$\limsup A_n = \bigcap_{k=1}^{\infty} \bigcup_{n=k}^{\infty} A_n$$

$$\liminf A_n = \bigcup_{k=1}^{\infty} \bigcap_{n=k}^{\infty} A_n .$$

EXERCISE 1.3 Prove that:

$$\cap A_n \subseteq \liminf A_n \subseteq \limsup A_n \subseteq \cup A_n .$$

If for the sequence $A_1, A_2, \ldots$ of sets we have $\limsup A_n = \liminf A_n$, then the sequence of sets is termed convergent, and we write $A = \lim A_n$ for their common value.

EXERCISE 1.4 A sequence of sets $A_1, A_2, \ldots$ is termed monotone increasing (monotone decreasing) iff $A_1 \subseteq A_2 \subseteq A_3 \subseteq \ldots$ $(A_1 \supseteq A_2 \supseteq A_3 \supseteq \ldots)$. Prove that a monotone sequence of sets is always convergent to $\lim A_n = \cup A_n$ in the increasing case and to $\lim A_n = \cap A_n$ in the decreasing case.

EXERCISE 1.5 Construct an example of a non-convergent sequence of sets of real numbers, and evaluate lim sup and lim inf for the sequence.

EXERCISE 1.6 (A Complementation Rule) Let $\mathcal{A}$ be any collection of sets, and let A be any set expressible as unions and/or intersections of sets (or their complements) belonging to $\mathcal{A}$. For instance, if $\mathcal{A} = \{A_1, A_2, \ldots\}$ we might consider:

$$A = \bigcup_{k\geq n}^{\infty} \bigcap_{n=1}^{\infty} A_k$$

or

$$A = \bigcup_{n=1}^{\infty} [A_n \backslash A_{n-1}] = \bigcup_{n=1}^{\infty} [A_n \cap A_{n-1}^c]$$

or

$$A = \bigcap_{n=1}^{\infty} A_n \quad .$$

Prove the following complementation rule: the complement A^c of A is obtained by modifying the expression for A as follows:

(i) replacing each $\cap$ symbol by $\cup$

(ii) replacing each $\cup$ symbol by $\cap$

(iii) replacing each set by its complement (note: $(B^c)^c = B$).

Accordingly, in the above illustrations, the complement of

$$A = \bigcup_{k\geq n}^{\infty} \bigcap_{n=1}^{\infty} A_k \quad \text{is} \quad A^c = \bigcap_{k\geq n}^{\infty} \bigcup_{n=1}^{\infty} A_k^c \quad ;$$

the complement of

$$A = \bigcup_{n=1}^{\infty} [A_n \cap A_{n-1}^c] \quad \text{is} \quad A^c = \bigcap_{n=1}^{\infty} [A_n^c \cup A_{n-1}] \quad ;$$

and the complement of

$$A = \bigcap_{n=1}^{\infty} A_n \quad \text{is} \quad A^c = \bigcup_{n=1}^{\infty} A_n^c \quad .$$

EXERCISE 1.7 (Binary Sequences) Let U consist of the 2^n different ordered n-tuples that can be constructed using only the integers 0 and 1. For each index i $(i = 0,1,2,\ldots,n)$ define A_i as the set of n-tuples in U possessing exactly i 1's, and for each index j $(j = 1,2,\ldots,n)$ define B_j as the set of n-tuples in U possessing a 1 in location j of the n-tuple.

(i) Let T be the subset of U consisting of n-tuples with alternating 0's and 1's only. Express T in terms of the B_j's (and/or their complements).

(ii) Express the A_i's similarly.

(iii) How many elements does A_i (B_j) contain?

(iv) Let S_r be that subset of T whose n-tuples contain at most r 1's. Obtain an expression for S_r.

(v) Verify that the A_i's are mutually exclusive but the B_j's are not. Relate this to the fact that summation of the number of elements in the A_i's yields 2^n whereas such is not the case for the B_j's.

(vi) Obtain an expression for any given element of U in terms of the B_j's (and/or their complements).

(vii) Obtain an expression for S_r^c (above). Interpret this set.

EXERCISE 1.8 (Infinite Binary Sequences) Let V consist of all ordered tuples of the form $(d_1, d_2, d_3, \ldots)$, where d_i, the element at the i-th location of the tuple, may be either 0 or 1. V is, by the way, an uncountable set.

(i) For any positive integer r, and for $n = 1,2,3,\ldots$, we define $A_n(r)$ as the set of points in V for which there are at least r occurrences of the pattern 1,1,1 amongst the first n locations (note: a 0 would necessarily precede the first 1 and follow the third 1 to form the stated pattern). Prove that this sequence of sets is convergent.

(ii) For any index n define B_n as that set of elements in V for which a 1 appears in position n. Evaluate $\liminf B_n$ and $\limsup B_n$. Is the sequence convergent?

Of particular importance are sets of real numbers (subsets of E_1). The usual (Euclidean) definition of the distance between two real numbers x and y is $|x-y|$. It then follows that for all such real numbers, $|x| \geq 0$, $|x+y| \leq |x| + |y|$ and $|x-y| = 0$ iff $x = y$. We shall now develop some properties of sets of real numbers that are a consequence of this definition of distance.

A non-empty set A of real numbers is termed bounded (bounded above, bounded below) iff there exists a finite real bound B (U, L) such that $|x| \leq B$ $(x \leq U,\ x \geq L)$ for all $x \in A$. Such numbers, if they exist, are not necessarily unique.

Otherwise, the set A is termed unbounded (unbounded above, unbounded below).

Sets of real numbers have the characteristic property that every set A that is bounded above (bounded below) has a smallest upper bound (greatest lower bound) denoted sup A (inf A), sometimes alternately denoted l.u.b. A (g.l.b. A). If a set A is unbounded above (unbounded below) we write $\sup A = +\infty$ ($\inf A = -\infty$). By convention, $\sup \emptyset = -\infty$ and $\inf \emptyset = +\infty$.

For a bounded set A of real numbers, an actual constructive proof of the existence of inf A and sup A is not given here, since it involves some theory of real numbers that is not assumed. However, the result usually appears "intuitively obvious" to most.

<u>EXAMPLE 1.2</u> (Some Properties of sup A and inf A)
For a bounded set A of real numbers, let:

$$a_{\cdot} = \inf A$$

and

$$a^{\cdot} = \sup A \quad .$$

Then for any real number a belonging to A:

$$a_{\cdot} \leq a \leq a^{\cdot} \quad .$$

Furthermore, given any $\epsilon > 0$ there can always be found points a' and a'' belonging to A such that:

$$a' - a_{\cdot} < \epsilon \quad \text{and} \quad a^{\cdot} - a'' < \epsilon \, .$$

(If $a_{\cdot}$ and $a^{\cdot}$ actually belong to A we may, of course, choose $a' = a_{\cdot}$ and $a'' = a^{\cdot}$. Such need not be the case, however.)

<u>EXERCISE 1.9</u> Evaluate the sup and inf of the following sets of real numbers, and illustrate the remarks of Example 1.2 for each: $A = \{x\colon 0 \le x < 1\}$, $A = \{x\colon x=1/n \ (n=1,2,\ldots,)\}$ and $A = \{x\colon x= 1-1/n \ (n=\pm 1, \pm 2,\ldots)\}$.

For any set A which is bounded above (bounded below), the number $\sup A$ $(\inf A)$ need not actually belong to A. If it does, however, it is further denoted by $\max A$ $(\min A)$.

<u>EXERCISE 1.10</u> The following result is sometimes of value. For any pair a,b of real numbers:

$$\min(a,b) = \frac{a+b}{2} - \frac{|a-b|}{2}$$

$$\max(a,b) = \frac{a+b}{2} + \frac{|a-b|}{2} \, .$$

Prove this result.

<u>EXERCISE 1.11</u> Construct a bounded set of real numbers for which the max and min do not exist.

<u>EXERCISE 1.12</u> Prove that for any non-empty set A of real numbers, $\inf A \le \sup A$, with equality iff A consists of a single point.

EXERCISE 1.13 Suppose that for a pair A,B of non-empty sets of real numbers we have $A \subseteq B$. Prove then that:

$$\inf A \geq \inf B$$

$$\sup A \leq \sup B \quad .$$

Illustrate these results with several examples and diagrams.

EXERCISE 1.14 Prove that for any two sets of real numbers A,B: $\sup(A \cup B) = \max(\sup A, \sup B)$ and $\inf(A \cup B) = \min(\inf A, \inf B)$. Are there analogous results for the sup and inf of $A \cap B$, $A\setminus B$ and $A \Delta B$? Generalize wherever possible to the case of $k > 2$ sets.

A real number c is termed a limit point of the set A iff given any $\epsilon > 0$ there exists a point $x \in A$ such that $0 < |x-c| < \epsilon$ (i.e. there are points of A distinct from and arbitrarily close to c). A limit point c may or may not belong to the set A.

EXAMPLE 1.3 $c = 0$ is a limit point of the set A defined as $A = \{x: x=1/n \ (n=1,2,\ldots)\}$, yet $c \notin A$. Every point of the set A defined as $A = \{x: 0 \leq x \leq 1\}$ is a limit point of A. However, $c = 1$ is a limit point of the set A defined as $A = \{x: 0 \leq x < 1\}$, but $c \notin A$.

EXERCISE 1.15 Prove that a finite set A has no limit points.

<u>EXERCISE 1.16</u> An ϵ-neighborhood $(\epsilon > 0)$ of a real number c is a set of the form: $\{x: |x-c| < \epsilon\}$. Prove that c is a limit point of the set A iff every ϵ-neighborhood of c contains infinitely many points of A.

The set A together with all limit points of A is termed the closure of A and written $\bar{A}$. Then, a set A is termed closed iff $A = \bar{A}$ (i.e. any limit point of A belongs to A).

<u>EXERCISE 1.17</u> Determine $\bar{A}$ for each of the following sets. Hence, determine which sets are closed:

$A = \{x: 0 < x \leq 1\}$, $A = \{x: x \geq 2\}$, $A = \{x: x=1/n (n=1,2,...)\}$, $A = \{$all rational numbers between zero and one inclusive$\}$.

The set A is termed open iff it is the complement of a closed set. Of course, a set A need be neither open nor closed.

<u>EXERCISE 1.18</u> (Alternate Definition of Open Set) Prove that a set A is open iff for every point $a' \in A$ there is some ϵ-neighborhood about a' completely contained in A. Illustrate the above result for some examples of open sets of real numbers. Also give several examples of sets of real numbers that are neither open nor closed.

<u>EXERCISE 1.19</u> (Combinations of Open or Closed Sets) Prove that the union (intersection) of a countable number of open (closed) sets is open (closed). Then, prove that the words union and intersection may be interchanged provided a finite number of sets is involved (but not in general an infinite

number). Illustrate the last point by two examples.

<u>EXERCISE 1.20</u> (Inverse images) Let f be any real-valued function of the real variable x. For any set R of real numbers define the inverse image $f^{-1}(R)$ of R under f as: $f^{-1}(R) = \{x: f(x) \in R\}$. Draw a diagram illustrating this definition. Then prove that:

(1) $R_1 \subseteq R_2 \Rightarrow f^{-1}(R_1) \subseteq f^{-1}(R_2)$

(2) $f^{-1}(R^c) = [f^{-1}(R)]^c$

(3) $f^{-1}(R_1 \cup R_2) = f^{-1}(R_1) \cup f^{-1}(R_2)$

(4) $f^{-1}(R_1 \cap R_2) = f^{-1}(R_1) \cap f^{-1}(R_2)$.

<u>EXERCISE 1.21</u> (Direct images) Let f be any real-valued function of the real variable x having domain D and range R. For any subset S of D we define the (direct) image $f(S)$ of S as follows: $f(S) = \{f(x): x \in S\}$. Accordingly, $R = f(D)$. Establish the following properties of direct images.

(1) $S_1 \subseteq S_2 \Rightarrow f(S_1) \subseteq f(S_2)$

(2) $f(S^c) = [f(S)]^c$ iff $f(S) \cap f(S^c) = \emptyset$

(3) $f(S_1 \cup S_2) = f(S_1) \cup f(S_2)$

(4) $f(S_1 \cap S_2) \subseteq f(S_1) \cap f(S_2)$.

<u>EXERCISE 1.22</u> (Continuation) State the obvious generalizations of (1), (3) and (4) of the preceding Exercise for

the case of $k > 2$ sets. Also, give a specific example proving that equality need not hold in (4). If the added condition in (2) is removed, can any general relationship be established? Compare these with the results for inverse images in Exercise 1.20.

<u>EXERCISE 1.23</u> (Continuous functions and open and closed sets) Prove that if f is continuous and R is an open (closed) set of reals then $f^{-1}(R)$ is an open (closed) set of reals. Thus conclude that $\{x: f(x) \geq c \text{ or } \leq c\}$ and $\{x: a \leq f(x) \leq b\}$ are closed sets of reals, and $\{x: f(x) > c \text{ or } < c\}$ and $\{x: a < f(x) < b\}$ are open sets of reals.

The following basic result asserts that every infinite, bounded set A possesses at least one limit point (that may or may not belong to A).

<u>THEOREM 1.1</u> (Bolzano-Weierstrass) Every infinite, bounded set A of real numbers possesses at least one limit point.

PROOF. If A is infinite and bounded then it is contained in some finite interval I, and at least one of the halves of I must contain an infinite number of points of A. Let I_1 be one. Next, bisect I_1 and again note that at least one of its halves, say I_2, must contain an infinite number of points of A. By continuing this process we obtain a monotone decreasing sequence of intervals, each containing an infinite number of points of A. The left endpoints of these intervals form a set of numbers which is bounded above and so possesses a finite supremum denoted s. We assert that s is a limit point of A. To prove this note that given any $\epsilon > 0$ there exists an n sufficiently large so that I_n is contained within an ϵ-neighborhood of s (because the lengths of the intervals tend to zero as $n \to \infty$). Then, because every I_n contains an infinite number of points of A, s is a limit point of A. This completes the proof.

Finally, a point d is termed a boundary point of the set A of real numbers iff every ϵ-neighborhood of d contains points of both A and A^c. The boundary of A is denoted $b(A)$.

EXERCISE 1.24 Prove that $b(A) = \bar{A} \cap \overline{(A^c)}$, whence conclude that $b(A)$ is a closed set.

EXERCISE 1.25 Prove that if A is a finite set then $b(A) = A$. Then evaluate $b(A)$ for the following sets:

$A = \{x: x=1/n \ (n=1,2,\dots)\}$, $A = \{x: 0 < x \le 1\}$,

$A = \{$all rational numbers between zero and one inclusive$\}$.

When is $b(A) = \emptyset$?

Another basic notion is that of a sequence $a_1, a_2, \dots$ or $\{a_n\}$ of real numbers. It is a collection of real numbers indexed on the positive integers. In contrast with a set, the individual members (terms) of a sequence need not all be different. If $1 \le n_1 < n_2 < \dots$ is an increasing sequence of positive integral indices then $a_{n_1}, a_{n_2}, \dots$ is termed a subsequence of $a_1, a_2, \dots$.

A real number a' is termed a limit point of the sequence $a_1, a_2, \dots$ iff given any $\epsilon > 0$ and any index N there exists an index $n \geq N$ for which $|a_n - a'| < \epsilon$. Note that a' itself need not occur as a term of the sequence.

<u>EXERCISE 1.26</u> (Limit Point of a Sequence vs. Set) Prove that a real number a' is a limit point of the sequence $a_1, a_2, \ldots$ iff either a' occurs infinitely often as a term of the sequence or a' is a limit point of the set of real numbers composed of the distinct terms of the sequence.

<u>EXERCISE 1.27</u> (Alternate Definition of Limit Point of a Sequence) Prove that a' is a limit point of $a_1, a_2, \ldots$ iff given any $\epsilon > 0$ there are infinitely many indices n for which $|a_n - a'| < \epsilon$.

<u>EXERCISE 1.28</u> Show that the sequence $1, -2, 3, -4, \ldots$ has no limit points whereas the sequence $1, 1/2, 1/3, \ldots$ has one, and $1, 2, 1/2, 2^{1/2}, 1/3, 2^{1/3}, \ldots$ has two. Give an example of a sequence having infinitely many limit points.

For any sequence $a_1, a_2, \ldots$ we denote by lim inf a_n (lim sup a_n) the inf (sup) of the set of real numbers consisting of the limit points of the sequence. Clearly, then, $\liminf a_n \leq \limsup a_n$.[1]

<u>EXERCISE 1.29</u> A sequence $a_1, a_2, \ldots$ is termed bounded (bounded below, bounded above) iff there exists a real number B (L, U) such that $|a_n| \leq B$ $(a_n \geq L,\ a_n \leq U)$

[1]Provided the sequence possesses at least one limit point. Otherwise, $\liminf a_n = +\infty$ and $\limsup a_n = -\infty$.

for $n = 1,2,\ldots$. First prove that every bounded sequence has at least one limit point. Then prove that for a sequence which is bounded below (bounded above): $L \le \liminf a_n$ ($\limsup a_n \le U$).

<u>EXERCISE 1.30</u> Let $a_1,a_2,\ldots$ be a sequence of real numbers and $ca_1, ca_2,\ldots$ the sequence obtained from it by multiplying each term by the real number c. Establish the relationship between: the lim inf's and lim sup's of the two sequences.

<u>EXERCISE 1.31</u> Let $a_1,a_2,\ldots$ and $b_1,b_2,\ldots$ be two sequences of real numbers and $a_1+ b_1, a_2+ b_2,\ldots$ the sequence obtained from them by adding them termwise. Establish the following relationships:

$$\liminf a_n + \liminf b_n \le \liminf (a_n + b_n) \le$$
$$\liminf a_n + \limsup b_n \le \limsup (a_n + b_n) \le$$
$$\limsup a_n + \limsup b_n .$$

A sequence $a_1,a_2,\ldots$ is termed convergent iff it possesses a single limit point a. In this case, $\liminf a_n = \limsup a_n = a$. We write $\lim a_n = a$ (or $a_n \to a$ as $n \to \infty$) and say that the sequence converges to a. Otherwise the sequence is termed divergent.

<u>EXERCISE 1.32</u> Prove that a convergent sequence is bounded.

<u>EXERCISE 1.33</u> A sequence $a_1,a_2,\ldots$ is said to diverge properly to $+\infty$ iff given any (arbitrarily large) positive constant C there exists an index N such that $a_n \ge C$ for all $n \ge N$. Provide a definition for divergence

to $-\infty$, then give an example of each type of sequence.

EXERCISE 1.34 (Alternate Definition of Convergence) Prove that a sequence $a_1, a_2, \ldots$ converges to a iff given any $\epsilon > 0$ there exists an index N (possibly depending upon ϵ) such that $|a_n - a| < \epsilon$ for all indices $n \geq N$. Find N and a explicitly for the sequence $1, 1/2, 1/3, \ldots$.

EXERCISE 1.35 Using the alternate definition of convergence, first prove that if $a_n \to a$ then $\bar{a}_n \to a$ where $\bar{a}_n = \frac{1}{n}(a_1 + \ldots + a_n)$. Next prove that if f is a continuous function at $x = a$ then $f(a_n) \to f(a)$. Finally prove that if $a_n > 0$ $(n = 1,2,\ldots)$ then $\sqrt[n]{a_1 \cdot a_2 \ldots a_n} \to a$.

EXERCISE 1.36 (Properties of Lim Inf and Lim Sup of a Sequence of Numbers) Let $\{a_n\}$ be a sequence of real numbers and set $\underline{a} = \liminf a_n$ and $\bar{a} = \limsup a_n$. Establish the following properties:

(i) The sequence $\{a_n\}$ is bounded iff both $\underline{a}$ and $\bar{a}$ are finite.

(ii) If $\{a_n\}$ is bounded, then given any $\epsilon > 0$ there can always be found an index N sufficiently large so that:

$a_n > \underline{a} - \epsilon$ for all indices $n \geq N$ and

$|a_n - \underline{a}| < \epsilon$ for infinitely many (but not necessarily all) indices $n \geq N$.

(iii) If $\{a_n\}$ is bounded, then given any $\epsilon > 0$ there can always be found an index N sufficiently large so that:

$a_n < \overline{a} + \epsilon$ for <u>all</u> indices $n \geq N$ and

$|a_n - \overline{a}| < \epsilon$ for infinitely many (but not necessarily all) indices $n \geq N$.

(iv) If $\{a_n\}$ is bounded, then given any $\epsilon > 0$ there can always be found an index N sufficiently large so that:

$\underline{a} - \epsilon < a_n < \overline{a} + \epsilon$ for <u>all</u> indices $n \geq N$.

(v) From $\{a_n\}$ can always be chosen a subsequence converging (or properly diverging, as the case may be) to either $\underline{a}$ or $\overline{a}$. Deduce from this that every subsequence of a convergent sequence is convergent (to the same limit, of course).

(vi) If $\{a_n\}$ is only assumed bounded below (above) can it be concluded that $\underline{a}$ $(\overline{a})$ is finite?

(vii) Illustrate the preceding results with various sequences. A diagram is often helpful.

<u>EXERCISE 1.37</u> (Monotone sequences) A sequence $a_1, a_2, \ldots$ is termed monotone increasing (decreasing) iff $a_1 \leq a_2 \leq a_3 \leq \ldots$ $(a_1 \geq a_2 \geq a_3 \geq \ldots)$. Prove that a monotone increasing (decreasing) sequence is convergent iff it is bounded above (below). Otherwise prove that the sequence is divergent to $+\infty$ $(-\infty)$, i.e. properly divergent.

EXERCISE 1.38 Using a previous Exercise, prove that if $\lim a_n = a$ and $\lim b_n = b$ then $\lim (ra_n + sb_n) = ra + sb$, where r,s are arbitrary real numbers. Generalize this result by considering functions $f(x,y)$ that are continuous at (a,b).

EXERCISE 1.39 Prove that a' is a limit point of $a_1, a_2, \ldots$ iff there exists a subsequence converging to a'. Thus prove that every subsequence of a convergent sequence is convergent to the limit.

An alternate criterion for convergence of a sequence is the so-called Cauchy Criterion. The often useful result is contained in the following Theorem.

THEOREM 1.2 (Cauchy Criterion for convergence) A necessary and sufficient condition that the sequence $a_1, a_2, \ldots$ be convergent is that given any $\epsilon > 0$ there exists an index N (possibly depending upon ϵ) such that $|a_m - a_n| < \epsilon$ for all indices $m,n \geq N$.

PROOF. Suppose $a_n \to a$. Then, given any $\epsilon > 0$ there exists an index N such that $|a_n - a| < \frac{1}{2}\epsilon$ for all indices $n \geq N$. Thus for all indices $m,n \geq N$ we have $|a_m - a_n| \leq |a_m - a| + |a_n - a| < \frac{1}{2}\epsilon + \frac{1}{2}\epsilon = \epsilon$, so the Cauchy Criterion is necessary.

Next, suppose the Cauchy Criterion is satisfied. Then for all indices $n \geq N$ we have $|a_N - a_n| < \epsilon$, or $a_N - \epsilon < a_n < a_N + \epsilon$. Thus the sequence is bounded and also $a_N - \epsilon \leq \liminf a_n \leq \limsup a_n \leq a_N + \epsilon$ whence

$0 \leq \lim\sup\ a_n - \lim\inf\ a_n \leq 2\epsilon$. Since ϵ is arbitrary, we conclude that $\lim\inf\ a_n = \lim\sup\ a_n = a$ (say), whence $a_n \to a$. Thus the Cauchy Criterion is sufficient and the proof is complete.

A sequence satisfying the Cauchy Criterion is usually termed a Cauchy Sequence.

EXERCISE 1.40 (Alternate Definition of a Closed Set) Prove that a set A of real numbers is closed iff the limit of every Cauchy (convergent) sequence, all of whose terms belong to A, also belongs to A.

EXERCISE 1.41 (Monotone Sequence of Closed Sets) Prove that if $A_1 \supseteq A_2 \supseteq A_3 \supseteq \cdots$ is a monotone decreasing sequence of non-empty, bounded, closed sets of real numbers then $\lim A_n = \bigcap A_n$ contains at least one point. Provide a counter-example for the case where the sets are either not assumed closed or not assumed bounded.

Recall that E_n consists of all ordered n-tuples $\underline{x} = (x_1, \ldots, x_n)$ of real numbers. Because this set cannot be linearly ordered for $n > 1$ (in contrast with the "line" of points in E_1), any notions dependent upon the ideas of "largest" or "smallest" have no meaning. Thus, for example, we cannot define a sup or inf of a subset of points in E_n, $n > 1$.

However, if we define the distance between two points in E_n, $\underline{x} = (x_1, \ldots, x_n)$ and $\underline{y} = (y_1, \ldots, y_n)$, as $\| \underline{x} - \underline{y} \| = \{\Sigma(x_i - y_i)^2\}^{\frac{1}{2}}$, then all preceding results of this section that depend only upon the notion of distance (in E_1) have immediate valid counterparts in E_n, $n > 1$. Thus, for example, the notions of limit point and limit (of a set or sequence), closed and open sets, bounded sets and sequences, closure, boundary, and so on have immediate parallels in E_n, $n > 1$ using the above definition of (Euclidean) distance in E_n.

EXERCISE 1.42 Generalize the appropriate preceding results of this section so as to be valid for E_n, $n > 1$. Review proofs to see where modifications are necessary.

EXERCISE 1.43 (Sup, Inf, Max, and Min as Applied to Functions) The notions of sup, inf, max, and min, defined only for sets of real numbers at this point, are also used in describing extrema properties of real-valued functions defined over a domain D. First, we establish the connection between the two concepts. For example, if S is a subset of D ($S \neq \emptyset$) then

$$\sup_{x \in S} f(x) \quad \text{means} \quad \sup f(S)$$

and

$$\inf_{x \in S} f(x) \quad \text{means} \quad \inf f(S)$$

where

$$f(S) = \{f(x): \ x \in S\} ,$$

that is, $f(S)$ is the set of values of $f(x)$ as x varies over S. In view of our knowledge of sets of real numbers, $\sup f(S)$ and/or $\inf f(S)$ need not belong to $f(S)$.

If either does, they are further denoted by $\max f(S)$ and $\min f(S)$ in which case the expressions:

$$\max_{x \in S} f(x) \quad \text{and} \quad \min_{x \in S} f(x)$$

are meaningful. Prove the following Classical and often useful result: if S is a closed, bounded set of real numbers, and f is continuous over S then $f(S)$ is closed and bounded, whence $\max_{x \in S} f(x)$ and $\min_{x \in S} f(x)$ exist. Further deduce that there exists at least one pair of points, say x^o and x_o, in S such that

$$f(x^o) = \max_{x \in S} f(x)$$

and

$$f(x_o) = \min_{x \in S} f(x) \quad ,$$

that is, both the maximum and minimum values of f over S are actually attained for at least one point within S.

<u>EXERCISE 1.44</u> (Some Properties of Sup and Inf of Functions) Let f and g be real-valued functions defined over a common domain D. Establish the following properties (all subsets are non-empty):

(i) If $S_1 \subseteq S_2$ are subsets of D then

$$\sup_{x \in S_1} f(x) \le \sup_{x \in S_2} f(x)$$

$$\inf_{x \in S_1} f(x) \ge \inf_{x \in S_2} f(x) \quad .$$

(ii) If $f(x) \le g(x)$ for all x in a subset S of D, then:

$$\sup_{x \in S} f(x) \le \sup_{x \in S} g(x)$$

$$\inf_{x \in S} f(x) \le \inf_{x \in S} g(x) \quad .$$

(iii) For any subset S of D:

$$\sup_{x \in S}[f(x) + g(x)] \le \sup_{x \in S} f(x) + \sup_{x \in S} g(x)$$

$$\inf_{x \in S}[f(x) + g(x)] \ge \inf_{x \in S} f(x) + \inf_{x \in S} g(x)$$

$$\sup_{x \in S} -f(x) = -\inf_{x \in S} f(x)$$

$$\inf_{x \in S} -f(x) = -\sup_{x \in S} f(x) \quad .$$

(iv) If $f(x)$ and $g(x)$ are non-negative over the subset S,

$$\inf_{x \in S} f(x)\, g(x) \ge \inf_{x \in S} f(x) \cdot \inf_{x \in S} g(x)$$

$$\sup_{x \in S} f(x)\, g(x) \le \sup_{x \in S} f(x) \cdot \sup_{x \in S} g(x)$$

$$\sup_{x \in S} [1/f(x)] = 1/\inf_{x \in S} f(x) \quad \text{and} \quad \inf_{x \in S} [1/f(x)] = 1/\sup_{x \in S} f($$

<u>EXERCISE 1.45</u> (Some Properties of Max and Min of Functions) Show that whenever the max or min of the indicated functions in Exercise 1.44 exist, we may replace sup by max and inf by min in any expression. Accordingly, a sufficient condition that all expressions in Exercise 1.44

be valid with max and min replacing sup and inf is that the functions involved be continuous over the sets, which are to be both closed and bounded sets of real numbers.

<u>EXERCISE 1.46</u> (Some Mathematical Problems in Game Theory) Problems related to iterated[1] extrema for functions of two real variables occur in such areas as the Theory of Games. Let $f(x,y)$ be a real-valued function defined over the rectangular domain $D = \{(x,y): x \in A,\ y \in B\}$ where A and B are sets of real numbers.

For each fixed $x \in A$ there is defined an ordinary function $f(x,y)$ of the real variable y, and the quantity:

$$\sup_{y \in B} f(x,y)$$

has previously been defined (see Exercise 1.43). Accordingly, then, as x is allowed to vary over A we obtain a function of the real variable x, in which case the quantities:

(a) $$\inf_{x \in A} \sup_{y \in B} f(x,y)$$

and

(b) $$\sup_{x \in A} \sup_{y \in B} f(x,y)$$

are well-defined. Analogously, we may define the quantities:

---

[1]To be distinguished from the 'sup', 'inf', 'max' or 'min' of $f(x,y)$ as (x,y) varies over the set D.

(c) $$\sup_{x \in A} \inf_{y \in B} f(x,y)$$

and

(d) $$\inf_{x \in A} \inf_{y \in B} f(x,y) \quad .$$

Similarly, the process may be reversed, and the quantities:

(e) $$\inf_{y \in B} \sup_{x \in A} f(x,y)$$

(f) $$\sup_{y \in B} \sup_{x \in A} f(x,y)$$

(g) $$\sup_{y \in B} \inf_{x \in A} f(x,y)$$

(h) $$\inf_{y \in B} \inf_{x \in A} f(x,y)$$

defined. In view of the results of Exercise 1.43, similar expressions with 'max' replacing 'sup' and/or 'min' replacing 'inf' may or may not be meaningful in either some or all cases.

Assume now that the conditions exist so that all of the above expressions are meaningful with 'max' and 'min' replacing 'sup' and 'inf'. The Theory of Games, for example, then considers the problem of determining conditions under which there exists a point $x_o \in A$ and a point $y_o \in B$ such that:

$$\max_{x \in A} \min_{y \in B} f(x,y) = \min_{y \in B} \max_{x \in A} f(x,y) = f(x_o,y_o) \quad .$$

Such a point (x_o,y_o) is termed a 'saddle point'.

Establish the following results related to this problem:

(i) $\max_{x\in A} \min_{y\in B} f(x,y) \le \min_{y\in B} \max_{x\in A} f(x,y)$ in general.

(ii) $\min_{x\in A} \min_{y\in B} f(x,y) = \min_{y\in B} \min_{x\in A} f(x,y)$ in general.

(iii) $\max_{x\in A} \max_{y\in B} f(x,y) = \max_{y\in B} \max_{x\in A} f(x,y)$ in general.

The next basic notion is that of an infinite series. An infinite series of real numbers is formally an 'infinite sum' of the form $a_1 + a_2 + a_3 + \ldots$, also denoted $\sum_1^\infty a_n$. The n-th partial sum of the series is defined as $s_n = a_1 + a_2 + \ldots + a_n$, or alternately $\sum_1^n a_i$, and the series is said to converge or diverge according as the sequence of partial sums $s_1, s_2, s_3, \ldots$ converges or diverges as a sequence. If $s_n \to s$ (finite) then we write $\sum_1^\infty a_n = s$ and call s the sum of the infinite series.

<u>EXAMPLE 1.4</u> The infinite series $\sum_1^\infty (\frac{1}{2})^{n-1} = 1+\frac{1}{2}+(\frac{1}{2})^2+\ldots$ converges to (has) a sum of 2 because $s_n = 1+(\frac{1}{2})+\ldots+(\frac{1}{2})^{n-1} = 2[1-(\frac{1}{2})^n]$, and clearly $s_n \to 2$ as $n \to \infty$.

<u>EXERCISE 1.47</u> Prove that $\sum_1^\infty 1/n(n+1) = 1$.

There is an alternate and equivalent criterion for convergence of an infinite series contained in the following Theorem. It stems from the Cauchy Criterion for convergence applied to the sequence of partial sums.

THEOREM 1.3 (General Principle of Convergence for a Series) Let $s_1, s_2, s_3, \ldots$ (where $s_n = a_1 + a_2 + \ldots + a_n$) be the sequence of partial sums corresponding to the infinite series $\sum_1^\infty a_n$. A necessary and sufficient condition that the series be convergent is that given any $\epsilon > 0$ there exists an index N (possibly depending upon ϵ) such that $|s_m - s_n| = |\sum_{n+1}^{m} a_i| < \epsilon$ for all indices $m > n \geq N$.

EXERCISE 1.48 Prove Theorem 1.3.

The preceding Theorem can now be applied to derive a well-known necessary condition for convergence of an infinite series.

THEOREM 1.4 (Necessary Condition for Convergence of a Series) If $\sum_1^\infty a_n$ converges then $a_n \to 0$ as $n \to \infty$.

EXERCISE 1.49 Prove Theorem 1.4.

EXAMPLE 1.5 From the preceding Theorem the series $\sum_1^\infty n/(n+1)$ diverges since $a_n \not\to 0$. Note however that the condition $a_n \to 0$ is not sufficient for convergence of $\sum_1^\infty a_n$. To see this, consider the series $\sum_1^\infty 1/n$ for which $a_n = 1/n \to 0$. Note that $s_{2n} - s_n = 1/(n+1) + \ldots + 1/2n > n/2n = \frac{1}{2}$, and thus by the General Principle of Convergence, the series diverges.

Convergent infinite series obey many of the familiar algebraic properties of ordinary finite sums. In particular, we have the following two results:

<u>THEOREM 1.5</u> If $\sum_1^\infty a_n = s$ then $\sum_1^\infty ka_n = ks$, where k is any constant independent of n.

PROOF. This result follows immediately from the results of Exercise 1.30, in the special case of convergence, applied to the sequence of partial sums.

<u>THEOREM 1.6</u> If $\sum_1^\infty a_n = s$ and $\sum_1^\infty b_n = t$ then $\sum_1^\infty (a_n \pm b_n) = s \pm t$.

<u>EXERCISE 1.50</u> Use the basic definition of convergence of an infinite series to prove Theorem 1.6.

If we bracket together (without altering order) consecutive groups of a finite number of terms of an infinite series, we obtain a new infinite series whose n-th term is the sum of the terms appearing in the n-th set of brackets. The following Theorem asserts that if the original series converges then the 'bracketed' series converges to the same sum. Certainly finite sums obey this property.

<u>THEOREM 1.7</u> (Bracketing) If $\sum_1^\infty a_n = s$ then the same is true for any series obtained from it by bracketing consecutive groups of a finite number of terms without altering order of terms.

PROOF. Let $\sum_1^\infty a'_n$ be the series obtained from $\sum_1^\infty a_n$ by bracketing consecutive groups of a finite number of terms without altering order, and let s'_n be its n-th partial sum. By construction, $s'_n = s_N$ for some index $N \geq n$ (where s_N is the N-th partial sum of the original series). Since $s_N \to s$ as $N \to \infty$, it follows that $s'_n \to s$ as $n \to \infty$ and the theorem is proved.

EXAMPLE 1.6 In general the converse of Theorem 1.7 is false, for consider the bracketed series $(1-1) + (1-1) + \ldots$ which converges to (has sum) zero. However, the unbracketed series $\sum_1^\infty (-1)^{n+1}$ clearly diverges since $a_n \not\to 0$.

To prove a general result concerning reordering of terms of an infinite series we require the added assumption that all (but perhaps a finite number) of the terms are non-negative. Such series occur frequently in Applied Mathematics. First we need the following result:

THEOREM 1.8 (Non-Negative Series) If $a_n \geq 0$ then either $\sum_1^\infty a_n$ is convergent or else it is properly divergent to $+\infty$, in the sense that its sequence of partial sums $s_1, s_2, \ldots$ is monotone increasing and unbounded.

EXERCISE 1.51 Prove Theorem 1.8 .

THEOREM 1.9 (Reordering Terms in Non-Negative Series) Suppose $a_n \geq 0$ and let $\sum_1^\infty b_n$ be any series whose terms are those of $\sum_1^\infty a_n$ but possibly in a different order. Then the two series either both converge to the same sum or else both properly diverge to $+\infty$.

PROOF. Let s_n be the n-th partial sum of $\sum_1^\infty a_n$ and s_n' the n-th partial sum of $\sum_1^\infty b_n$. Suppose $b_1 = a_{m_1}, b_2 = a_{m_2}, \ldots$ and let $p = \max(m_1, m_2, \ldots, m_n)$. Now $s_n' = \sum_1^\infty a_{m_j} \le \sum_1^p a_i = s_p$ where $p \ge n$. Reversing the roles of the a's and b's we can then find an index q such that $s_n = s_q'$ where $q \ge n$. Thus if $\sum_1^\infty a_n = s$ then both $\lim s_n' \le s$ and $\lim s_n' \ge s$ whence equality must hold and $\sum_1^\infty b_n = s$. The case of proper divergence to $+\infty$ follows parallel lines and the proof is complete.

In summary then, we have the following result stating that, as far as bracketing and/or alteration of order of terms is concerned, infinite series of non-negative terms behave like ordinary finite sums.

THEOREM 1.10 (Bracketing and/or Reordering Non-negative Series) Suppose $a_n \ge 0$. Let $\sum_1^\infty c_n$ be any series obtained from $\sum_1^\infty a_n$ by reordering and/or bracketing. If $\sum_1^\infty a_n$ converges to s (diverges to $+\infty$), then the same is true of $\sum_1^\infty c_n$.

EXERCISE 1.52 Prove Theorem 1.10.

It is useful to be able to express the sum of a convergent infinite series in 'closed form', i.e. as a single number. However, a first step is to develop methods for determining which infinite series are convergent to begin with. We now consider some tests for convergence of infinite series.

First we consider the notions of absolute and conditional convergence. The series $\sum_1^\infty a_n$ is termed absolutely convergent iff the series $\sum_1^\infty |a_n|$ is convergent. It may happen that a series is convergent but not absolutely convergent. In this case the series is termed conditionally convergent. However, an absolutely convergent series is always convergent, as the following Theorem proves.

<u>THEOREM 1.11</u> (Absolute Convergence Implies Convergence)
If $\sum_1^\infty a_n$ is absolutely convergent it is convergent.

PROOF. Suppose $\sum_1^\infty |a_n|$ converges. By the General Principle of Convergence, given any $\epsilon > 0$ there exists an index N such that $\big||a_{n+1}|+\ldots+|a_m|\big| < \epsilon$ for all indices $m > n \geq N$. Thus, if s_n is the n-th partial sum of $\sum_1^\infty a_n$ we have $|s_m - s_n| = |\sum_{n+1}^m a_i| \leq |a_{n+1}|+\ldots+|a_m| < \epsilon$ for all indices $m > n \geq N$. So, the same principle guarantees that $\sum_1^\infty a_n$ converges. This completes the proof.

Later we shall give an example of a series that is conditionally convergent but not absolutely convergent.

The results of Theorem 1.10 can be extended from non-negative series alone to absolutely convergent series. We already know that any convergent series may be bracketed without altering the sum. Thus, all that is necessary to prove is that the terms of an absolutely convergent series may be reordered in any manner without altering the sum. This result is contained in the following Theorem.

THEOREM 1.12 (Reordering Terms of Absolutely Convergent Series) The terms of an absolutely convergent series may be rearranged in any manner without altering the convergence or sum.

PROOF. Suppose $\sum_1^\infty a_n$ is absolutely convergent and let $\sum_1^\infty b_n$ be any series obtained from it by rearranging terms. Define:

$$u_n = \begin{cases} a_n & \text{if } a_n \geq 0 \\ 0 & \text{if } a_n < 0 \end{cases} \qquad v_n = \begin{cases} -a_n & \text{if } a_n < 0 \\ 0 & \text{if } a_n \geq 0 \end{cases}$$

$$u_n' = \begin{cases} b_n & \text{if } b_n \geq 0 \\ 0 & \text{if } b_n < 0 \end{cases} \qquad v_n' = \begin{cases} -b_n & \text{if } b_n < 0 \\ 0 & \text{if } b_n \geq 0 \end{cases} .$$

Now absolute converges of $\sum_1^\infty a_n$ guarantees that the two non-negative series $\sum_1^\infty u_n$ and $\sum_1^\infty v_n$ are convergent. But $\sum_1^\infty u_n'$ and $\sum_1^\infty v_n'$ are obtained from them by reordering only, hence it follows from Theorem 1.9 that the latter are convergent and also $\sum_1^\infty u_n = \sum_1^\infty u_n'$, $\sum_1^\infty v_n = \sum_1^\infty v_n'$. Thus, via the algebraic properties of convergent series, $\sum_1^\infty b_n = \sum_1^\infty (u_n' - v_n')$ converges, and to the sum of the series

$\sum_1^\infty (u_n - v_n) = \sum_1^\infty a_n$. This completes the proof.

In what follows, we shall be considering a variety of well-known tests for convergence of infinite series. Some tests apply to series in general; some are restricted to non-negative series. Because absolute convergence implies convergence, we see that a test restricted to non-negative series can be used to establish absolute convergence, perhaps, of an arbitrary series, which in turn would establish convergence of the series. Thus, a test for convergence of non-negative series might not be as restrictive as it might first appear.

The first general test for convergence of an infinite series is based upon the familiar improper Riemann integral; it is a test restricted to non-negative series.

THEOREM 1.13 (Integral Test for Convergence) Let $\sum_{1}^{\infty} a_n$ be non-negative series, and $f(x)$ a non-negative function of x satisfying:

(i) $f(x)$ is continuous for $x \geq 0$ [1]

(ii) $f(x)$ is non-increasing

(iii) $f(x) \to 0$ as $x \to \infty$

(iv) $f(n) = a_n$, $n = 1,2,\ldots$ [1]

Then the infinite series $\sum_{1}^{\infty} a_n$ and the improper Reimann integral $\int_0^{\infty} f(x)\,dx$ converge or properly diverge to $+\infty$ together.

PROOF. The series $\sum_{1}^{\infty} a_n$ either converges or else properly diverges to $+\infty$. Now because $f(x)$ is non-increasing it follows that for any positive integer m, $\int_{m-1}^{m} f(x)\,dx \geq f(m) = a_m$. Thus for any positive integer n we have that $0 \leq a_1 + a_2 + \ldots + a_n \leq \int_0^n f(x)\,dx \leq \int_0^{\infty} f(x)\,dx$ so that the sequence of partial sums is bounded, hence the

[1] These first two conditions need only hold for all x sufficiently large, say $x \geq c$; condition (iv) need only hold for all n sufficiently large, say $n \geq N$. Then, the convergence test depends upon convergence of the improper Reimann Integral $\int_K^{\infty} f(x)\,dx$, where $K \geq \max(c,N)$.

series is convergent if the improper integral converges. Next note that for any positive integer m we have $\int_m^{m+1} f(x)\,dx \le f(m) = a_m$, so that for any positive integer n, $\int_1^{n+1} f(x)\,dx \le a_1 + a_2 + \ldots + a_n$ so that if the improper integral diverges then so does the infinite series. This completes the proof.

EXAMPLE 1.7 The infinite series $\sum_1^\infty 1/n^p$ converges iff $p > 1$ by the Integral Test since the Improper Riemann Integral $\int_1^\infty (1/x)^p\,dx$ converges iff $p > 1$.

EXERCISE 1.53 Test the following series for convergence:

$$\sum_1^\infty 1/(n+1)^{3/2}\ ,\quad \sum_1^\infty 2n/(1+n^2)\ ,\quad \sum_2^\infty 1/(n\log n)\ .$$

Once we know the convergence properties of a basic set of non-negative series, it may be possible to determine the (absolute) convergence properties of other series by means of a suitable comparison. This method is set forth in the following Theorem.

THEOREM 1.14 (Comparison Tests) Let $\sum_1^\infty a_n$ and $\sum_1^\infty b_n$ be two non-negative series;[1]

(1) if $0 \le a_n \le b_n$ then $\sum_1^\infty b_n$ convergent $\Rightarrow \sum_1^\infty a_n$ convergent

(2) if $0 \le a_n \le b_n$ then $\sum_1^\infty a_n$ divergent $\Rightarrow \sum_1^\infty b_n$ divergent

(3) if $0 \le k_n \le K < \infty$ then $\sum_1^\infty a_n$ convergent $\Rightarrow \sum_1^\infty k_n a_n$ convergent

[1] The conditions of parts (1) through (4) obviously need hold only for all n sufficiently large.

(4) if $0 < K \le k_n$ then $\sum_1^\infty a_n$ divergent $\Rightarrow \sum_1^\infty k_n a_n$ divergent

(5) if $a_n/b_n \to c > 0$ then $\sum_1^\infty a_n$ and $\sum_1^\infty b_n$ converge or diverge[1] together

(6) if $a_n/b_n \to 0$ then $\sum_1^\infty b_n$ convergent $\Rightarrow \sum_1^\infty a_n$ convergent[1]

(7) if $a_n/b_n \to \infty$ then $\sum_1^\infty b_n$ divergent $\Rightarrow \sum_1^\infty a_n$ divergent.[1]

<u>EXERCISE 1.54</u> Prove parts (1) through (4) of Theorem 1.14.

PROOF. Parts (5), (6), (7). To prove part (5) it suffices to consider the case $c=1$ where $a_n \sim b_n$ as $n \to \infty$.[2] For any fixed ϵ, $0 < \epsilon < 1$, and all indices n sufficiently large we then have $(1-\epsilon)b_n \le a_n \le (1+\epsilon)b_n$. Thus, if $\sum_1^\infty b_n$ converges then $\sum_1^\infty a_n$ converges by comparison with the convergent series $\sum_1^\infty (1+\epsilon)b_n$; on the other hand, if $\sum_1^\infty a_n$ converges then $\sum_1^\infty b_n$ converges by comparison with the convergent series $\sum_1^\infty a_n/(1-\epsilon)$. Similar reasoning establishes the result in case of divergence.

To prove part (6) note that for any ϵ, $0 < \epsilon < 1$, and for all indices n sufficiently large we have $0 \le a_n \le \epsilon\, b_n$ so that convergence of $\sum_1^\infty b_n$ implies convergence of $\sum_1^\infty a_n$ by comparison with the convergent series $\sum_1^\infty \epsilon\, b_n$.

Finally, to prove part (7) note that for any constant K, $0 < K < \infty$, and all indices n sufficiently large we have $a_n \ge Kb_n$ so that $\sum_1^\infty a_n$ diverges whenever $\sum_1^\infty b_n$

---

[1] Stronger results exist. See References.

[2] Section 6 treats the $\sim$ notation.

diverges by comparison with the divergent series $\sum_1^\infty Kb_n$. This completes the proof.

<u>EXERCISE 1.55</u> Determine if any of the comparison tests of Theorem 1.14 can be modified so as to apply to series in general, not necessarily non-negative.

<u>EXAMPLE 1.8</u> The series $\sum_1^\infty 1/n^{\log n}$ converges since $0 \le 1/n^{\log n} \le 1/n^2$ for all n sufficiently large, and the series $\sum_1^\infty 1/n^2$ is known convergent. Also the series $\sum_1^\infty (n^2+ 5n + 3)^{-\frac{1}{2}}$ diverges because $(n^2+ 5n + 3)^{-\frac{1}{2}} \sim 1/n$ as $n \to \infty$, and the series $\sum_1^\infty 1/n$ is already known to be divergent. Also, the series $\sum_1^\infty [(n^4+ 4n^3+ 1)/(7n^7+ 5n^4+ 8)]^{2/5}$ converges since its general term has the same order as $K/n^{6/5}$ (K a constant) as $n \to \infty$, and of course the series $\sum_1^\infty 1/n^{6/5}$ converges.

<u>EXERCISE 1.56</u> Test the following series for convergence:

$$\sum_1^\infty n^2/(1+n^3), \quad \sum_1^\infty 1/n!, \quad \sum_1^\infty (n-1)\sin^2 n/n^3 \quad .$$

<u>EXAMPLE 1.9</u> (Geometric Series) The series $\sum_1^\infty ar^{n-1} (a\neq 0)$ is termed the Geometric Series with common ratio r. It converges to the sum $a/(1-r)$ for $|r| < 1$ and diverges for $|r| \ge 1$. To prove this note that the n-th partial sum of the Geometric Series is exactly $s_n= a(1-r^n)/(1-r)$ which for $|r| < 1$ converges to $a/(1-r)$ as $n \to \infty$. For any $|r| \ge 1$ the n-th term of the series fails to tend to zero, whence for these values of r the Geometric Series diverges.

A test of wide applicability for absolute convergence of a series $\sum_1^\infty a_n$ $(a_n \neq 0;\ n=1,2,\ldots)$ is based upon the limiting behavior of the absolute values of the ratios forming the sequence of ratios of successive terms, $\{a_{n+1}/a_n\}$, as $n \to \infty$. It is known as D'Alembert's Ratio Test, and is presented in the following Theorem.

<u>THEOREM 1.15</u> (D'Alembert's Ratio Test) For any series $\sum_1^\infty a_n$ $(a_n \neq 0;\ n=1,2,\ldots)$ [1] define:

$\underline{r} = \liminf |a_{n+1}/a_n|$ and $\overline{r} = \limsup |a_{n+1}/a_n|$. Then

(1) $\overline{r} < 1 \Rightarrow \sum_1^\infty a_n$ absolutely convergent

(2) $\underline{r} > 1 \Rightarrow \sum_1^\infty a_n$ divergent

(3) no conclusion if $\underline{r} \le 1 \le \overline{r}$.

PROOF. First suppose $\overline{r} < 1$ and let $a = (1+\overline{r})/2$. By properties of limit supremum we then have $|a_{n+1}/a_n| \le a < 1$ for all indices $n \ge N$, say. Thus, $|a_{N+p}| \le |a_N| a^p$ $(p=1,2,\ldots)$ so that $\sum_1^\infty |a_n|$ converges by comparison with the Geometric Series.

Next suppose $\underline{r} > 1$ [2] and let $b = (1+\underline{r})/2$. By properties of limit infimum we then have $|a_{n+1}/a_n| \ge b > 1$ for all indices $n \ge N$, say. Thus $|a_{N+q}| \ge |a_N| b^q$ $(q=1,2,\ldots)$ so that $a_n \not\to 0$ as $n \to \infty$ whence the series diverges. This completes the proof.

---

[1] This condition need only hold for all n sufficiently large.

[2] Assume $\underline{r}$ finite. If $\underline{r} = +\infty$ the proof can be modified easily.

It should be pointed out that if the ratios $|a_{n+1}/a_n|$ actually have a limit r as $n \to \infty$ then $\underline{r} = \overline{r} = r$, and D'Alembert's Ratio Test is based upon this limit. However, the test does not require existence of the limit; it is more general.

EXAMPLE 1.10 Consider the series $1 + b + bc + b^2c + b^2c^2 + \ldots + b^nc^{n-1} + b^nc^n + \ldots$ where $0 < |b| < |c|$. Since $|a_{n+1}/a_n|$ ratios alternate between $|b|$ and $|c|$, it follows that $\underline{r} = |b|$ and $\overline{r} = |c|$ so that the series converges absolutely if $|c| < 1$, diverges if $|b| > 1$, with no conclusion if $|b| < 1 < |c|$.

EXAMPLE 1.11 The series $\sum_1^\infty x^n/n!$ converges absolutely for every real x since $|a_{n+1}/a_n| = |x|/(n+1) \to 0$ $(= r)$ as $n \to \infty$. However, the series $\sum_1^\infty x^n/n$ has $|a_{n+1}/a_n| = |x|n/(n+1)$ so it is absolutely convergent for $|x| < 1$, divergent for $|x| > 1$. The series is known convergent for $x=1$; it will later be shown conditionally convergent for $x= -1$.

EXERCISE 1.57 Test for convergence $\sum_1^\infty a_n$ where a_n equals: $c^n/n!$ $(c > 0)$, $n!/n^n$, $(-1)^n/n!$, $(2^n + 1)/(3^n + n)$, n^2/c^n $(c > 0)$, $[5-(-1)^n]/(3 \cdot 2^n)$.

EXERCISE 1.58 (Comparison of Ratios) Let $\sum_1^\infty a_n$ and $\sum_1^\infty b_n$ be two series. Prove that if for all n sufficiently large $a_n \neq 0$, $b_n \neq 0$ and

(1) $|a_{n+1}/a_n| \le |b_{n+1}/b_n|$ then $\sum_1^\infty b_n$ absolutely convergent $\Rightarrow \sum_1^\infty a_n$ absolutely convergent

(2) $|a_{n+1}/a_n| \ge |b_{n+1}/b_n|$ then $\sum_1^\infty b_n$ divergent $\Rightarrow \sum_1^\infty a_n$ divergent.

Another important test for absolute convergence of a series $\sum_1^\infty a_n$ is based upon the behavior of the n-th root of $|a_n|$ as $n \to \infty$. This test, which will be shown later to be related to the previous test, is known as Cauchy's Root Test. It is set forth in the following Theorem.

<u>THEOREM 1.16</u> (Cauchy's Root Test) For any series $\sum_1^\infty a_n$ define $\bar{c} = \limsup \sqrt[n]{|a_n|}$. Then

(1) $\bar{c} < 1 \Rightarrow \sum_1^\infty a_n$ absolutely convergent

(2) $\bar{c} > 1 \Rightarrow \sum_1^\infty a_n$ divergent

(3) no conclusion if $\bar{c} = 1$.

PROOF. Suppose first that $\bar{c} < 1$ and set $a = (1+\bar{c})/2$. By properties of limit supremum of a sequence of numbers we shall have $\sqrt[n]{|a_n|} \le a$ for all indices $n \ge N$, say. Because $a < 1$ and so $|a_n| \le a^n$ for $n \ge N$, we see that the series converges absolutely by comparison with the Geometric Series.

Next suppose $\bar{c} > 1$ and set $b = (1+\bar{c})/2$ so $b > 1$.[1] By properties of limit supremum of a sequence of numbers we know that $\sqrt[n]{|a_n|} \ge b$ for infinitely many (arbitrarily large) values of the index n. Because $b > 1$ it is impossible for a_n to tend to zero as $n \to \infty$ so the series must be divergent. This completes the proof.

[1]Assume $\bar{c}$ finite; the proof can be modified easily to account for the case where $\bar{c}$ is not finite.

Note that if $\sqrt[n]{|a_n|}$ actually has a limit c as $n \to \infty$ then Cauchy's Root Test is based upon this limit since $\bar{c} = c$. Furthermore, it is difficulty to find an infinite series for which this limit does not exist. Nevertheless, the test as stated does not require the limit to exist.

EXAMPLE 1.12 (Comparing the Ratio and Root Tests) Consider again the series of Example 1.10, namely, $1 + b + bc + b^2c + b^2c^2 + \ldots + b^nc^{n-1} + b^nc^n + \ldots$. Results of the Ratio Test were (assuming $0 < |b| < |c|$): absolute convergence if $|c| < 1$, divergence if $|b| > 1$, no conclusion if $|b| < 1 < |c|$. The Root Test does better. Note that $a_{2n} = b^nc^{n-1}$ and $a_{2n-1} = b^{n-1}c^{n-1}$ $(n=1,2,\ldots)$ so that $\sqrt[n]{|a_n|} \to |bc|^{\frac{1}{2}}$ as $n \to \infty$. Thus we have absolute convergence if $|bc| < 1$, and divergence if $|bc| > 1$, since in that case $a_n \not\to 0$.

The preceding Example might suggest that there is some connection between the Ratio and Root tests, and that, in some sense, the Root test is superior. This is rigorously set forth in the following Theorem.

THEOREM 1.17 (Connecting Ratio and Root Tests) For any series $\sum_1^\infty a_n$ ($a_n \neq 0$ for all indices n sufficiently large) let

$$\underline{r} = \liminf |a_{n+1}/a_n|, \quad \bar{r} = \limsup |a_{n+1}/a_n|$$

$$\underline{c} = \liminf \sqrt[n]{|a_n|}, \quad \bar{c} = \limsup \sqrt[n]{|a_n|} .$$

Then these quantities are related as follows:

$$\underline{r} \le \underline{c} \le \bar{c} \le \bar{r} .$$

EXERCISE 1.59 Using References if necessary, prove at least one part of the preceding inequalities.

EXERCISE 1.60 Deduce from the preceding Theorem that if a series can be proved either absolutely convergent or divergent by the Ratio Test, then the same can also be done with the Root Test. Then, show that the converse is false in general by treating the series $\sum_1^\infty a_n$ where $a_n = 2^{-n-(-1)^n}$. Thus the Root Test is superior in general.

EXAMPLE 1.13 The series $\sum_2^\infty 1/(\log n)^n$ is convergent since $\sqrt[n]{|a_n|} = 1/\log n \to 0$ as $n \to \infty$; so also is the series $\sum_1^\infty [n/(1+n^2)]^{n/2}$ since $\sqrt[n]{|a_n|} = \sqrt{n/(1+n^2)} \to 0$.

EXERCISE 1.61 Test for convergence $\sum_1^\infty a_n$ where a_n equals $1/n^n$, $[n/(n+1)]^{n^2}$, $n!/n^n$, $\log n/n^2$, $2^n/n!$, $2^n/1\cdot 3\cdots(2n+1)$, $\log[n/(n+1)]$.

Finally, there is a test for convergence that is sometimes valuable in treating 'boundary' cases (i.e. $r = 1$ or $\bar{c} = 1$) left inconclusive by the Ratio or Root Tests; it is valuable in establishing conditional convergence. The result is contained in the following Theorem.

THEOREM 1.18 (Leibnitz' Test for Alternating Series) If $\sum_1^\infty a_n$ is an alternating-sign series whose terms, in absolute value, form a monotone, null sequence, then the series is convergent (and the remainder is numerically less than the absolute value of the first neglected term.)

PROOF. The series can be written as $\sum_1^\infty (-1)^{n-1}|a_n|$ where $|a_n| \downarrow 0$ as $n \to \infty$. The even partial sums $s_2, s_4, \ldots$ are monotone increasing because $s_{2n+2} - s_{2n} = |a_{2n+1}| - |a_{2n+2}| > 0$. Similarly the odd partial sums are monotone decreasing. Thus the intervals $[s_2, s_1]$, $[s_4, s_3], \ldots, [s_{2n}, s_{2n-1}], \ldots$ form a monotone decreasing sequence of closed sets that determine a unique real number s which is the common limit of the sequence of odd and even partial sums. Thus the series converges to s. Furthermore, the remainder $r_n = |s - s_n| \le |s_{n+1} - s_n| = |a_{n+1}|$, as asserted. This completes the proof.

EXAMPLE 1.14 The Harmonic Series $\sum_1^\infty (-1)^{n-1}/n = 1-1/2+1/3-\ldots$ is (conditionally) convergent by Leibnitz' test; so is the series $\sum_2^\infty (-1)^n/\log n$.

EXAMPLE 1.15 For any series $\sum_1^\infty a_n$ the condition $a = o(1/n)$[1] as $n \to \infty$ (that is, $na_n \to 0$ as $n \to \infty$) is neither necessary nor sufficient for convergence. The series with $a_n = 1/n\log n$ $(n = 2,3,\ldots)$ proves that the condition is not sufficient for convergence; either of the following two series proves that the condition is not necessary for convergence: $a_n = (-1)^n/n$ $(n = 1,2,\ldots)$ or $b_n = (\frac{1}{2})^n$ $(n \ne 2^m;\ m = 0,1,2,\ldots)$, $b_{2^m} = m/2^m$ $(m = 0,1,2,\ldots)$.

EXERCISE 1.62 Prove that the condition $a_n = o(1/n^{1+a})$ $(a > 0)$ is sufficient for convergence of $\sum_1^\infty a_n$. Can o be replaced by 0?

[1]A brief look at Section 6 will introduce the o, 0 notation.

<u>EXERCISE 1.63</u> Construct an example to prove that the condition $a_n = 0(1/n)$ as $n \to \infty$, (that is, na_n bounded as $n \to \infty$), is not necessary even for a non-negative series to converge (it is obviously not sufficient for convergence, why?).

<u>Hints and Answers to Exercises: Section 1.</u>

1.1 Hint: The standard "Cantor Diagonalization Technique" uses the real numbers x within $[0,1]$ in their binary expansion form. Assume, to the contrary, that this set <u>is</u> countable. Then it would be possible to enumerate the set as follows:

$$x_1 = .a_{11}a_{12}a_{13}\cdots\cdots$$

$$x_2 = .a_{21}a_{22}a_{23}\cdots\cdots$$

$$x_3 = .a_{31}a_{32}a_{33}\cdots\cdots$$

$$\vdots$$

where $a_{ij} = 0$ or 1 as the case may be. However, if we construct the real number x^* as follows:

$$x^* = .a_1^*a_2^*a_3^*\cdots\cdots$$

where $a_i^* = 0$ or 1 according as $a_{ii} = 1$ or 0, then the fact that this number within $[0,1]$ differs from every number on the presumably exhaustive list in at least one position establishes a contradiction, thereby proving uncountability of the set of points within $[0,1]$.

1.2 Hint: Prove first that $\limsup A_n \subseteq \bigcap_1^\infty \bigcup_k^\infty A_n$ by showing that if $x \in \limsup A_n$ then $x \in \bigcap_1^\infty \bigcup_k^\infty A_n$. Next show that $\bigcap_1^\infty \bigcup_k^\infty A_n \subseteq \limsup A_n$ by establishing that if $x \in \bigcap_1^\infty \bigcup_k^\infty A_n$ then $x \in \limsup A_n$. These two results combined establish equality of the two sets. Analogous reasoning establishes the result involving $\liminf A_n$.

1.3 Hint: It suffices to establish that $x \in \bigcap A_n \Rightarrow x \in \liminf A_n \Rightarrow x \in \limsup A_n \Rightarrow x \in \bigcup A_n$.

1.4 Hint: Prove that $\bigcap_1^\infty \bigcup_k^\infty A_n = \bigcup_1^\infty \bigcap_k^\infty A_n = \bigcup A_n$ in the increasing case and equals $\bigcap A_n$ in the decreasing case.

1.5 Hint: Make the example interesting.

1.6 Hint: The simplest cases, namely:

(i) $(A \cup B)^c = A^c \cap B^c$

(ii) $(A \cap B)^c = A^c \cup B^c$

(known as DeMorgan's Laws) are easy to prove directly. For the remainder of the problem, it suffices first to establish the following two sets of laws:

(a) Associative Laws:

$A \cup (B \cup C) = (A \cup B) \cup C$

$A \cap (B \cap C) = (A \cap B) \cap C$

(b) Distributive Laws:

$$A \cup (B \cap C) = (A \cup B) \cap (A \cup C)$$

$$A \cap (B \cup C) = (A \cap B) \cup (A \cap C) \quad .$$

Next, use DeMorgan's Laws to prove the validity of the statement of the Exercise for the combination of sets appearing on the left-hand sides in (a) and (b). Finally, note that a proof for more complex expressions is simply based upon the preceding four cases and DeMorgan's Laws.

1.7 Answer: (i) $T = R \cup S$ where $R = B_1 \cap B_2^c \cap B_3 \cap B_4^c \cap \ldots$ and $S = B_1^c \cap B_2 \cap B_3^c \cap B_4 \cap \ldots$.

(ii) Consider the intersection of n sets, where the k-th set in the intersection is either B_k or B_k^c . There are $\binom{n}{i}$ different such intersections in which exactly i complemented sets occur. A_i is the union of these $\binom{n}{i}$ sets.

(iii) A_i contains $\binom{n}{i}$ elements; each B_j contains 2^{n-1} elements

(iv) $S_r = T \cap [\bigcup_{i=0}^{r} A_i]$.

(vi) Suppose the element is $(d_1, d_2, \ldots, d_n)$. The expression for this element in terms of the B_j's is an intersection of n sets, the i-th set in the intersection being B_i or B_i^c according as $d_i = 1$ or 0.

(vii) $S_r^c = T \cup [\bigcap_{i=0}^{r} A_i^c]$.

1.8 Hint: For part (i) establish that, for sufficiently large n, the sequence of sets is monotone. Then use the results of Exercise 1.4. In part (ii), define Z_n as the subset of sequences in V that have <u>0 elements only</u> from location n outward, and define W_n as the subset of sequences in V that have <u>1 elements only</u> from location n outward. Then, $\liminf B_n = \cup W_n$ and $\limsup B_n = \cup Z_n$. The sequence is not convergent.

1.9 Hint:

(i) For $A = \{x: 0 \le x < 1\}$, $a_{\cdot} = 0$ and $a^{\cdot} = 1$. Clearly $a_{\cdot} \in A$ whereas $a^{\cdot} \notin A$.

(ii) For $A = \{x: x = 1/n \ (n = 1,2,\ldots)\}$, $a_{\cdot} = 0$ and $a^{\cdot} = 1$. Only $a_{\cdot}$ belongs to A.

(iii) For $A = \{x: x = 1 - 1/n \ (n = \pm 1, \pm 2, \ldots)\}$ we have $a_{\cdot} = -1$ and $a^{\cdot} = 1$. Neither belongs to A.

1.10 Hint: If $a \le b$ then $|a-b| = b-a$ whereas if $a \ge b$ then $|a-b| = a-b$.

1.11 Hint: Any bounded interval of the form (a,b) will suffice.

1.12 Hint: For the first part, assume to the contrary that $\inf A > \sup A$ and establish a contradiction. For the second part the "if" statement is obvious; the "only if" statement is established by showing that if A contains (at least) two distinct points, say a and b, then: $\sup A - \inf A \ge |a-b| > 0$.

1.14 Hint: For instance, for the first part prove that any constant k satisfying: $k \geq \max(\sup A, \sup B)$ is an upper bound for $A \cup B$ whereas no k satisfying the inequality: $k < \max(\sup A, \sup B)$ can be an upper bound for $A \cup B$.

1.15 Hint: For instance, prove that if c is a limit point of S then given any positive integer n there can always be found an integer $N \geq n$ and corresponding point $x_N \in S$ such that $1/(N+1) \leq |c-x_N| \leq 1/N$. This, in turn, proves that S cannot be a finite set.

1.16 Hint: This would follow upon expanding the method suggested for the preceding Exercise.

1.17 Answer: Respectively, $\bar{A} = \{x: 0 \leq x \leq 1\}$, $\{x: x \geq 2\}$, $\{x: x = 0, x = 1/n \ (n=1,2,\ldots)\}$, {all real numbers between zero and one inclusive}.

1.18 Hint: If, to the contrary, for some point $a \in A$ this were not true, then this in turn would contradict the fact that A^c is closed.

1.19 Hint: For the first part, use the alternate definition of an open set contained in Exercise 1.18 then obtain the result for closed sets by complementation. The second part could be established by showing, for example, that $(0,1] = \bigcup_1^\infty A_n$ is not closed where $A_n = [1/n,1]$ or that the singleton point $\{0\} = \bigcap_1^\infty A_n$ is not an open set where $A_n = (-1/n,1/n)$.

1.20 Hint: For Part (1) it suffices to prove that if x' is a member of $\{x: f(x) \in R_1\}$ and $R_1 \subseteq R_2$, then x' is also a member of $\{x: f(x) \in R_2\}$. The three remaining parts can be established by showing that the set(s) on the right-hand sides are contained in the set(s) on the left-hand sides and vice versa.

1.21 Hint: Part (1) follows upon noting that if $x \in S_1$ then $x \in S_2$, so that $f(x) \in f(S_1)$ implies $f(x) \in f(S_2)$. In part (2) the Universal set is R. Use part (3) with $S_1 = S$ and $S_2 = S^c$. Parts (3) and (4) are straightforward.

1.22 Hint: The fact that $S_1 \cap S_2 = \emptyset$ may hold true while at the same time $f(S_1) \cap f(S_2) \neq \emptyset$ should lead to an example.

1.23 Hint: For the first part it suffices to show that if R is open then for any $x' \in f^{-1}(R)$ there can always be found an $\epsilon > 0$ such that all points x satisfying: $|x-x'| < \epsilon$ also belong to $f^{-1}(R)$. The 'closed' follows by complementation, and the final parts form special cases.

1.24 Hint: Equivalently prove that d is a limit point of both A and A^c iff $d \in \bar{A} \cap \bar{A^c}$.

1.25 Hint: For the first part, use the result of the preceding Exercise.
Answer: Respectively, $b(A) = \{x: x = 0, x = 1/n \ (n=1,2,\ldots)\}$, $\{x: x = 0,1\}$, $\{x: 0 \le x \le 1\}$.

1.26 Hint: Note that a two-part proof is required, both an "if" and "only if" part.

1.27 Hint: The "if" part follows immediately from the Definition. For the "only if" part, a straightforward proof by contradiction will work.

1.28 Hint: For the first sequence show that no real number a' can satisfy the definition of a limit point. For the second sequence show that no other point except zero satisfies the definition of a limit point. For the third sequence note that $2^0 = 1$. Any enumeration of the rational numbers is a sequence having infinitely many limit points, for example.

1.29 Hint: For the first part, use the results of Exercise 1.25 together with Theorem 1.1 (Bolzano-Weierstrass). The last part follows from the assumptions together with the definition of lim sup and lim inf.

1.30 Hint: Consider the cases $c < 0$, $c = 0$, and $c > 0$ separately. See also the Hint to Exercise 1.31.

1.31 Hint: Establish the following equivalent definitions of $\liminf a_n$ and $\limsup a_n$:

$$\liminf a_n = \sup_n \inf_{k \geq n} a_k$$

$$\limsup a_n = \inf_n \sup_{k \geq n} a_k \quad .$$

Then, to prove the first part of the chain of inequalities, all that needs proof is that:

$$\inf_{k\geq n} (a_k + b_k) \geq \inf_{k\geq n} a_k + \inf_{k\geq n} b_k .$$

The remaining parts follow analogously.

1.32 Hint: It suffices to establish that if c is the limit of the (convergent) sequence, then for any $\epsilon > 0$, all but a finite number of terms of the sequence lie outside of the interval from $c - \epsilon$ to $c + \epsilon$.

1.33 Answer: A sequence $a_1, a_2, \ldots$ is said to diverge to $-\infty$ iff given any (arbitrarily large) positive constant C there exists an index N such that $a_n \leq -C$ for all indices $n \geq N$. Clearly the sequence $1,2,3,\ldots$ diverges to $+\infty$ and $-1,-2,-3,\ldots$ diverges to $-\infty$. A more interesting example would be instructive.

1.34 Hint: This follows immediately from the definition of a limit point of a sequence since a convergent sequence possesses a single limit point.
Answer: Clearly $a = 0$ and given any $\epsilon > 0$, $|1/n - 0| < \epsilon$.

1.35 Answer: For the first part, assume $a = 0$. Now given any $\epsilon > 0$ we can always find an index M sufficiently large so that $|a_n| < \epsilon/2$ for all indices $n \geq M$. Next note that for any such index n:

$$|\bar{a}_n - 0| \leq \frac{|a_1 + \ldots + a_M|}{n} + \frac{|a_{M+1} + \ldots + a_n|}{n} \leq \frac{k}{2} + \frac{\epsilon}{2}$$

where $k = |a_1+\ldots+a_M|$. Since k is fixed, we can then choose an index $N \geq M$ sufficiently large so that $k/n < \varepsilon/2$ for all indices $n \geq N$. Accordingly, we have $|\bar{a}_n - 0| < \varepsilon$ whenever $n \geq N$ whence $\bar{a}_n \to 0$ as $n \to \infty$. If $a \neq 0$, apply similar reasoning to the $\bar{a}_n$'s obtained from the sequence whose n-th term is $a_n - a$.

Hint: For the second part, since f is continuous at $x = a$, given any $\varepsilon > 0$ there exists a $\delta > 0$ such that: $|f(a_n) - f(a)| < \varepsilon$ whenever $|a_n - a| < \delta$. Because $a_n \to a$ as $n \to \infty$ we can then choose an index N sufficiently large so that $|a_n - a| < \delta$ whenever $n \geq N$.

Hint: For the third part, apply the results of the two preceding parts and work with the sequence whose n-th term is:

$$\log \sqrt[n]{a_1 a_2 \ldots a_n} = 1/n(\log a_1 + \log a_2 + \ldots + \log a_n).$$

1.36 Hint: (i) If $\{a_n\}$ is bounded then both $\underline{a}$ and $\bar{a}$ are finite via Exercise 1.29 and the fact that $\underline{a} \leq \bar{a}$. On the other hand, if $\underline{a}$ and $\bar{a}$ are both finite, first prove that $\underline{a}$ and $\bar{a}$ are both actually limit points of $\{a_n\}$, and accordingly, <u>any</u> limit point a' of $\{a_n\}$ must satisfy: $\underline{a} \leq a' \leq \bar{a}$. Next, use Exercise 1.26 to prove that, given any $\varepsilon > 0$, no more than a finite number of terms of the sequence can lie outside of the interval from $\underline{a} - \varepsilon$ $\bar{a} + \varepsilon$. Otherwise, there would exist a limit point a'' of $\{a_n\}$ for which either $a'' < \underline{a}$ or $a'' > \bar{a}$, which is impossible. Boundedness of $\{a_n\}$ then follows immediately.

(ii) The first part follows from (i) and the second part follows from Exercise 1.26.

(iii) Same as for (ii).

(iv) This follows upon a simple combination of the results in the first parts of (ii) and (iii).

(v) If $\underline{a}$ and $\bar{a}$ are finite, then they are limit points of $\{a_n\}$. Let $\{d_n\}$ be a sequence of positive real numbers converging to zero. In accordance with Exercise 1.27, for each positive integer n there exists an index i_n such that $|a_{i_n} - \underline{a}| < d_n$ (and the indices may be chosen in an increasing order). Accordingly, the subsequence $\{a_{i_n}\}$ of $\{a_n\}$ converges to $\underline{a}$. Analogously for $\bar{a}$. A slight modification in reasoning is necessary if either $\underline{a}$ or $\bar{a}$ is not finite. The second part follows from the fact that convergence of $\{a_n\}$ is equivalent to the condition: $\underline{a} = \bar{a}$.

(vi) No. For example, the sequence $\{1,2,3,...\}$ has no limit points, whence $\underline{a} = +\infty$.

1.37 Hint: One possible method of proof is to use the alternate definitions of $\liminf a_n$ and $\limsup a_n$ contained in the Hint to Exercise 1.31 and the definition of convergence requiring that $\liminf a_n = \limsup a_n$.

1.38 Hint: For the first part, note that $|ra_n + sb_n - ra - sb| \leq |r||a_n - a| + |s||b_n - b|$. For the second part, the assumed continuity of f guarantees that given any $\epsilon > 0$ there can always be found a $\delta > 0$ such that whenever $\sqrt{(a_n-a)^2 + (b_n-b)^2} < \delta$ then $|f(a_n,b_n) - f(a,b)| < \epsilon$.

Accordingly, choose an index N sufficiently large so that both $(a_n - a)^2 < \frac{1}{2}\delta^2$ and $(b_n - b)^2 < \frac{1}{2}\delta^2$ for all indices $n \geq N$, and combine the two results.

1.39 Hint: If a' occurs infinitely often as a term in the sequence then the result follows immediately; otherwise a' is a limit point of the set consisting of the distinct terms of the sequence. In this case, the Hint for Exercise 1.15 is applicable. The second part can be established using proof by contradiction.

1.40 Hint: Show that the assumptions imply that $A = \bar{A}$.

1.41 Hint: (This is a generalization of Cantor's Nested Set Theorem). Let a_n be a point chosen at will within A_n $(n=1,2,\ldots)$. Although not necessarily convergent, the sequence $\{a_n\}$ is bounded, and so possesses at least one limit point, say a^*, which can be shown to belong to $\bigcap_1^\infty A_n$ because there exists at least one subsequence $\{a_{n_i}\}$ of $\{a_n\}$ converging to a^*.

1.43 Answer: We first prove that the set $f(S)$ is bounded. To do this it is sufficient to prove that any sequence $\{f(x_n)\}$ of values of f in $f(S)$ is bounded.

First set $f_* = \lim\inf f(x_n)$ and $f^* = \lim\sup f(x_n)$. By properties of the lim inf, there exists a subsequence, say $\{f(x_{i_n})\}$ of $\{f(x_n)\}$ converging (or properly diverging, as the case may be) to f_*. Now we prove that f_* is finite. Accordingly, set $x_* = \lim\inf x_{i_n}$ and let $\{x_{i_{j_n}}\}$ be a subsequence of $\{x_{i_n}\}$ converging to x_* (finite).

Since S is a closed set, x_* belongs to S. Furthermore, f is continuous at x_* whence given any $\epsilon > 0$ there can always be found an index N sufficiently large so that $|f(x_{i_{j_n}}) - f(x_*)| < \epsilon$ whenever $n \geq N$. This, of course, implies that the sequence $\{f(x_{i_{j_n}})\}$ converges to $f(x_*)$. But $\{f(x_{i_{j_n}})\}$ is a subsequence of the convergent (to f_*) sequence $\{f(x_{i_n})\}$, which implies that $f_* \equiv f(x_*)$ is finite. A similar argument establishes that f^* is finite, and accordingly the results of Exercise 1.36 guarantee that the sequence $\{f(x_n)\}$ is bounded.

Now we prove that $f(S)$ is closed. Accordingly, let s be a limit point (finite) of the set $f(S)$, and $\{f(x_n)\}$ any sequence of points in $f(S)$ converging to s. According to the results of Exercise 1.40, it is sufficient to prove that s belongs to $f(S)$. Set $\underline{x} = \liminf x_n$ and let $\{x_{i_n}\}$ be a subsequence of $\{x_n\}$ converging to $\underline{x}$, which belongs to S since this set is assumed closed. By continuity of f and boundedness of S it follows that $f(x_{i_n}) \to f(\underline{x})$ as $n \to \infty$ and, since $\{x_{i_n}\}$ is a subsequence of $\{x_n\}$, it follows that $f(x_n) \to f(\underline{x}) \equiv s$ as $n \to \infty$, thereby establishing that s belongs to $f(S)$.

Finally, by properties of a closed, bounded set of real numbers, and the very definition of the set $f(S)$, it follows that there exists at least one pair of points, say x^o and x_o, belonging to S such that:

$$f(x^o) = \max_{x \in S} f(x)$$

and

$$f(x_o) = \min_{x \in S} f(x) \quad .$$

(Could this result be generalized to include functions of n real variables defined over a closed, bounded subset of E_n ?)

1.44 Hint: (i) Note that $f(S_1) \subseteq f(S_2)$, and accordingly the result follows from Exercise 1.13.

(ii) From Example 1.2 it follows that for all $x \in S$

$$\inf_{x \in S} f(x) \leq f(x) \leq \sup_{x \in S} f(x)$$

and

$$\inf_{x \in S} g(x) \leq g(x) \leq \sup_{x \in S} g(x) \quad .$$

Since $g(x) - f(x) \geq 0$ for all $x \in S$, the two results follow upon subtraction, first using the right-hand sides then the left-hand sides of the pair of inequalities.

(iii) For the first inequality, note that for any fixed point $x \in S$ we have $f(x) \leq \sup_{x \in S} f(x)$, and similarly $g(x) \leq \sup_{x \in S} g(x)$. Upon simple addition we have:

$$f(x) + g(x) \leq \sup_{x \in S} f(x) + \sup_{x \in S} g(x) \quad .$$

The function on the left is bounded above by the constant (function) on the right, so the result follows from the first part of (ii). Analogous reasoning applies to the second part. The third and fourth parts follow upon proof and application of the following Lemma: if A is a set of real numbers, and "$-A$" denotes the set of real numbers obtained from A by changing the sign of each number of A, then: $\sup -A = -\inf A$ and $\inf -A = -\sup A$.

(iv) For the first inequality, note that for any fixed point $x \in S$ we have $f(x) \geq \inf_{x \in S} f(x) \geq 0$ and $g(x) \geq \inf_{x \in S} g(x) \geq 0$, accordingly,

$$f(x)\, g(x) \geq \inf_{x \in S} f(x) \cdot \inf_{x \in S} g(x) \geq 0 \quad .$$

Thus the function on the left is bounded below by the constant (function) on the right; accordingly the required result follows upon application of the second part of (ii). Analogous reasoning establishes the second inequality.

(v) If $\inf_{x \in S} f(x) = 0$ the first result follows directly from the identity " $+ \infty = + \infty$ ". Assume, therefore, that $\inf_{x \in S} f(x) = m > 0$. Since for any $x \in S$ we have:

$$1/f(x) \leq 1/\inf_{x \in S} f(x)$$

it follows immediately from the first part of (ii) that:

$$\sup_{x \in S} 1/f(x) \leq 1/\inf_{x \in S} f(x) \quad .$$

Assume to the contrary that equality does <u>not</u> hold; that is:

$$\sup_{x \in S} 1/f(x) < 1/\inf_{x \in S} f(x) = 1/m \quad .$$

Accordingly there can be found an $\epsilon > 0$ sufficiently small such that for all $x \in S$ we have:

$$1/f(x) \le \sup_{x \in S} [1/f(x)] = 1/(m+\epsilon) < 1/\inf_{x \in S} f(x) = 1/m \quad .$$

This, in turn, implies that $f(x) \ge m+\epsilon$ for all $x \in S$, thereby contradicting the fact that $\inf_{x \in S} f(x) = m$. Therefore, equality must hold. The remaining part follows by similar reasoning.

1.46 Hint: (i) For an arbitrary selection of $x \in A$ and $y \in B$ we have:

$$f(x,y) \le \max_{x \in A} f(x,y)$$

whence

$$\min_{y \in B} f(x,y) \le \max_{x \in A} f(x,y) \quad .$$

Because the preceding inequality is true for an arbitrary selection of $x \in A$ and $y \in B$ it follows that in general:

$$\max_{x \in A} \min_{y \in B} f(x,y) \le \min_{y \in B} \max_{x \in A} f(x,y) \quad .$$

(ii) First prove that:

$$\min_{x \in A} \min_{y \in B} f(x,y) \le \min_{y \in B} \min_{x \in A} f(x,y)$$

then

$$\min_{y \in B} \min_{x \in A} f(x,y) \le \min_{x \in A} \min_{y \in B} f(x,y) \quad .$$

(iii) Use a technique analogous to that used in (ii).

1.47 Hint: Rewrite $1/n(n+1)$ as $1/n - 1/(n+1)$ and observe the 'telescoping' that takes place upon addition of the terms in the n-th partial sum s_n .

1.48 Hint: This follows immediately from Theorem 1.2, the Cauchy Criterion for Convergence, applied to the sequence $\{s_n\}$ of partial sums.

1.49 Hint: Apply the Cauchy Criterion with $m = n + 1$.

1.51 Hint: Apply the results of Exercise 1.37 in the special case of convergence, to the appropriate sequences of partial sums.

1.52 Hint: This is a straightforward combination of Theorem 1.8 and Theorem 1.9.

1.53 Answer: Convergent, Properly Divergent, Properly Divergent.

1.54 Hint: Using the assumptions, apply the results of Exercise 1.37 to the appropriate sequences of partial sums along with the basic definition of convergence and proper divergence.

1.56 Answer: Properly Divergent, Convergent, Convergent.

1.57 Answer: Convergent, Convergent, Convergent, Convergent, $c > 1$ - Convergent; $c < 1$ - Properly Divergent, Convergent.

1.59 Hint: See, for example, the book "Infinite Series" by J. M. Hyslop, Oliver and Boyd, New York: Interscience Publishers, Inc., pp. 45-46.

1.60 Answer: If $\bar{r} < 1$ then $\bar{c} < 1$, and if $\underline{r} > 1$ then $\underline{c} > 1$. Note that for the particular case given, $\sqrt[n]{a_n} \to \frac{1}{2}$ as $n \to \infty$, whence $\underline{c} = \bar{c} = \frac{1}{2}$ whereas $\underline{r} = 1/8$ and $\bar{r} = 2$.

1.61 Answer: Convergent, Properly Divergent, Convergent, Convergent, Convergent, Convergent, Properly Divergent.

1.62 Hint: Use the definitions of o and 0 together with parts (5) and (a slight generalization of) (6) and the known convergence of the series $\sum_1^\infty 1/n^{1+a}$ for $a > 0$.

1.63 Answer: For the first part, the series $\Sigma\, a_m$ where:

$$a_m = (\tfrac{1}{2})^m \quad \text{if} \quad m \neq 2^n \quad (n=1,2,\ldots)$$

$$a_m = m(\tfrac{1}{2})^m \quad \text{if} \quad m = 2^n \quad (n=1,2,\ldots)$$

suffices, and for the second part the series $\sum_1^\infty 1/n$ suffices.

References to Additional and Related Material: Section 1

1. Boas, R., "A Primer of Real Functions", Carus Mathematical Monograph 13, Mathematical Association of America (1961).

2. Bromwich, T., "An Introduction to the Theory of Infinite Series", Macmillan and Co. (1926).

3. Ford, W., "Divergent Series", Chelsea Publishing Co., Inc. (1960).

4. Francis, E., "Examples in Infinite Series, with Solutions", Deighton, Bell and Co. (1953).

5. Goldberg, S., "Probablity: an Introduction", Prentice-Hall, Inc. (1964).

6. Green, J., "Sequences and Series", Glencoe, Ill. Free Press (1958).

7. Halberstam, H. and K. Roth, "Sequences", Clarendon Press (1966).

8. Hirschman, I., "Infinite Series", Holt, Reinhart and Winston, Inc. (1962).

9. Hobson, E., "Theory of Functions of a Real Variable", Vol. I, Dover Publications, Inc.

10. Hyslop, J., "Infinite Series", Oliver and Boyd, Ltd. (1959).

11. Jolley, L., "Summation of Series", Second Edition, Dover Publications Inc. (1961).

12. Knopp, K., "Infinite Sequences and Series", Dover Publications, Inc.

13. Lipschutz, S., "Schaum's Outline of Theory and Problems of Set Theory and Related Topics", Schaum Publishing Co. (1964).

14. Rainville, E., "Infinite Series", Macmillan and Co. (1963).

15. Royden, H., "Real Analysis", Macmillan and Co. (1963).

16. Stanaitis, D., "An Introduction to Sequences, Series, and Improper Integrals", Holden-Day, Inc. (1967).

17. Thielman, H., "Theory of Functions of Real Variables", Prentice-Hall, Inc. (1959).

2. Doubly Infinite Sequences and Series

A natural generalization of the notion of an infinite sequence of real numbers is that of a doubly (or, more generally, multiply) infinite sequence of real numbers $\{a_{m,n}\}_{m,n=1}^{\infty}$, briefly $\{a_{m,n}\}$.

The terms of a doubly infinite sequence can be viewed conveniently when arranged in the doubly infinite array:

$$\begin{matrix} a_{11} & a_{12} & a_{13} & \cdots \\ a_{21} & a_{22} & a_{23} & \cdots \\ a_{31} & a_{32} & a_{33} & \cdots \\ \vdots & \vdots & \vdots & \ddots \end{matrix}$$

Figure 2-1

Each term of such a sequence might arise naturally as the value of a function $f(m,n)$ of the two positive integer variables (m,n): $m,n = 1,2,\ldots$, in which case the array would appear as:

$$\begin{matrix} f(1,1) & f(1,2) & f(1,3) & \cdots \\ f(2,1) & f(2,2) & f(2,3) & \cdots \\ f(3,1) & f(3,2) & f(3,3) & \cdots \\ \vdots & \vdots & \vdots & \ddots \end{matrix}$$

Figure 2-2

Each row, column or "Path" through the array (for which at least one of the indices of the terms constituting such a "Path" becomes infinite) amounts to an ordinary sequence of real numbers with which we have already dealt. Thus, for example, we may discuss convergence properties of any such sequence. Now we formulate a definition of convergence for the doubly infinite sequence, and relate it to previous results.

Accordingly, the doubly infinite sequence $\{a_{m,n}\}$ of real numbers is said to converge to the real number a iff for any given $\epsilon > 0$ there can always be found an index P sufficiently large so that:

$$|a_{m,n} - a| < \epsilon \quad \text{for all indices} \quad m \geq P \quad \text{and} \quad n \geq P.$$

In this case we write

$$\lim_{\substack{m\to\infty \\ n\to\infty}} a_{m,n} = a \quad \text{or} \quad \lim a_{m,n} = a$$

<u>EXAMPLE 2.1</u> The sequence $\{1/mn\}$ clearly converges to zero, whereas the sequence $\{m/n\}$ does not converge because for $m = n$ the sequence $\{m/n\}$ converges to 1 whereas for $m = 2n$ the sequence $\{m/n\}$ converges to 2.

<u>EXERCISE 2.1</u> The limit (when existent) of a doubly infinite sequence is clearly unique. Prove that if the sequence $\{a_{m,n}\}$ converges to a, then any ordinary sequence of real numbers obtained by extracting a "Path" from the doubly infinite array (for which <u>both</u> indices of the terms constituting the members of the "Path" tend to infinity) also converges to a.

<u>EXERCISE 2.2</u> Demonstrate by examples that the limit of various ordinary sequences obtained from the doubly infinite array representing $\{a_{m,n}\}$ may exist without $\{a_{m,n}\}$ being convergent.

<u>EXERCISE 2.3</u> Formulate a definition of proper divergence of a doubly infinite sequence $\{a_{m,n}\}$ to either $+\infty$ or $-\infty$. Apply the definitions to the two sequences whose general terms are: $a_{m,n} = m^r n^s$ $(r,s > 0)$ and $a_{m,n} = m \cdot \text{Cos}(2n+1)\pi$.

More generally, a real number a' is termed a limit point of the doubly infinite sequence $\{a_{m,n}\}$ iff given any $\varepsilon > 0$ and any index P' there are always infinitely many terms of the sequence $\{a_{m,n}\}$ with indices $m \geq P'$ and $n \geq P'$ for which $|a_{m,n} - a'| < \varepsilon$. The inf and sup of the set consisting of the limit points of the sequence $\{a_{m,n}\}$ are denoted by:

$$\lim\inf a_{m,n} \quad \text{and} \quad \lim\sup a_{m,n}$$

respectively. These quantities always exist, though are not necessarily finite.

<u>EXERCISE 2.4</u> State and prove the double sequence versions of the results proved for ordinary sequences in Exercise 1.36. See Example 2.3.

<u>EXERCISE 2.5</u> Prove that $\lim a_{m,n} = a$ (finite) if and only if $\lim\inf a_{m,n} = \lim\sup a_{m,n} = a$.

EXERCISE 2.6 Formulate and prove a "Cauchy Criterion" for convergence of a doubly infinite sequence of real numbers.

Of frequent practical occurrence in Applied Mathematics are the so-called iterated limits associated with a doubly infinite sequence of real numbers. These iterated limits do not always exist; however, when they do exist their values, and their relationship to the doubly infinite sequence as a whole, may be of general importance.

Using the function notation for convenience, consider a doubly infinite sequence $\{f(m,n)\}$ of real numbers (not necessarily convergent). For each (any) fixed value m of the first index we may consider the ordinary sequence of real numbers $\{f(m,n)\}_{n=1}^{\infty}$ whose elements consist of the members of the m-th row of the doubly infinite array associated with the sequence. Naturally, such a sequence need not converge for any value of m,[1] e.g. $f(m,n) = (-1)^m (-1)^n$.

Suppose, however, for the doubly infinite sequence $\{f(m,n)\}$ under consideration, each such sequence does in fact converge. Then we may write, say,

$$\lim_{n\to\infty} f(m,n) = h(m) \quad \text{for} \quad m = 1,2,\ldots \quad .$$

Accordingly, another ordinary sequence $\{h(m)\}$ of real numbers is determined which also, in turn, may or may not be convergent.[1]

[1] It is clear that in a general treatment, we must consider the lim inf and lim sup of the appropriate sequences as n and/or m tend to infinity. We restrict ourselves here to the simpler situation. Further results may be found in the References.

EXAMPLE 2.2 Consider the doubly infinite sequence $\{f(m,n)\}$ with $f(m,n) = (-1)^m n/(n+1)$. Clearly $h(m)$ exists for each value of m and equals $(-1)^m$. Obviously the sequence $\{h(m)\}$ does not converge. Can any conclusion be drawn concerning the possible existence of $\lim f(m,n)$ in this case?

Suppose, however, that the sequence $\{h(m)\}$ is convergent, with limit f_1. In this case we write:

$$\lim_{m\to\infty} \lim_{n\to\infty} f(m,n) = f_1$$

and call f_1 the iterated limit of the sequence $\{f(m,n)\}$ (n followed by m).

By reversing the roles of m and n in the limiting processes (and assuming that the limits involved all exist), we may also obtain the iterated limit f_2 of the sequence $\{f(m,n)\}$ (m followed by n) written as follows:

$$\lim_{n\to\infty} \lim_{m\to\infty} f(m,n) = f_2 \quad .$$

EXERCISE 2.7 Establish the following results:

(a) Show that both iterated limits may exist yet be unequal by considering $f(m,n) = m/(m+n)$.

(b) Show that both iterated limits may exist and be equal without the doubly infinite sequence being convergent by considering $f(m,n) = 2mn/(m^2+ n^2)$.

For a given doubly infinite sequence $\{f(m,n)\}$, it may already be apparent that computation of iterated limits may be 'easier' than attempting to evaluate (assuming it exists) $\lim f(m,n)$ directly from the definition. Accordingly, the following results do establish a connection between the two types of limits under certain conditions (but not in general; review Exercise 2.7(b)).

<u>EXERCISE 2.8</u> Prove that if the double limit:

$$f^* = \lim_{\substack{m\to\infty \\ n\to\infty}} f(m,n)$$

of the doubly infinite sequence $\{f(m,n)\}$ exists, <u>and</u> for each (any) fixed value m of the first index, the limit:

$$\lim_{n\to\infty} f(m,n) = h(m) \quad \text{exists for} \quad m = 1,2,\ldots,$$

then the iterated limit (n followed by m) also exists and equals f^*, that is,

$$\lim_{m\to\infty} \lim_{n\to\infty} f(m,n) = f^* .$$

Provide an example to prove that the second part of the hypothesis cannot be dropped in general.

Our next result is, in a sense, a partial converse to the preceding Exercise. It deals with a single iterated limit, and its relationship to the double limit. It is contained in the following Theorem.

<u>THEOREM 2.1</u> If, for a given doubly infinite sequence $\{f(m,n)\}$,

(i) $\lim_{n\to\infty} f(m,n) = h(m)$ exists, uniformly in n for each (every) $m = 1,2,\ldots,$ [1]

and

(ii) $\lim_{m\to\infty} h(m) = f^*$ exists,

then

$$\lim_{\substack{m\to\infty \\ n\to\infty}} f(m,n) = f^* \quad .$$

PROOF. Given $\epsilon > 0$ there exists an index N sufficiently large so that

$$|f(m,n) - h(m)| < \epsilon/2$$

for all $n \geq N$ and for $m = 1,2,\ldots$. Next, from the assumed convergence of $\{h(m)\}$ to f^*, it follows that there exists an index M sufficiently large so that

$$|h(m) - f^*| < \epsilon/2$$

for all $m \geq M$. Accordingly,

$$|f(m,n)-f^*| \leq |f(m,n)-h(m)| + |h(m)-f^*| < \epsilon/2+\epsilon/2 = \epsilon$$

for all indices $m \geq P'$ and $n \geq P'$ with P' chosen as $P' = \max(M,N)$. This establishes the required result.

The final result of this type deals with sufficient conditions under which both iterated limits exist, are equal, and

[1]Given any $\epsilon > 0$ there can always be found an index N such that $|f(m,n)-h(m)| < \epsilon$ for all $n \geq N$ and all $m = 1,2,\ldots$.

imply that the double limit exists and equals their common value.

<u>THEOREM 2.2</u> If, for a doubly infinite sequence $\{f(m,n)\}$:

(i) $\lim_{n\to\infty} f(m,n) = h(m)$ exists, uniformly in n, for each (every) $m = 1,2,\ldots,$

and

(ii) $\lim_{m\to\infty} f(m,n) = g(n)$ exists for each $n = 1,2,\ldots,$

then both iterated and double limits exist and are equal:

$$\lim_{\substack{m\to\infty \\ n\to\infty}} f(m,n) = \lim_{m\to\infty}\lim_{n\to\infty} f(m,n) = \lim_{n\to\infty}\lim_{m\to\infty} f(m,n).$$

PROOF. Given $\epsilon > 0$ there exists an index N sufficiently large so that:

$$|f(m,n) - h(m)| < \epsilon/6$$

for all $n \geq N$ and all $m = 1,2,\ldots$. Consider now a <u>fixed</u> index $n^* \geq N$. Accordingly, then, there can be found an index M sufficiently large so that

$$|f(m,n^*) - g(n^*)| < \epsilon/6$$

for all $m \geq M$. Finally, let (m',n') and (m'',n'') be any pair of indices for which $m',m'' \geq M$ and $n',n'' \geq N$.

Then:

$$
\begin{aligned}
|f(m',n') - f(m'',n'')| \le\ & |f(m',n') - h(m')| + \\
& |h(m') - f(m',n^*)| + \\
& |f(m',n^*) - g(n^*)| + \\
& |g(n^*) - f(m'',n^*)| + \\
& |f(m'',n^*) - h(m'')| + \\
& |h(m'') - f(m'',n'')| < \\
& \epsilon/6 + \epsilon/6 + \epsilon/6 + \epsilon/6 + \epsilon/6 + \epsilon/6 = \epsilon .
\end{aligned}
$$

Therefore $\{f(m,n)\}$ satisfies the obvious two-dimensional generalization of the Cauchy Criterion (Exercise 2.6) and is accordingly convergent. The remaining parts follow directly from Exercise 2.8. (Note: the common value of the respective limits may be obtained by considering $\lim_{m\to\infty} h(m)$ or $\lim_{n\to\infty} g(n)$.)

A logical next step is to consider the generalization of an ordinary (singly) infinite series, namely, a doubly (or, more generally, multiply) infinite series. Since such series frequently occur in Applied Mathematics, it is necessary to consider the Mathematical problems related to them.

By a doubly infinite series:

$$\sum_{m,n=1}^{\infty} a_{m,n}$$

of real numbers is meant, in effect, the 'sum' of the real

numbers in the doubly infinite array:

$$\begin{array}{cccccccc}
a_{11} & + & a_{12} & + & a_{13} & + & \dots \\
+ & & + & & + & & \\
a_{21} & + & a_{22} & + & a_{23} & + & \dots \\
+ & & + & & + & & \\
a_{31} & + & a_{32} & + & a_{33} & + & \dots \\
+ & & + & & + & + & \\
\vdots & & \vdots & & \vdots & & \ddots
\end{array}$$

Figure 2-3

Provided a 'sum' can be defined and exists, there appear to be many different methods for obtaining a 'sum'. For example,

(i) by rows: $\sum_{m=1}^{\infty} [\sum_{n=1}^{\infty} a_{m,n}]$

(ii) by columns: $\sum_{n=1}^{\infty} [\sum_{m=1}^{\infty} a_{m,n}]$

(iii) by diagonals: $\sum_{t=2}^{\infty} [\sum_{m+n=t} a_{m,n}]$.

These are known as iterated sums.

There are many other different ways. However, even if it were possible to obtain a 'sum' by two or more of the above (or other) methods, there is no a-priori guarantee that their values would (or should) be equal.

Accordingly, to resolve the above difficulties, we are forced to approach the problem of summing doubly infinite series by resorting to the basic notions of partial sums and doubly infinite sequences of partial sums. Therefore, to each doubly infinite series:

$$\sum_{m,n=1}^{\infty} a_{m,n} \quad \text{briefly} \quad \Sigma\, a_{m,n}$$

we associate a corresponding doubly infinite sequence $\{s_{m,n}\}_{m,n=1}^{\infty}$ of partial sums defined as follows:

$$s_{m,n} = \sum_{i=1}^{m} \sum_{j=1}^{n} a_{ij} \quad .$$

As before, this doubly infinite sequence $\{s_{m,n}\}$ can be viewed conveniently when arranged in the doubly infinite array:

$$\begin{matrix} s_{11} & s_{12} & s_{13} & \cdots \\ s_{21} & s_{22} & s_{23} & \cdots \\ s_{31} & s_{32} & s_{33} & \cdots \\ \vdots & \vdots & \vdots & \ddots \end{matrix}$$

Figure 2-4

Convergence of the doubly infinite series will then be based upon the convergence properties of this corresponding doubly infinite sequence of partial sums.

Thus, the doubly infinite series $\Sigma\, a_{m,n}$ is said to converge to (or have sum) s according as the corresponding doubly infinite sequence $\{s_{m,n}\}$ of partial sums converges <u>as a sequence</u> to the (finite) real number s. Proper divergence of a doubly infinite series is then defined by applying the results of Exercise 2.3. Convergent doubly infinite series (and sequences, too) obey the same algebraic properties as their singly infinite counterparts (see, e.g. Theorems 1.5 and 1.6 for series, and Exercises 1.30, 1.31 and 1.38 for sequences).

EXAMPLE 2.3 (Bounded and Totally Bounded Sequences) In general, the doubly infinite sequence $\{a_{m,n}\}$ is termed bounded if there exists some index P and a positive constant B such that $|a_{m,n}| \le B$ for all indices $m \ge P$ and $n \ge P$. Accordingly, the following sequence would be considered bounded by 1:

$$\begin{matrix} 1 & 2 & 3 & 4 & 5 & \cdots\cdots \\ 2 & 1 & 1 & 1 & 1 & \cdots\cdots \\ 3 & 1 & 1 & 1 & 1 & \cdots\cdots \\ 4 & 1 & 1 & 1 & 1 & \cdots\cdots \\ 5 & 1 & 1 & 1 & 1 & \cdots\cdots \\ \vdots & \vdots & \vdots & \vdots & \vdots & \ddots \end{matrix}$$

even though the terms $a_{1,n} = n$ and $a_{m,1} = m$ are unbounded. The sequence converges to 1 as well; convergence properties of a doubly infinite sequence are determined by behavior of terms with indices m,n sufficiently large. Total boundedness would require the existence of a constant B' such that $|a_{m,n}| \le B'$ for all indices m and n. Note, however, that when the terms of the doubly infinite sequence are partial sums $\{s_{m,n}\}$ corresponding to a doubly infinite series, then total boundedness of of this sequence is a necessary condition (prerequisite) for convergence of the doubly infinite series.

The first result concerning doubly infinite series essentially states that if a doubly infinite series is convergent, then it makes no difference how the terms are grouped when performing the process of summation. This important result is contained in the following Theorem.

THEOREM 2.3 If $\Sigma\, a_{m,n} = s$ (finite), then regardless of the manner in which the terms of this series are grouped during the process of summation, any such process will always yield a sum of s.

PROOF. EXERCISE 2.9.

Accordingly, if the doubly infinite series $\Sigma\, a_{m,n} = s$ converges, then both 'iterated sums'

$$\sum_{m=1}^{\infty} \sum_{n=1}^{\infty} a_{m,n} = s \quad \text{and} \quad \sum_{n=1}^{\infty} \sum_{m=1}^{\infty} a_{m,n} = s$$

and the diagonal sum

$$\sum_{t=2}^{\infty} \sum_{m+n=t} a_{m,n} = s$$

yield this same value.

Accordingly, if a doubly infinite series is known to be convergent, any method of summation will yield the same 'sum'. Tests for convergence, therefore, will provide the means for making use of Theorem 2.3.

Before proceeding with some tests for convergence of doubly infinite series, however, we obtain the following special case of Theorem 2.3 that applies to series of non-negative terms.

THEOREM 2.4 (Series with Non-Negative Terms Only)
If for $m,n = 1,2,\ldots$ $a_{m,n} \geq 0$, then the three series:[1]

$$\sum_{m,n=1}^{\infty} a_{m,n}\ , \quad \sum_{m=1}^{\infty}\sum_{n=1}^{\infty} a_{m,n}\ , \quad \sum_{n=1}^{\infty}\sum_{m=1}^{\infty} a_{m,n}$$

either all converge to the same finite sum s, or else all properly diverge to $+\infty$.

PROOF. Consider first the series $\Sigma\, a_{m,n}$. Its corresponding doubly infinite sequence $\{s_{m,n}\}$ of partial sums has the property that if $m' \geq m$ and $n' \geq n$ then $s_{m',n'} \geq s_{m,n}$. Accordingly, the sequence of partial sums converges to a finite real number, say s, if it is bounded above (that is, if there exists a positive constant B such $s_{m,n} \leq B$ for all indices m and n). Otherwise, the sequence of partial sums and the series diverge to $+\infty$. Equality in value of the three series then follows from Theorem 2.3.

One practical consequence of Theorem 2.4 is demonstrated by the following Example.

[1] In view of Theorem 2.3, it is clear that any other summation method could be included in the statement.

EXAMPLE 2.4 Consider the doubly infinite series $\Sigma\,(\tfrac{1}{2})^{(m+n)}$. Using the 'diagonal' method of summation, observe:

$$\sum_{t=2}^{\infty} \sum_{m+n=t} (\tfrac{1}{2})^{(m+n)} = \sum_{t=2}^{\infty} (t-1)(\tfrac{1}{2})^{t} = 1 \; .$$

Accordingly, the doubly infinite series is convergent with sum 1, and any other method of summation would have resulted in the same sum. ('Rows' or 'columns' could have been used as well.) By the way, the latter (singly) infinite series was evaluated using the well-known result that for $|r| < 1$,

$$\sum_{n=1}^{\infty} nr^{n-1} = 1/(1-r)^2 \quad ,$$

which is obtained in Section 5 (Power Series). (In addition, it is to be noted that tabulations of 'standard' convergent series are to be found in many of the References.)

EXAMPLE 2.5 Had we been given the series $\Sigma\, 1/(m+n-1)^3$ (which is known convergent), summation by 'rows' or 'columns' would have been impossible. However,

$$\sum_{t=2}^{\infty} \sum_{m+n=t} 1/(m+n-1)^3 = \sum_{s=1}^{\infty} 1/s^2 = \pi^2/6 \; .$$

Accordingly, Theorem 2.4 guarantees that this is the sum of the given doubly infinite series. This demonstrates the value of the "choice of method" of summation that Theorem 2.4 establishes.

<u>EXAMPLE 2.6</u> Theorem 2.4 may also be used as a test for convergence of a doubly infinite series, even though a 'closed form' need not exist for the sum itself. Consider the series $\Sigma\, 1/(m^t + n^t)$. We choose to examine the 'diagonal form' of the series, namely,

$$\sum_{p=2}^{\infty} \sum_{m+n=p} 1/(m^t + n^t) \; .$$

It is easy to show that the series properly diverges for $t \le 1$. Accordingly, assume that $t > 1$ and note that for $m + n = p$ we have:

$$2p^t > m^t + n^t > 2(\tfrac{1}{2}p)^t \; .$$

and so

$$\frac{p-1}{2p^t} < \sum_{m+n=p} \frac{1}{m^t+n^t} < \frac{p-1}{2(\frac{1}{2}p)^t} \; .$$

Therefore the original doubly infinite series and the (singly) infinite series $\sum_{1}^{\infty} 1/p^{t-1}$ converge or properly diverge together. It is known that the latter series is convergent for $t > 2$ and properly divergent for $t \le 2$. With this the Example is completed.

It is apparent, then, that we need consider some tests for convergence at this point. Before doing so, it should be pointed out that tests for convergence for series with non-negative terms also apply to series in general (as in the case of singly-infinite series) by virtue of the following result.

<u>THEOREM 2.5</u> (Absolute Convergence Implies Convergence) If the doubly infinite series $\Sigma\ |a_{m,n}|$ converges, then $\Sigma\ a_{m,n}$ converges.

PROOF. <u>EXERCISE 2.10</u>.

Accordingly, a series such as $\Sigma\ (-1)^{m+n}/(m^t+n^t)$ is convergent, in view of the results in Example 2.6, at <u>least</u> for $t > 2$.

The first test, in effect, enables one to eliminate some series quite easily from being convergent.

<u>THEOREM 2.6</u> (Necessary Condition for Convergence) If either

$$\lim_{n\to\infty} a_{m,n} \neq 0 \quad \text{or} \quad \lim_{m\to\infty} a_{m,n} \neq 0$$

then the doubly infinite series $\Sigma\ a_{m,n}$ diverges.

PROOF. <u>EXERCISE 2.11</u>.

Accordingly, series such as $\Sigma\ m/(m+n^5)$ or $\Sigma\ (m^r+ n^s)/m^r n^s$, $r,s > 0$, will fail to converge.

<u>EXERCISE 2.12</u> Theorem 2.6 essentially states that in the doubly infinite sequence $\{a_{m,n}\}$ associated with the series, each row and column sequence must have limit 0 before $\Sigma\ a_{m,n}$ can converge. How might this be generalized?

<u>THEOREM 2.7</u> (A Comparison Test for Convergence) If the doubly infinite series $\Sigma\, a_{m,n}$ of non-negative terms is convergent, and for all indices m and n sufficiently large, $0 \le b_{m,n} \le a_{m,n}$, then the doubly infinite series $\Sigma\, b_{m,n}$ converges.

PROOF. <u>EXERCISE 2.13</u>.

<u>EXERCISE 2.14</u> (Extension) It will be noted that Theorem 2.7 is closely related to the results of Theorem 1.14 for (singly) infinite series. Generalize other results so as to provide tests for doubly infinite series.

<u>EXERCISE 2.15</u> Apply a suitable comparison test to examine convergence properties of $\Sigma\, m^r n^s/(m^t + n^t)$.

Another test for convergence is a straightforward generalization of the Integral Test for (singly) infinite series. This result follows.

<u>THEOREM 2.8</u> (An Integral Test for Convergence) Let $\Sigma\, a_{m,n}$ be a doubly infinite series of non-negative terms, and $f(x,y)$ a function of the two real variables (x,y) that is non-negative and satisfies: [1]

(i) $f(x,y)$ is continuous for all $x,y > 0$

(ii) $f(x,y)$ is non-increasing in x and y

(iii) $f(m,n) = a_{m,n}$ for $m,n = 1,2,\ldots$.

[1]The conditions need hold only for all values of the variables sufficiently large.

Then, the series $\Sigma a_{m,n}$ and the improper Riemann Integral $\int_0^\infty \int_0^\infty f(x,y)dydx$ both converge or properly diverge together.

EXERCISE 2.16 Prove Theorem 2.8.

EXERCISE 2.17 Apply the Integral Test to determine convergence properties of the series $\Sigma 1/m^r n^s$ and $\Sigma p^m q^n$, $p,q > 0$.

Other tests for convergence may be found in the References.

EXERCISE 2.18 Using whatever method is appropriate, examine the following series for convergence:

(a) $\sum_{m,n=1}^{\infty} 1/(m^r + n^s)$

(b) $\sum_{m,n=2}^{\infty} 1/(am^t + bn^t)(\log mn)^s$

(c) $\sum_{m,n=2}^{\infty} 1/(am^r + bn^s)(\log mn)^t$

We conclude this Section with some results closely related to convergence properties of doubly infinite series. It has to do with reversal (or interchange) of the order of summation in a doubly infinite series, and we have touched upon this topic earlier.

It often occurs in Applied Mathematics that one is required to evaluate an iterated sum, say $\sum_{m=1}^{\infty} \sum_{n=1}^{\infty} a_{m,n}$, which proves too difficult to accomplish, yet it is observed that the iterated sum with m and n summation interchanged, viz. $\sum_{n=1}^{\infty} \sum_{m=1}^{\infty} a_{m,n}$, is relatively 'easy' to evaluate. The question arises as to the conditions under which we may be assured the two are indeed equal. The results are contained in what follows. Further information is found in the References.

<u>THEOREM 2.9</u> If any one of the three series:

$$\sum_{m,n=1}^{\infty} a_{m,n}, \quad \sum_{m=1}^{\infty} \sum_{n=1}^{\infty} a_{m,n}, \quad \sum_{n=1}^{\infty} \sum_{m=1}^{\infty} a_{m,n}$$

is absolutely convergent, then all three series are absolutely convergent (hence convergent), and the sums are equal.

We shall not prove this strong result here (see Hyslop, pp. 114-5). Instead, we shall prove a weaker version that suffices for many practical purposes. It is contained in the following result.

<u>THEOREM 2.10</u> If $\Sigma\, a_{m,n}$ is a doubly infinite series for which one of the iterated sums is absolutely convergent, then $\sum_{m=1}^{\infty} \sum_{n=1}^{\infty} a_{m,n} = \sum_{n=1}^{\infty} \sum_{m=1}^{\infty} a_{m,n}$ (i.e. the order of summation may be interchanged without effecting the value).

PROOF. By Theorem 2.4, the hypotheses imply that both iterated sums, and the doubly infinite series, are absolutely convergent to the same value. Now we apply Theorem 2.2. Accordingly, we have the following quantities, all of which exist (finite) since absolute convergence of $\Sigma\, a_{m,n}$

guarantees its convergence:

$$f(m,n) = s_{m,n} = \sum_{i=1}^{m} \sum_{j=1}^{n} a_{i,j} \qquad m,n = 1,2,...$$

$$h(m) = \sum_{i=1}^{m} \sum_{j=1}^{\infty} a_{i,j} \qquad m = 1,2,...$$

$$g(n) = \sum_{i=1}^{\infty} \sum_{j=1}^{n} a_{i,j} \qquad n = 1,2,...$$

Clearly convergence guarantees that:

$$\lim_{m\to\infty} f(m,n) = g(n) \quad \text{for} \quad n = 1,2,...$$

and

$$\lim_{n\to\infty} f(m,n) = h(m) \quad \text{for} \quad m = 1,2,... \quad .$$

In view of the results of Theorem 2.2, the present Theorem will be established if we can prove that the latter of the two preceding limits is uniform in n.

To do this, note first that:

$$|f(m,n) - h(m)| = \left| \sum_{i=1}^{m} \sum_{j=n+1}^{\infty} a_{i,j} \right|$$

$$\leq \sum_{i=1}^{m} \sum_{j=n+1}^{\infty} |a_{i,j}|$$

$$\leq \sum_{i=1}^{\infty} \sum_{j=n+1}^{\infty} |a_{i,j}| = r(n+1) \quad .$$

It is clear that $0 \le r(n+2) \le r(n+1)$ for $n = 1,2,\ldots,$ and accordingly the following limit exists:

$$\lim_{n\to\infty} r(n) = r \ge 0 \quad .$$

Assuming $r > 0$ leads to a contradiction, so that uniformity of the required limit follows directly. This completes the proof.

<u>EXERCISE 2.19</u> Verify that, in the preceding proof, the assumption that $r > 0$ leads to a contradiction.

<u>EXAMPLE 2.7</u> In Theorem 2.10, the requirement of absolute convergence of one of the iterated sums cannot be dropped. The doubly infinite series $\Sigma\, a_{m,n}$ whose terms appear in the following array demonstrates this point.

$$\begin{array}{cccccc}
2 & -2 & 0 & 0 & 0 & \cdots\cdots \\
0 & 2 & -2 & 0 & 0 & \cdots\cdots \\
0 & 0 & 2 & -2 & 0 & \cdots\cdots \\
0 & 0 & 0 & 2 & -2 & \cdots\cdots \\
\vdots & \vdots & \vdots & \vdots & \vdots & \ddots
\end{array}$$

Clearly

$$\sum_{n=1}^{\infty} \sum_{m=1}^{\infty} a_{m,n} = 0 \qquad \text{(by rows)}$$

whereas

$$\sum_{m=1}^{\infty} \sum_{n=1}^{\infty} a_{m,n} = 2 \qquad \text{(by columns)} \quad .$$

The iterated sums were not <u>absolutely</u> convergent.

EXAMPLE 2.8 The following illustration demonstrates the value of being able to interchange order of summation (validly). Suppose we are required to evaluate the (iterated) sum:

$$\sum_{m=0}^{\infty} \sum_{n=1}^{\infty} \frac{me^{-1/n^2}(1/n^2)^m}{m!} .$$

This manner of summation presents us with an extremely difficult task. However, interchanging the order of summation, observe:

$$\sum_{n=1}^{\infty} \sum_{m=0}^{\infty} \frac{me^{-1/n^2}(1/n^2)^m}{m!} = \sum_{n=1}^{\infty} 1/n^2 = \pi^2/6 .$$

There is no question about the absolute convergence of the latter iterated sum. Hence, both are equal. In fact, Theorem 2.4 would yield a stronger conclusion, since the series is non-negative.

EXERCISE 2.20 Prove that if the (singly) infinite series Σa_m is absolutely convergent, and the non-negative (singly) infinite series Σb_n is convergent, then the conditions of Theorem 2.10 are satisfied for the doubly infinite series $\Sigma a_m b_n$, so that the two iterated sums would be equal.

In practice, the stronger result, stated in the paragraphs preceding Theorem 2.10, is most often applied. Here, however, we have restricted ourselves to use only of the result proved in Theorem 2.10.

Hints and Answers to Exercises: Section 2

2.1 Hint: Any "Path" through the array defining $\{a_{m,n}\}$ is determined by a sequence (m_k, n_k); $k = 1,2,\ldots$ of distinct pairs of indices. By the assumption of convergence of the doubly infinite sequence to s, given any $\epsilon > 0$ there can always be found an index P sufficiently large so that $|a_{m,n} - a| < \epsilon$ for all indices $m,n \geq P$. Since, by hypothesis, both $m_k \to \infty$ and $n_k \to \infty$ as $k \to \infty$, there can always be found a K sufficiently large so that $m_k, n_k \geq P$ for all $k \geq K$.

2.2 Hint: Consider, for instance, the doubly infinite sequence $\{(-1)^{m+2n}\}$ with m fixed and $n = 1,2,\ldots$ or the doubly infinite sequence $\{(-1)^{m+n}\}$ with $m \to \infty$ and $n \to \infty$ in such a manner that $m + n$ remains even (or odd). It should be clear that neither doubly infinite sequence is convergent. Construct further examples.

2.3 Hint: Consider, for instance, proper divergence to $+\infty$. The doubly infinite sequence $\{a_{m,n}\}$ is said to properly diverge to $+\infty$ iff given any (arbitrarily large) fixed positive constant C there can always be found an index P sufficiently large so that $a_{m,n} \geq C$ for all indices $m,n \geq P$. For the case where $a_{m,n} = m^r n^s$ $(r,s > 0)$ we can choose P as the first integer exceeding $C^{1/(r+s)}$. Note, as with the case of an ordinary (singly) infinite sequence, a doubly infinite sequence need be neither convergent nor properly divergent. This is illustrated by the two doubly infinite sequences given in the Hint for Exercise 2.2.

2.4 Hint: Begin by defining a doubly infinite sequence $\{a_{m,n}\}$ to be <u>bounded</u> iff there exists some positive constant B and some index P such that $|a_{m,n}| \le B$ for all indices $m,n \ge P$. This definition may, at first, seem rather odd since according to it, the doubly infinite sequence defined by the array:

$$\begin{matrix} 1 & 2 & 3 & 4 & 5 & \cdots \\ 2 & 1 & 1 & 1 & 1 & \cdots \\ 3 & 1 & 1 & 1 & 1 & \cdots \\ 4 & 1 & 1 & 1 & 1 & \cdots \\ 5 & 1 & 1 & 1 & 1 & \cdots \\ \vdots & \vdots & \vdots & \vdots & \vdots & \ddots \end{matrix}$$

would be considered bounded (by 1) even though the terms $a_{1,n} = n$ and $a_{m,1} = m$ are unbounded as m and n tend to $+\infty$. However, it is to be noted that convergence of a doubly infinite sequence is determined by the behavior of terms $a_{m,n}$ only for all m and n sufficiently large, and accordingly convergence is uneffected by terms $a_{m,n}$ where $m \le P$ or $n \le P$ for any fixed index P. The sequence defined above, for example, converges to 1.

The first part of the analogue to 1.36(i) is easy to establish. Concerning the second part, prove that if both $\underline{a} = \liminf a_{m,n}$ and $\overline{a} = \limsup a_{m,n}$ are finite, then given any $\epsilon > 0$ there can always be found an index P sufficiently large so that $\underline{a} - \epsilon < a_{m,n} < \overline{a} + \epsilon$ for all indices $m,n \ge P$. The doubly infinite analogues of the remaining parts of Exercise 1.36 follow in a straightforward manner.

2.5 Hint: Apply the basic definitions of $\liminf a_{m,n}$ and $\limsup a_{m,n}$.

2.6 Hint: A doubly infinite sequence $\{a_{m,n}\}$ is termed a Cauchy sequence if, given any $\epsilon > 0$, there can always be found a pair of indices M and N (possibly depending upon ϵ) such that $|a_{m,n} - a_{r,s}| < \epsilon$ for all indices $m,r \geq M$ and $n,s \geq N$.

2.7 Hint: (a) For each fixed value of m, $\lim_{n\to\infty} f(m,n) = 1$, whereas for each fixed value of n we have $\lim_{m\to\infty} f(m,n) = 0$. (b) Clearly $\lim_{n\to\infty} \lim_{m\to\infty} f(m,n) = \lim_{m\to\infty} \lim_{n\to\infty} f(m,n) = 0$, yet the doubly infinite sequence $\{f(m,n)\}$ cannot be convergent since, for example, given any index P, $f(P,P) = 1$ whereas for, say, $m = 2P$ and $n = 3P$, we have $f(2P, 3P) = 1/3$. Accordingly, no limit can exist.

2.8 Hint: If $f^* = \lim f(m,n)$ exists, then given any $\epsilon > 0$ there can always be found an index P such that $|f(m,n)-f^*| < \epsilon$ for all indices $m,n \geq P$. Upon taking limits of the preceding expression as $n \to \infty$ (which are assumed to exist for $m = 1,2,\ldots$) we obtain that $|h(m)-f^*| \leq \epsilon$ for all $m \geq P$. This, in turn, implies that the sequence $\{h(m)\}$ converges to f^* as $m \to \infty$, which is equivalent to the statement:

$$\lim_{m\to\infty} \lim_{n\to\infty} f(m,n) = f^* ,$$

which was to be established.

As an illustration of an example required in the second part, we could take the doubly infinite sequence $\{f(m,n)\}$ where $f(m,n) = \text{Cos } n\pi/m$. Although it is easy to prove that $\lim f(m,n) = 0$ exists, the quantities $\lim_{n\to\infty} f(m,n) = h(m)$ do not exist for any $m = 1,2,\ldots$ and the iterated limit $\lim_{m\to\infty} \lim_{n\to\infty} f(m,n)$ does not exist. (Note, however, with the roles of m and n reversed, the hypotheses of the Exercise are satisfied, and accordingly the iterated limit $\lim_{n\to\infty} \lim_{m\to\infty} f(m,n) = 0$ exists.)

2.9 Hint: Consider first the special case where $a_{m,n} \geq 0$ for $m,n = 1,2,\ldots$ and let $G_1, G_2, \ldots$ be any (finite or infinite) collection of mutually exclusive subsets of indices which taken together constitute the entire set of indices, viz. $\{(m,n): m,n = 1,2,\ldots\}$. For each G_k $(k = 1,2,\ldots)$ define a doubly infinite series as follows:

$$\sum_{m,n=1}^{\infty} a_{m,n}^{(k)} \quad \text{where} \quad a_{m,n}^{(k)} = \begin{cases} a_{m,n} & \text{if } (m,n) \in G_k \\ 0 & \text{if } (m,n) \notin G_k \end{cases}.$$

It is not difficult to establish that the doubly infinite sequence $\{s_{m,n}^{(k)}\}$ of partial sums corresponding to the k-th such series is monotone non-decreasing in each index, and all are bounded above by s. Accordingly, it can then be proved that this k-th sequence of partial sums converges to, say, s_k (finite). Thus we have that

$$\sum_{G_k} a_{m,n} = \sum_{m,n=1}^{\infty} a_{m,n}^{(k)} = s_k .$$

Upon an obvious term-by-term addition of the doubly infinite series for $k = 1,2,\ldots$, it takes little additional argument to establish that:

$$\sum_{k=1} \sum_{G_k} a_{m,n} = \sum_{k=1} \sum_{m,n=1}^{\infty} a_{m,n}^{(k)} = \sum_{k=1} s_k = \sum_{m,n=1}^{\infty} a_{m,n} = s ,$$

as was to be proved. In the general case where the terms of the convergent doubly infinite series may be positive or negative, a proof can be based upon the above results after such a series is 'split' into two doubly infinite series, one containing the positive terms and the other containing the negative terms along with 'dummy' zero terms.

2.10 Hint: (Method 1) Generalize the proof of Theorem 1.11 using the generalization of the Cauchy Criterion introduced in Exercise 2.6. (Method 2) If $\Sigma|a_{m,n}|$ is convergent set:

$$b_{m,n} = \begin{cases} a_{m,n} & \text{if } a_{m,n} \geq 0 \\ 0 & \text{if } a_{m,n} < 0 \end{cases}$$

and

$$c_{m,n} = \begin{cases} -a_{m,n} & \text{if } a_{m,n} < 0 \\ 0 & \text{if } a_{m,n} \geq 0 \end{cases}$$

Both $\Sigma\, b_{m,n}$ and $\Sigma\, c_{m,n}$ converge by comparison with $\Sigma|a_{m,n}|$. Furthermore, since $a_{m,n} \equiv b_{m,n} - c_{m,n}$, it follows from the algebraic properties of convergent doubly infinite series that the series:

$$\Sigma\, a_{m,n} = \Sigma\, (b_{m,n} - c_{m,n}) \quad \text{is convergent} .$$

2.11 Hint: Let "P" be any path through the doubly infinite array $\{a_{m,n}\}$. Accordingly, "P" is determined by a sequence (m_k, n_k); $k = 1,2,\ldots$ of distinct pairs of indices so that either $m_k \to \infty$ or $n_k \to \infty$ (or both) as $k \to \infty$. Now establish that $\Sigma\, a_{m,n}$ cannot converge unless $a_{m_k,n_k} \to 0$ as $k \to \infty$ for any such path.

2.13 Hint: If $\{s_{m,n}\}$ is the doubly infinite sequence of partial sums corresponding to the doubly infinite series $\Sigma\, a_{m,n}$ and $\{s'_{m,n}\}$ is the corresponding sequence for $\Sigma\, b_{m,n}$, the proof can be based upon the fact that the hypotheses imply that $0 \le s'_{m,n} \le s_{m,n}$ for $m,n = 1,2,\ldots$.

2.14 Hint: Note that in extending parts (5), (6) and (7) of Theorem 1.14 the double limit:

$$\lim_{\substack{m\to\infty \\ n\to\infty}} a_{m,n}/b_{m,n}$$

is to be used, not iterated limits.

2.16 Hint: Generalize the proof of Theorem 1.13.

2.17 Answer: Convergence for $r,s > 1$ and divergence otherwise. Note that the function $f(x,y) = 1/x^r y^s$ factors into a function of x alone and y alone, and accordingly the 2-dimensional Riemann Integral involved in the Integral Test also factors.

2.18 Hint: (a) Integral Test: Convergence if $r, s > 1$ and $s > r/(r-1)$. Divergence otherwise.

(b) Convergent if $ab > 0$ (i.e., a and b have the same sign) and $t > 2$. Divergent otherwise.

2.19 Hint: Note that

$$\sum_{i=1}^{\infty} \sum_{j=1}^{\infty} |a_{i,j}| = \sum_{i=1}^{\infty} \sum_{j=1}^{n} |a_{i,j}| + r(n+1) .$$

2.20 Hint: Let $\Sigma\, b_n = s$ and observe that

$$\sum_{m=1}^{\infty} \sum_{n=1}^{\infty} |a_m b_n| \le s \sum_{m=1}^{\infty} |a_m| < \infty .$$

References to Additional and Related Material: Section 2

1. Boas, R., "A Primer of Real Functions", Carus Mathematical Monograph 13, Mathematical Association of America (1961).

2. Bromwich, T., "An Introduction to the Theory of Infinite Series", Macmillan and Co. (1926).

3. Ford, W., "Divergent Series", Chelsea Publishing Co., Inc. (1960).

4. Francis, E., "Examples in Infinite Series, with Solutions", Deighton, Bell and Co. (1953).

5. Goldberg, S., "Probability: An Introduction", Prentice-Hall, Inc. (1964).

6. Green, J., "Sequences and Series", Glencoe, Ill. Free Press (1958).

7. Halberstam, H. and K. Roth, "Sequences", Clarendon Press (1966).

8. Hirschman, I., "Infinite Series", Holt, Reinhart and Winston, Inc. (1962).

9. Hobson, E., "Theory of Functions of a Real Variable", Vol, I, Dover Publications, Inc.

10. Hyslop, J., "Infinite Series", Oliver and Boyd, Ltd. (1959).

11. Jolley, L., "Summation of Series", Second Edition, Dover Publications Inc. (1961).

12. Knopp, K., "Infinite Sequences and Series", Dover Publications, Inc.

13. Lipschutz, S., "Schaum's Outline of Theory and Problems of Set Theory and Related Topics", Schaum Publishing Co. (1964).

14. Rainville, E., "Infinite Series", Macmillan and Co. (1963).

15. Royden, H., "Real Analysis", Macmillan and Co. (1963).

16. Stanaitis, D., "An Introduction to Sequences, Series, and Improper Integrals", Holden-Day, Inc. (1967).

17. Theilman, H., "Theory of Functions of Real Variables", Prentice-Hall, Inc. (1959).

3. Sequences and Series of Functions

The natural and useful outgrowths of sequences and series of numbers are the parallel concepts of sequences and series of functions. In many areas of Applied Mathematics we must deal with these two notions.

A sequence $f_1, f_2, \ldots, \ldots$ or $\{f_n\}$ of real functions is a collection of real-valued functions $f_n = f_n(x)$ of the real variable x which are indexed on the positive integers $n = 1, 2, \ldots$. Each function (term) in such a sequence is assumed to be defined over the same domain D of values of x.

Accordingly, given any sequence of functions, and any specific value of x in D, there is determined a sequence $f_1(x), f_2(x), \ldots$ of real numbers of the type considered in Section 1. Since such a sequence need not be convergent, we must begin by dealing with its lim inf and lim sup which are always defined. For any specific x in D we define $\underline{f(x)}$ and $\overline{f(x)}$ as follows:

$$\underline{f(x)} = \liminf f_n(x)$$

$$\overline{f(x)} = \limsup f_n(x) \quad .$$

As x varies over D, it follows that two functions $\underline{f}$ and $\overline{f}$ are determined which are naturally termed the lim inf and lim sup of the sequence of functions. We denote this as follows:

$$\underline{f} = \liminf f_n$$

$$\overline{f} = \limsup f_n \quad .$$

<u>EXERCISE 3.1</u> (Indicator Functions) Suppose A is a subset of some fixed "Universal" set U. The function I_A defined for each point x of U as follows:

$$I_A(x) = \begin{cases} 1 & \text{if } x \in A \\ 0 & \text{if } x \notin A \end{cases}$$

is called the <u>indicator function</u> of the set A. Prove the following results concerning indicator functions of subsets of U:

(i) $A \subseteq B$ iff $I_A(x) \leq I_B(x)$ for all x in U

(ii) $A = B$ iff $I_A(x) = I_B(x)$ for all x in U

(iii) $I_{A^c} = 1 - I_A$

(iv) $I_{A \cap B} = I_A \cdot I_B$

(v) $I_{A \cup B} = I_A + I_B - I_{A \cap B}$

(vi) $I_{\bigcup_1^n A_k} = I_{A_1} + I_{A_1^c} \cdot I_{A_2} + \ldots + I_{A_1^c} \cdot I_{A_2^c} \cdots I_{A_{n-1}^c} \cdot I_{A_n}$.

In addition, reformulate the notions of lim inf, lim sup, and lim for a sequence of sets in terms of equivalent notions involving the indicator functions of the sets in the given sequence.

If, for a specific value of x in D, we have $\underline{f}(x) = \overline{f}(x) = f(x)$, then of course the sequence $f_1(x), f_2(x), \ldots$ of real numbers is convergent (to $f(x)$). Should this be the case for <u>every</u> value of x in D then the sequence of functions is termed <u>pointwise convergent</u> (or simply convergent) over D. Naturally, the <u>limit</u> of this convergent sequence of functions is the function f which is the common value of the two functions $\underline{f}$ and $\overline{f}$. In such a case we write:

$$f = \lim f_n \quad \text{or} \quad f_n \to f \quad \text{over} \quad D \ .$$

<u>EXAMPLE 3.1</u> Let $r_1, r_2, \ldots$ be an enumeration of the rational numbers, and define a related sequence of functions as follows:

$$f_n(x) = \begin{cases} 1 & \text{if } x = r_1, r_2, \ldots, r_n \\ 0 & \text{otherwise} \end{cases}$$

for $n = 1,2,\ldots$. It is not difficult to see that the sequence converges (over all real x) to the function f defined as:

$$f(x) = \begin{cases} 1 & \text{if } x \text{ is rational} \\ 0 & \text{if } x \text{ is irrational} \end{cases} \ .$$

Note, by the way, that each term in the sequence is continuous at all but a finite number of points, whereas the limit function is discontinuous everywhere.

<u>EXERCISE 3.2</u> (Bounded Sequences of Functions) Formulate a definition of what would be meant by a <u>bounded</u> sequence of functions. Then prove that the limit function of a bounded sequence of functions is bounded whenever such a sequence converges. Can this be generalized when the sequence does not converge?

In certain practical applications of the notion of convergence of a sequence of functions, a stronger (than pointwise) type of convergence is required. The following Example may suggest why this is the case.

EXAMPLE 3.2 (Shortcoming of Pointwise Convergence) Suppose that $\{f_n\}$ is a sequence of Riemann Integrable functions converging over a domain D to a limit function f. Then it is not necessarily true that f is even Riemann Integrable, or, even if it is, it is not necessarily true that the sequence of integrals $\{\int_D f_n dx\}$ converges to $\int_D f dx$. To illustrate this point, consider the sequence of functions of Example 3.1. Each function in the sequence is Riemann-Integrable with value 0. However, the limit (pointwise) function of the sequence, being everywhere discontinuous, is not even Riemann-Integrable.

EXERCISE 3.3 (Further Shortcomings of Pointwise Convergence) Provide examples that demonstrate that the limit of a (pointwise) convergent sequence of continuous (differentiable) functions need not be continuous (differentiable).

The stronger type of convergence that is required to overcome, at least partially, the difficulties demonstrated above is termed uniform convergence. It is set forth in the following definition.

DEFINITION 3.1 (Uniform convergence) A convergent sequence $\{f_n\}$ of functions is termed uniformly convergent over D to the limit f, written $f_n \twoheadrightarrow f$ over D, iff given any $\epsilon > 0$ there can be found an index N, possibly depending upon ϵ but not upon x, such that for all indices $n \geq N$

we have $|f_n(x) - f(x)| < \epsilon$ for <u>all</u> x in D.

The following diagram may help to illustrate the notion of uniform convergence.

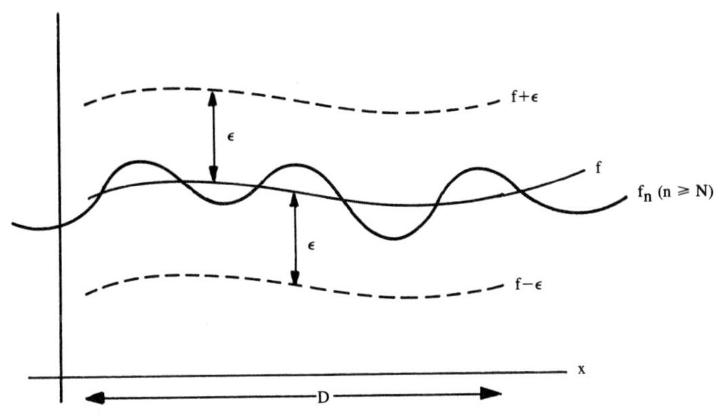

Figure 3-1

<u>EXERCISE 3.4</u> Consider the sequence $\{f_n\}$ where $f_n(x) = x^n$. Prove that this sequence converges uniformly to the function $f(x) \equiv 0$ over $D = [0,C]$, $0 < C < 1$. Find the index N for any given $\epsilon > 0$. Draw a figure illustrating why convergence is <u>not</u> uniform (but only pointwise) over $D = [0,1)$. Actually prove that the required index N cannot be found in this latter case.

As with other definitions, it is often useful to have alternate but equivalent formulations. Several alternate criteria for uniform convergence are given in the following Exercises.

EXERCISE 3.5 (Cauchy Criterion for Uniform Convergence) A sequence $\{f_n\}$ of functions is termed uniformly Cauchy over D iff for any $\epsilon > 0$ there can be found an index N, possibly depending upon ϵ but not on x, such that for all indices $m,n \geq N$ we have $|f_m(x) - f_n(x)| < \epsilon$ for all x in D. Prove that a sequence which is uniformly Cauchy is uniformly convergent, and identify the limit function.

EXERCISE 3.6 (Null Sequence Criterion for Uniform Convergence) Prove that $f_n \twoheadrightarrow f$ over D iff there exists some sequence of numbers, say $\{b_n\}$, not depending upon x and converging to zero, such that $|f_n(x) - f(x)| < b_n$ for $n = 1,2,...$ and all x in D.

EXAMPLE 3.3 (Illustration of Null Sequence Criterion) The sequence $\{f_n\}$ with $f_n(x) = x - x^n/n$ converges uniformly to $f(x) = x$ over the domain $[-1,1]$ since for all x in this domain $|f_n(x) - f(x)| = |x^n/n| \leq 1/n$ and clearly $b_n = 1/n \to 0$.

The next result is a first application of the notion of uniform convergence. It states a sufficient condition under which a convergent sequence of continuous functions will be guaranteed to have a continuous limit function.

Before proceeding with the result itself, we must first understand a 'Mathematical fine point' that might be overlooked otherwise. When it is stated that a function f is continuous (differentiable) over a closed, bounded interval $[a,b]$, what is meant precisely is that:

(i) at any point c interior to the interval, the function is continuous (differentiable) in the usual sense,

(ii) at the left endpoint a the function is continuous from the right (the derivative from the right exists), and

(iii) at the right endpoint b the function is continuous from the left (the derivative from the left exists).[1]

The reason for this precaution is, of course, found in the very definition of continuity (differentiability). The definitions involve two-sided limits. For any point c interior to the interval $[a,b]$ the function(s) involved are presumably defined and the appropriate limits can be computed to check for either continuity or differentiability at this point. However, at and endpoint, say the left endpoint a of the interval, we cannot assume that the function is even defined for any value of $x < a$; analogously at the right endpoint. Accordingly, two sided limits are not appropriate at these points.

In most of the following results that involve this fine point, we shall usually only prove case (i) of the three cases.

[1]See Section 5 for a discussion of two-sided (ordinary), left-hand, and right-hand limits.

The other two cases will usually occur in Exercises. In these Exercises it will be discovered that only slight changes in the proof for case (i) will be needed.

<u>THEOREM 3.1</u> (Continuity of Limit Function) If $f_n \twoheadrightarrow f$ over the closed, bounded interval $[a,b]$, and each f_n is continuous over $[a,b]$, then the limit function f is continuous over $[a,b]$.

PROOF. Let c be any point interior to $[a,b]$. We shall show that f is continuous at $x = c$. Accordingly, it suffices to establish that, given any $\epsilon > 0$ there can always be found a corresponding $\delta > 0$ such that $|f(x) - f(c)| < \epsilon$ whenever $|x-c| < \delta$. Note first that for any index n we have:

$$\begin{aligned} |f(x) - f(c)| &= |f(x) - f_n(x) + f_n(x) - f_n(c) + f_n(c) - f(c)| \\ &\le |f(x) - f_n(x)| + |f_n(x) - f_n(c)| + |f_n(c) - f(c)|. \end{aligned}$$

Now because $f_n \twoheadrightarrow f$ over $[a,b]$ we can always find an index N such that for all (and any fixed) indices $n \ge N$ and all x in $[a,b]$ we have $|f(x) - f_n(x)| < \epsilon/3$. Furthermore, for any fixed index $n' \ge N$ we can choose a $\delta_{n'} > 0$ such that if $|x-c| < \delta_{n'}$ then $|f_{n'}(x) - f_{n'}(c)| < \epsilon/3$ (by continuity of $f_{n'}$). Combining these results we see that for such a fixed index n':

$$|f(x) - f(c)| < \epsilon/3 + \epsilon/3 + \epsilon/3 = \epsilon ,$$

whenever $|x-c| < \delta_{n'}$. With this the proof is completed.

EXERCISE 3.7 Describe the modifications needed in the preceding proof to complete the result for the endpoints a and b.

In passing, it should be noted that the preceding Theorem gives sufficient conditions under which the following statement is true:

$$\lim_{x \to c} \lim_{n \to \infty} f_n(x) = \lim_{n \to \infty} \lim_{x \to c} f_n(x)$$

(with proper care at the endpoints of the closed, bounded interval). Thus, Theorem 3.1 actually gives sufficient conditions under which the two limit operations: $\lim_{n \to \infty}$ and $\lim_{x \to c}$ may validly be interchanged.

EXERCISE 3.8 Verify that, in fact, the conclusion of Theorem 3.1 amounts to the above statement.

Although we already have several equivalent criteria for establishing uniform convergence of a sequence of functions, the next result applies to a special case where the sequence of functions involved is monotone.[1] It will be seen that this result could be quite convenient to use (as compared with the previous criteria for establishing uniform convergence) in this very special case.

---

[1]The sequence $\{f_n\}$ of functions is monotone increasing (decreasing) over a given set of values of x iff $f_n(x) \le f_{n+1}(x)$ $(f_n(x) \ge f_{n+1}(x))$ for $n = 1,2,\ldots$ and all x in the given set.

<u>THEOREM 3.2</u> (Monotone Sequences of Continuous Functions) If $\{f_n\}$ is a monotone sequence of continuous functions converging pointwise to a continuous limit function f over a closed, bounded interval $[a,b]$, then convergence is uniform.

PROOF. (By Contradiction). Suppose that the sequence is monotone increasing, whence $f_n(x) \le f_{n+1}(x) \le f(x)$ for all x in $[a,b]$ and $n = 1,2,\ldots$.

Assume, to the contrary, that convergence is <u>not</u> uniform. Then, for some $\epsilon > 0$ the condition: "$f(x) - f_n(x) < \epsilon$ for all indices $n \ge N$" cannot be satisfied for all x in $[a,b]$ by <u>any</u> index N.

Now let $N_1 < N_2 < N_3 < \ldots$ be any increasing sequence of indices. Accordingly, for each N_i our assumption guarantees the existence of an index $n_i \ge N_i$ along with an associated point x_i within $[a,b]$ such that $f(x_i) - f_{n_i}(x_i) \ge \epsilon$.

By Theorem 1.1 (Bolzano-Weierstrass), the bounded sequence $x_1, x_2, x_3, \ldots$ possesses at least one limit point, say c, within $[a,b]$, and so there exists a subsequence $x_1', x_2', x_3', \ldots$ (with associated indices $n_1'\ n_2'\ n_3'\ \ldots$) converging to c.

Now fix an index N <u>as large as desired.</u> Then for all indices $n_i' \ge N$ we must have:

$$f_N(x_i') \le f_{n_i'}(x_i') \le f(x_i') - \epsilon ,$$

whence, upon allowing $n_i' \to \infty$ (so $x_i' \to c$), we conclude that:

$$f_N(c) \le f(c) - \epsilon \quad ,$$

because all functions involved are assumed to be continuous. But this contradicts pointwise convergence of the original sequence $\{f_n\}$ at the point $x = c$, because the indicated index N could be chosen as large as desired. With this contradiction, we are forced to conclude that convergence _is_ uniform, thereby completing the proof.

EXERCISE 3.9 Review the proof of the preceding Theorem, and indicate what changes would be necessary to establish uniform convergence in case the sequence of functions was monotone decreasing.

EXAMPLE 3.4 We shall demonstrate that the sequence of continuous functions $\{f_n\}$, where $f_n(x) = (1+x/n)^n$, which converges pointwise to the continuous limit function $f(x) = e^x$ for $x \ge 0$, does so in a monotone increasing manner. Therefore, from the preceding Theorem we can conclude that for any positive constant B, the convergence is uniform over $[0,B]$. Two methods of proof will be used; observe that only monotonicity of convergence needs proof.

Method A: It suffices to show that $r_n(x) = \dfrac{f_{n+1}(x)}{f_n(x)} \ge 1$ for $n = 1,2,\ldots$ and $x \ge 0$, or, alternately, that $s_n(x) = r_n(x) - 1 \ge 0$. Now because $s_n(0) = 0$, it suffices to establish that $s_n'(x) \geq 0$ for $x \geq 0$. Upon algebraic simplification:

$$s_n'(x) = \frac{xn^n(n+x)^{n-1}(n+x+1)^n}{(n+1)^{n+1}(n+x)^{2n}} \geq 0$$

for $n = 1,2,\ldots,$ thereby completing the proof.

Method B: The inequality: $\log z \geq 1 - 1/z$ for all real $z \geq 1$, is established easily by showing that the function of z: $s(z) = \log z + 1/z - 1$, satisfies the conditions $s(1) = 0$ and $s'(z) \geq 0$ for all real $z \geq 1$. We shall have use for this later.

First consider $f_n(x) = (1+x/n)^n$ as a single function of the real variable n $(n > 0)$ with x being some fixed non-negative real number. This function of n is monotone increasing in n iff the function $g_n(x) = \log f_n(x)$ is monotone increasing in n.

Clearly $g_n(x) = n \log(1+x/n)$, and accordingly $g_n'(x) = -x/(x+n) + \log(1+x/n)$. Therefore, $g_n(x)$ is a monotone increasing function of n, since for $n > 0$, $g_n(x) > 0$ and $g_n'(x) \geq 0$. The latter follows from the opening inequality with $z = (x+n)/n$. From this it now follows that the <u>sequence of functions</u> $\{f_n\}$ is monotone increasing.

<u>EXERCISE 3.10</u> Prove the last statement made in 'Method B'.

<u>EXERCISE 3.11</u> Use Theorem 3.2 to prove (easily) that the sequence of functions $\{f_n\}$, where $f_n(x) = x^n$, converges uniformly to $f(x) \equiv 0$ over $[0,B]$ for any $0 < B < 1$.

Next we consider the problems related to term-by-term integration of a sequence of (Riemann-Integrable) functions. It should not be surprising that a condition of uniform convergence becomes involved.

The first result deals with the term-by-term integration of a uniformly convergent sequence of Riemann-Integrable functions, and the relationship between the integrals of the individual terms and the integral of the limit function.

<u>THEOREM 3.3</u> (Term-by-Term Integration) Let $\{f_n\}$ be a sequence of Riemann-Integrable functions converging uniformly to the limit function f over a closed, bounded interval $[a,b]$. Then the limit function itself is Riemann-Integrable, and over $[a,b]$:

$$\int_a^x f_n(z)\,dz \twoheadrightarrow \int_a^x f(z)\,dz \quad .$$

PROOF. A direct proof that f is in fact Riemann-Integrable will not be given here.[1] We shall only establish the uniform convergence.

Accordingly, let $\epsilon > 0$ be given. Now since $f_n \twoheadrightarrow f$ over $[a,b]$ there can be found an index N such that $|f(x) - f_n(x)| < \epsilon/(b-a)$ for all x in $[a,b]$, and all indices $n \geq N$. Thus, for any specific value of x

[1] In view of Theorem 3.1, integrability of f would follow immediately if each f_n were continuous over $[a,b]$.

in $[a,b]$ and all $n \geq N$ we have:

$$\left|\int_a^x f_n(z)\,dz - \int_a^x f(z)\,dz\right| \leq \int_a^x |f_n(x)-f(z)|\,dz < \varepsilon(x-a)/(b-a) < \varepsilon$$

thereby establishing uniform convergence.

Note that we could restate the conclusion of the preceding Theorem in the alternate form:

$$\lim_{n\to\infty} \int_a^x f_n(z)\,dz = \int_a^x \lim_{n\to\infty} f_n(z)\,dz$$

uniformly for all x in $[a,b]$. Thus, the preceding Theorem actually provides sufficient conditions (including uniform convergence) under which the two operations:

$$\lim_{n\to\infty} \quad \text{and} \quad \int_a^x \quad \text{(Riemann)}$$

may validly be interchanged.

<u>EXERCISE 3.12</u> Verify that the above statement is, in fact, true.

<u>EXERCISE 3.13</u> Prove that:

$$\int_0^x (1+z/n)^n\,dz \to e^x - 1$$

as $n \to \infty$ (in fact, uniformly for all x in any interval of the form $[0,B]$ with $B > 0$).

The final result in this sequence of results involving uniform convergence deals with term-by-term differentiation of a sequence of differentiable functions, and the relationship between the derivatives of the terms of the sequence and the derivative (assuming existence) of the limit function. This result is contained in the following Theorem.

THEOREM 3.4 (Term-by-Term Differentiation) Let $\{f_n\}$ be a sequence of differentiable functions converging to the limit function f over a closed, bounded interval $[a,b]$. If the derivatives $\{\frac{d}{dx} f_n\}$ are continuous and converge uniformly over $[a,b]$, then f is itself differentiable and $\frac{d}{dx} f_n \twoheadrightarrow \frac{d}{dx} f$ over $[a,b]$.

PROOF. Suppose $\frac{d}{dx} f_n \twoheadrightarrow g$. Then g must be continuous over $[a,b]$, and furthermore,

$$\int_a^x \frac{d}{dz} f_n(z)dz = f_n(x) - f_n(a) \twoheadrightarrow \int_a^x g(z)dz$$

over $[a,b]$. But (pointwise) convergence of $\{f_n\}$ to f guarantees that $f_n(x) - f_n(a) \to f(x) - f(a)$ as $n \to \infty$; accordingly:

$$f(x) = f(a) + \int_a^x g(z)dz \quad .$$

Since g is continuous, the right-hand side of the preceding expression has as its derivative $g(x)$. Since this is true for any x within $[a,b]$ we conclude both that f is differentiable and indeed $g = \frac{d}{dx} f$.

Accordingly, $\frac{d}{dx} f_n \twoheadrightarrow \frac{d}{dx} f$ over $[a,b]$ and the proof is completed.

<u>EXERCISE 3.14</u> Restate the conclusion of Theorem 3.4 so as to emphasize the point that the Theorem provides conditions under which the operations of $\lim_{n\to\infty}$ and $\frac{d}{dx}$ may be interchanged. The conditions are sufficient but not necessary.

<u>EXERCISE 3.15</u> Construct an example illustrating Theorem 3.4. Does $\frac{d}{dx} f_n \twoheadrightarrow \frac{d}{dx} f$ imply $f_n \twoheadrightarrow f$?

The preceding results dealing with infinite <u>sequences</u> of real functions can now be applied to the study of infinite <u>series</u> of real functions. Such series, and their generalizations, occur frequently in the various areas of Applied Mathematics.

An <u>infinite series of real functions</u> is a 'sum' of the form

$$\sum_1^\infty f_n = f_1 + f_2 + \ldots + \ldots$$

where each term f_n in this 'sum' is a real-valued function $f_n = f_n(x)$ of the real variable x. Each such function (term) is assumed to be defined over the same domain D of values of x.

Accordingly, for each fixed x in D there is determined an ordinary infinite series of real numbers:

$$\sum_1^\infty f_n(x) = f_1(x) + f_2(x) + \ldots + \ldots$$

which may or may not be convergent. We denote the n-th partial sum of this series by:

$$S_n(x) = f_1(x) + f_2(x) + \ldots + f_n(x) .$$

In accordance with the results of Section 1, the infinite series $\sum_1^\infty f_n(x)$ converges according as the sequence of partial sums $\{S_n(x)\}$ converges as a sequence of real numbers.

If, for a given value of x in D, we have $S_n(x) \to S(x)$, say, then the series $\sum_1^\infty f_n(x)$ is said to converge to (or have sum) $S(x)$, written

$$\sum_1^\infty f_n(x) = S(x) \quad .$$

Should the series above converge for <u>all</u> values of x in D then we say that the series of functions $\sum_1^\infty f_n$ converges (pointwise) over D. The function s whose value $s(x)$ represents the sum of such a convergent series for each x in D is termed the <u>sum function</u> of the infinite series.

In dealing with infinite series of functions $\sum_{n=1}^\infty f_n(x)$ converging pointwise to the sum function $s(x)$ for all x in D, one often encounters the following problem: each term (function) $f_n(x)$ of the series possesses a limit, say c_n, as x approaches some value c (possibly $+\infty$ or $-\infty$).[1] That is,

$$\lim_{x \to c} f_n(x) = c_n \quad \text{for} \quad n = 1,2,\ldots \quad .$$

[1] In general, a two-sided (ordinary), left-hand, or right-hand limit might be involved. These topics are covered in Section 5.

In such a case, it is often of practical importance to inquire <u>whether or not</u> the following statement is true:

$$\sum_{n=1}^{\infty} [\lim_{x\to c} f_n(x)] = \lim_{x\to c} [\sum_{n=1}^{\infty} f_n(x)]$$

equivalently, <u>whether or not</u>

$$\sum_{n=1}^{\infty} c_n = \lim_{x\to c} S(x) \quad .$$

Observe:

(a) there is no a-priori guarantee that $S(x)$ possesses a limit as x approaches c, and

(b) there is no guarantee of equality in any event.

The following Example illustrates one such situation.

<u>EXAMPLE 3.5</u> Consider the infinite series $\sum_{n=1}^{\infty} f_n(x)$, where for $0 \le x \le 1$, $f_n(x) = x^n(1-x)$, $n = 2,3,\ldots$. Clearly $\lim_{x\to 1} f_n(x) = 0 = c_n$ for $n = 1,2,\ldots$ and the sum function $S(x)$ is given by

$$S(x) = \begin{cases} 1 & \text{if } 0 \le x < 1 \\ 0 & \text{if } x = 1 \end{cases} \quad .$$

Convergence is pointwise; however, it <u>is</u> obvious that $\lim_{x\to 1} S(x) = 1$ <u>is not</u> equal to $\sum_{n=1}^{\infty} c_n = 0$.

It appears, then, that conditions <u>additional to</u> pointwise; convergence are needed to assure equality, or else weaker conclusions (than equality) are required. We illustrate two such results below. Another related result is found in Section 5.

The first result is one in which additional conditions are imposed which guarantee equality.

<u>THEOREM 3.5</u> (Dominated Convergence) Suppose $\sum_{n=1}^{\infty} f_n(x)$ converges pointwise to the sum function $S(x)$ for all x in D and that $\lim_{x\to c} f_n(x) = c_n$ exists for $n = 1,2,\ldots$. If, in addition,

(a) $|f_n(x)| \le b_n$ for all x in D

and

(b) $\sum_{n=1}^{\infty} b_n$ converges,

then

$$\sum_{n=1}^{\infty} [\lim_{x\to c} f_n(x)] = \lim_{x\to c} [\sum_{n=1}^{\infty} f_n(x)],$$

equivalently,

$$\sum_{n=1}^{\infty} c_n = \lim_{x\to c} S(x) \quad .$$

PROOF. See Exercise 3.18.

A slight modification of Example 3.5 serves to illustrate use of Dominated Convergence.

<u>EXAMPLE 3.6</u> Consider the infinite series $\sum_{n=1}^{\infty} f_n(x)$, where for $0 \le x \le 1$ the terms are defined as follows:

$$f_n(x) = \begin{cases} x^n(1-x)^2 & \text{for } 0 \le x < n/(n+2) \\ 0 & \text{for } n/(n+2) \le x \le 1 \end{cases} \quad .$$

It is not difficult to show that the series converges pointwise for all $0 \le x \le 1$, but the sum function $S(x)$ cannot be put in a simple "closed form".

Observe that $\lim_{x \to 1} f_n(x) = 0 = c_n$ for $n = 1,2,\ldots$ and, even though $S(x)$ may be unknown, we intend to use Dominated convergence to establish that

$$\lim_{x \to 1} S(x) = \sum_{n=1}^{\infty} c_n = 0 \quad .$$

First note that Calculus procedures applied to the function $g_n(x) = x^n(1-x)^2 \quad 0 \le x \le 1$ establish that its maximum occurs at $x = n/(n+2)$, and this maximum value is $[n/(n+2)]^n \cdot 4/(n+2)^2 = b_n$. Now it is clear from the definitions that $|f_n(x)| \le b_n$ for $0 \le x \le 1$ and $n = 1,2,\ldots,$ and it is straightforward to prove that the series $\sum_{n=1}^{\infty} b_n$ converges. Accordingly, the conditions of Dominated Convergence are satisfied and the required equality established.

Note that the Dominated Convergence Theorem automatically applies to the case where the terms $f_n(x)$ of the infinite series are functions of an integer variable m, in which case the term $f_n(x)$ might be written $f(m,n)$, and the hypotheses and conclusion of the Theorem could be modified to read as follows:

<u>EXAMPLE 3.7</u> (Dominated Convergence: Integer Variable Functions) Let $\sum_{n=1}^{\infty} f(m,n)$ converge to the sum $g(m)$ for $m = 1,2,\ldots$ and suppose that $\lim_{m\to\infty} f(m,n) = c_n$ exists for $n = 1,2,\ldots$. If, in addition,

(a) $|f(m,n)| \le b_n$ for all $m = 1,2,\ldots$

and

(b) $\sum_{n=1}^{\infty} b_n$ converges,

then $\sum_{n=1}^{\infty} [\lim_{m\to\infty} f(m,n)] = \lim_{m\to\infty}[\sum_{n=1}^{\infty} f(m,n)]$, equivalently,

$$\sum_{n=1}^{\infty} c_n = \lim_{m\to\infty} g(m) \quad .$$

The situation illustrated by the preceding Example occurs often enough to deserve the special attention it has been given.

The second of the two results considers the case where (strong) conditions in addition to pointwise convergence of a series of functions $\sum_{n=1}^{\infty} f_n(x)$ are not added, and, in turn, a weaker conclusion (than equality) is drawn. This result follows.

<u>THEOREM 3.6</u> Suppose $\sum_{n=1}^{\infty} f_n(x)$ is a series of <u>non-negative</u> functions converging pointwise to the sum function $S(x)$ for all x in D. If $\lim_{x\to c} f_n(x) = c_n$ exists for $n = 1,2,\ldots,$ then:

$$\sum_{n=1}^{\infty} [\lim_{x\to c} f_n(x)] \le \lim_{x\to c} [\sum_{n=1}^{\infty} f_n(x)] \text{ , equivalently,}$$

$$\sum_{n=1}^{\infty} c_n \leq \lim_{x\to c} S(x) \quad .$$

PROOF. For any (fixed) index N,

$$s_N = \sum_{n=1}^{N} [\lim_{x\to c} f_n(x)] = \lim_{x\to c} [\sum_{n=1}^{N} f_n(x)]$$

$$\leq \lim_{x\to c} [\sum_{n=1}^{\infty} f_n(x)] = \lim_{x\to c} S(x) \quad .$$

Since the sequence $\{s_N\}$ is non-decreasing, the result follows upon taking limits as $N \to \infty$ throughout the above sequence of relationships.[1]

It is to be noted that if, in the preceding Theorem, the series $\sum_{n=1}^{\infty} b_n$ diverges, then we may conclude that the sum function becomes unbounded as x approaches c. Accordingly, the weakened conclusion may be of practical use. Similarly, if it is known that $\lim_{x\to c} S(x) = s$ (finite), then it may be concluded that $\sum_{n=1}^{\infty} b_n \leq s$.

Since c may be $+\infty$ or $-\infty$, and the terms $f_n(x)$ of the series $\sum_{n=1}^{\infty} f_n(x)$ may be functions of an integer variable, we may state an immediate version of Fatou's Lemma for this case. Again, this situation occurs often enough in Applied Mathematics so as to deserve special attention.

---

[1]In its many forms, this result is known as Fatou's Lemma. Note also that $\lim_{x\to c} S(x)$ exists since the f_n's are assumed non-negative.

EXAMPLE 3.8 (Fatou's Lemma: Integer variable functions) Suppose $\sum_{n=1}^{\infty} f(m,n)$ is a series of non-negative functions converging to the sum $g(m)$ for $m = 1,2,\ldots$. If $\lim_{m\to\infty} f(m,n) = c_n$ exists for $n = 1,2,\ldots$, then:

$$\sum_{n=1}^{\infty} [\lim_{m\to\infty} f(m,n)] \le \lim_{m\to\infty} [\sum_{n=1}^{\infty} f(m,n)], \text{ equivalently,}$$

$$\sum_{n=1}^{\infty} b_n \le \lim_{m\to\infty} g(m) \quad .$$

As noted earlier, a result such as this (an inequality, not an equality) may prove of practical value in a variety of situations.

Although many useful results can be obtained for pointwise convergent (infinite) series of functions, additional assumptions are required and/or weakened conclusions are drawn in general. The two preceding Theorems illustrate this point.

It is not surprising, then, that there is a 'stronger' type of convergence that is more frequently dealt with when considering infinite series of functions.

DEFINITION 3.2 (Uniform Convergence: Series) The infinite series $\sum_{n=1}^{\infty} f_n(x)$ of functions $f_n(x)$ is said to converge uniformly to the sum function $S(x)$ over D iff the corresponding sequence of partial sums, $S_n(x) = \sum_{m=1}^{n} f_m(x)$, $n = 1,2,\ldots$ converges uniformly (as a sequence) to the limit function $S(x)$ over D.

EXAMPLE 3.9 The series of functions[1] $\sum_{0}^{\infty} x^n$ converges uniformly to the sum function S defined by: $S(x) = 1/(1-x)$ for all x in $[-\frac{1}{2},\frac{1}{2}]$, for example. To see this, note that $S_n(x) = 1 + x + x^2 + \ldots + x^{n-1} = (1-x^n)/(1-x)$, accordingly,

$$|S_n(x) - S(x)| = \frac{|x^n|}{|1-x|} \leq (\tfrac{1}{2})^{n-1}$$

for all x in $[-\frac{1}{2},\frac{1}{2}]$. Thus, the Null Sequence Criterion (Exercise 3.6) guarantees the asserted uniform convergence.

There are alternate criteria and tests for uniform convergence of an infinite series of functions. These and related results are contained in the following Theorems and Exercises. The first result gives a 'Cauchy Criterion' for uniform convergence of an infinite series.

THEOREM 3.7 (Cauchy Criterion for Uniform Convergence: Series) The infinite series of functions $\sum_{1}^{\infty} f_n$ converges uniformly over the domain D of values of x iff given any $\epsilon > 0$ there can always be found an index N, depending possibly upon ϵ but not upon x, such that:

$$|S_m(x) - S_n(x)| = \left|\sum_{n+1}^{m} f_i(x)\right| < \epsilon$$

for all indices $m > n \geq N$ and all x in D.

PROOF. EXERCISE 3.16

---

[1] As with sequences, the convention that the index n begin with the integer 1 is arbitrary. All previous results apply by a simple re-indexing.

EXERCISE 3.17 (Algebra of Uniformly Convergent Series)
Let $\sum_1^\infty f_n$ and $\sum_1^\infty g_n$ be uniformly convergent series of functions over D. Suppose h is any bounded, continuous function over D. Prove that the following series of functions are then also uniformly convergent over D:

(i) $\sum_1^\infty (af_n + bg_n)$ a,b constants

(ii) $\sum_1^\infty h \cdot f_n$

Can these results be generalized?

The next result is a well-known and useful test for uniform convergence known as the Weierstrass M-Test.

THEOREM 3.8 (Weierstrass M-Test for Uniform Convergence)
Let $\sum_1^\infty f_n$ be a (pointwise) convergent series of functions over D. If there can be found a convergent series $\sum_1^\infty M_n$ of positive numbers such that $|f_n(x)| \le M_n$ for all x in D $(n = 1,2,\ldots)$, then the series of functions is uniformly convergent over D.

PROOF. Since $\sum_1^\infty M_n$ converges, given any $\epsilon > 0$ there can always be found an index N such that $M_{n+1}+\ldots+M_m < \epsilon$ for all indices $m > n \ge N$. (This follows from Theorem 1.2). Accordingly, for all such indices and all x in D we have:

$$|S_m(x) - S_n(x)| = |f_{n+1}(x) + \ldots + f_m(x)|$$
$$\le |f_{n+1}(x)| + \ldots + |f_m(x)|$$
$$\le M_{n+1} + \ldots + M_m$$
$$< \varepsilon$$

and uniform convergence follows from the Cauchy Criterion of Theorem 3.7. This completes the proof.

<u>EXERCISE 3.18</u> Prove that Dominated Convergence, Theorem 3.5, is an immediate consequence of the Weierstrass M-Test in the preceding Theorem.

<u>EXAMPLE 3.10</u> The series $\sum_1^\infty x^n/n^2$ is uniformly convergent over $[-1,1]$ since $|x^n/n^2| \le 1/n^2$ for all x in $[-1,1]$ and the series $\sum_1^\infty 1/n^2$ is known to be convergent. Analogously, the series $\sum_1^\infty \sin(nx)/n^3$ is uniformly convergent over any closed, bounded interval $[a,b]$ since for all x in such an interval, we have $|\sin(nx)/n^3| \le 1/n^3$, and the series $\sum_1^\infty 1/n^3$ is convergent.

<u>EXERCISE 3.19</u> Prove that the following series of functions converge uniformly over the indicated domains:

(i) $\sum_1^\infty (\tanh x)^n/n!$ for all real x

(ii) $\sum_1^\infty e^{nx}/2^n$ for $x \le \log(3/2)$

(iii) $\sum_1^\infty nx^n$ for all x in $[-a,a]$, $0 < a < 1$.

EXERCISE 3.20 (A Comparison Test for Uniform Convergence) Prove that if $\sum_1^\infty g_n$ is a uniformly, absolutely convergent series of functions over D and for all x in D $|f_n(x)| \le |g_n(x)|$ $(n = 1,2,\ldots)$[1], then the series of functions $\sum_1^\infty f_n$ is uniformly, absolutely convergent over D. Generalize this result.

We shall now consider, for series of functions, results parallel to those for sequences of functions that involve the interchange of various limit operations (associated with continuity, derivative and integral). It will be seen that uniform convergence of a series of functions plays an important part in these results. Before we can proceed, however, we have need for the following preparatory result.

THEOREM 3.9 (Lemma on Uniformly Convergent Series) Suppose $\sum_1^\infty f_n$ converges uniformly to the sum function S over a closed, bounded interval $[a,b]$. If c is any point in $[a,b]$ and for each $n = 1,2,\ldots$ we have $f_n(x) \to a_n$ as $x \to c$, then $\sum_1^\infty a_n$ is convergent and $S(x) \to \sum_1^\infty a_n$ as $x \to c$.

PROOF. Let $S_n(x) = \sum_1^n f_j(x)$ and $s_n = \sum_1^n a_j$. First observe that the series $\sum_1^\infty a_n$ does indeed converge since given any $\epsilon > 0$ there can always be found an index N, possibly depending upon ϵ but not upon x, such that for all indices $m > n \ge N$ and all x in $[a,b]$ we have: $|f_{n+1}(x) + \ldots + f_m(x)| < \epsilon$, so, upon permitting $x \to c$, we

[1]This condition need hold only for all indices n sufficiently large.

then have: $|a_{n+1}+\ldots+a_m| \leq \epsilon$ for all indices $m > n \geq N$, so convergence follows from Theorem 1.2 (Cauchy Criterion).

Next, given any $\delta > 0$ we can always find indices N_1 and N_2 say, so that $|\sum_1^\infty a_n - s_n| < \delta/3$ for all $n \geq N_1$ and $|S(x) - S_n(x)| < \delta/3$ for all x in $[a,b]$ and all indices $n \geq N_2$ (this includes the point $x = c$).

Now choose an index N^* with $N^* \geq \max(N_1,N_2)$. Corresponding to this index we can find a λ sufficiently small so that for any x in $[a,b]$ satisfying $|x - c| < \lambda$ we have: $|S_{N^*}(x) - s_{N^*}| < \delta/3$.

Combining the above results, we conclude that if $|x - c| < \lambda$ then

$$|S(x) - \sum_1^\infty a_n| = |S(x) - S_{N^*}(x) + S_{N^*}(x) - s_{N^*} + s_{N^*} - \sum_1^\infty a_n|$$

$$\leq |S(x) - S_{N^*}(x)| + |S_{N^*}(x) - s_{N^*}| + |s_{N^*} - \sum_1^\infty a_n|$$

$$< \delta/3 + \delta/3 + \delta/3 = \delta .$$

Thus $S(x) \to \sum_1^\infty a_n$ as $x \to c$ as was to be proved.

EXERCISE 3.21 (Extension of Theorem 3.9) Prove that the conclusion of the preceding Theorem remains valid if the domain of uniform convergence is replaced by $x \geq b$ $(x \leq a)$ and $f_n(x) \to a_n$ $(n = 1,2,\ldots)$ as $x \to \infty$ $(x \to -\infty)$.

We shall now use the preceding Theorem to establish a result concerning continuity of the sum function of a convergent infinite series of functions.

THEOREM 3.10 (Continuity of Sum Function) If the series $\sum_1^\infty f_n$ of continuous functions converges uniformly to the sum function S over the closed, bounded interval $[a,b]$, then S is a continuous function over that interval.

PROOF. Let c be any point interior to $[a,b]$.[1] Since each term f_n is continuous at c it follows that $f_n(x) \to f_n(c)$ as $x \to c$ $(n = 1,2,\ldots)$. Furthermore, $S(c) = \sum_1^\infty f_n(c)$ by its very definition. The proof is completed by a simple appeal to the preceding Theorem with $a_n = f_n(c)$.

Observe that we may alternately state the conclusion of the preceding Theorem as:

$$\lim_{x\to c} \sum_1^\infty f_n(x) = \sum_1^\infty \lim_{x\to c} f_n(x) \quad .$$

Thus the preceding Theorem essentially gives sufficient conditions on series of functions under which the two limit operations: $\lim_{x\to c}$ and $\sum_1^\infty$ may validly be interchanged.

The next result deals with convergent series of Riemann-Integrable functions.

THEOREM 3.11 (Term-by-Term Integration of Series) If the series $\sum_1^\infty f_n$ of Riemann-Integrable functions converges uniformly over the closed, bounded interval $[a,b]$, then its sum function S is also Riemann-Integrable and for

[1]Actually, the endpoints a and b require the special attention mentioned earlier in this Section.

any x in [a,b]:

$$\int_a^x S(z)\,dz = \sum_1^\infty \int_a^x f_n(z)\,dz \quad .$$

PROOF. The Riemann-Integrability of S will not be proved here (see footnote to Theorem 3.3). Concerning the second part of the result, let $\epsilon > 0$ be given. By uniform convergence it is always possible to find an index N, possibly depending upon ϵ but not upon x, such that for all indices $n \geq N$ and all x in [a,b] we have: $|S(x) - S_n(x)| < \epsilon/(b-a)$. Accordingly,

$$\left|\int_a^x S(z)\,dz - \sum_1^n \int_a^x f_i(z)\,dz\right| = \left|\int_a^x [S(z) - S_n(z)]\,dz\right|$$

$$\leq \int_a^x |S(z) - S_n(z)|\,dz$$

$$\leq \epsilon(x-a)/(b-a) \leq \epsilon$$

for $n \geq N$ and all x in [a,b]. This completes the proof.

Notice that Theorem 3.11 essentially gives conditions under which:

$$\int_a^x \sum_1^\infty f_n(z)\,dz = \sum_1^\infty \int_a^x f_n(z)\,dz$$

and so, it provides conditions under which the two limit operations

$$\sum_{1}^{\infty} \quad \text{and} \quad \int_{a}^{x} \quad \text{(Riemann)}$$

may validly be interchanged.

EXAMPLE 3.11 $\sum_{0}^{\infty} x^n = 1/(1-x)$ uniformly for all x in $[-a,a]$ $(0 < a < 1)$. Therefore,

$$\log 1/(1-x) = \int_{0}^{x} 1/(1-z)\, dz = \sum_{0}^{\infty} \int_{0}^{x} z^n dz = x + x^2/2 + x^3/3 + \ldots$$

(in fact, convergence of the series is uniform over the interval $[-a,a]$).

The final result in this Section deals with term-by-term differentiation of a convergent series of differentiable functions. It is contained in the following (somewhat cumbersome) Theorem.

THEOREM 3.12 (Term-by-Term Differentiation of Series) If the series $\sum_{1}^{\infty} f_n$ of differentiable functions converges to the sum function S over a closed, bounded interval $[a,b]$ and if the derivative functions $\frac{d}{dx} f_n$ $(n = 1,2,\ldots)$ are continuous over $[a,b]$, and if the series of derivatives $\sum_{1}^{\infty} \frac{d}{dx} f_n$ converges uniformly over $[a,b]$, then S is differentiable and $\frac{d}{dx} S = \sum_{1}^{\infty} \frac{d}{dx} f_n$ for all x in $[a,b]$.

PROOF. EXERCISE 3.22

Accordingly, the preceding Theorem gives conditions under which the two limit operations:

$$\frac{d}{dx} \text{ and } \sum_{1}^{\infty}$$

may validly be interchanged.

EXERCISE 3.23 Restate the conclusion of the preceding Theorem to illustrate the statement made above.

EXERCISE 3.24 Provide an illustration of an application of Theorem 3.12.

Finally, information regarding Mathematical results and problems related to either doubly (or, more generally, multiply) infinite sequences

$$\{f_{m,n}(x)\}_{m,n=1}$$

or series

$$\sum_{m,n=1} f_{m,n}(x)$$

of functions defined over a common domain D can be found in the References. These results are natural extensions of the results obtained in Section 2 for doubly infinite sequences and series of real numbers, and those of the present Section for ordinary (singly) infinite sequences and series of functions.

Hints and Answers: Section 3

3.1 Hint: (i) Suppose first that $A \subseteq B$. Then,

if $x \in A$ and $x \in B$ then $I_A(x) = I_B(x) = 1$,

if $x \notin A$ and $x \in B$ then $I_A(x) = 0 \leq I_B(x) = 1$,

if $x \notin A$ and $x \notin B$ then $I_A(x) = I_B(x) = 0$.

Next assume that $I_A(x) \leq I_B(x)$ for all x in U yet, to the contrary, A is <u>not</u> a subset of B. Accordingly, there would exist at least one point x in U such that $x \in A$ yet $x \notin B$. In such a case we would then have $I_A(x) = 1$ and $I_B(x) = 0$ which is a contradiction. Thus $A \subseteq B$.

(ii) Note that $A = B$ if and only if $A \subseteq B$ and $B \subseteq A$.

(iii) Note that $x \in A^c$ if and only if $x \notin A$.

(iv) Note that $x \in A \cap B$ if and only if $x \in A$ and $x \in B$.

(v) If $A \cap B = \emptyset$ the result is immediate. Note, however, if $A \cap B \neq \emptyset$ and $x \in A \cap B$ then $I_A(x) + I_B(x) = 2$.

(vi) First establish that:

$$A_1 \cup [A_1^c \cap A_2] \cup \ldots \cup [A_1^c \cap A_2^c \cap \ldots \cap A_{n-1}^c \cap A_n]$$

is a decomposition of $\bigcup_1^n A_k$ into n mutually exclusive sets. Finally, generalize (v) by showing that if $B_1, B_2, \ldots, B_n$ are mutually exclusive sets, then

$$I_{\bigcup_1^n B_k} = \sum_1^n I_{B_k} .$$

Finally, if $\underline{A} = \liminf A_n$ and $\overline{A} = \limsup A_n$, note that

$$I_{\underline{A}} = \sum_{n=1}^{\infty} \prod_{k=n}^{\infty} I_{A_k}$$

and

$$I_{\overline{A}} = \prod_{n=1}^{\infty} [I_{B_n} + I_{B_n^c \cap B_{n+1}} + I_{B_n^c \cap B_{n+1}^c \cap B_{n+2}} + \cdots]$$

where

$$B_n = \bigcup_{k=1}^{n} A_k \quad .$$

The sequence $A_1, A_2, \ldots$ of sets is convergent if and only if

$$I_{\underline{A}}(x) = I_{\overline{A}}(x) \quad \text{for all} \quad x \in U \quad .$$

Note that:

$$I_{\underline{A}} = \liminf I_{A_n}$$

and

$$I_{\overline{A}} = \limsup I_{A_n} \quad .$$

This requires some additional reasoning.

3.2 Answer: A sequence $\{f_n\}$ of functions is termed <u>bounded</u> over a domain D iff there exists a positive constant B such that $|f_n(x)| \le B$ for all x in D $(n = 1,2,\ldots)$. If $\{f_n\}$ is bounded over D then (generalization) clearly $-B \le \underline{f(x)} \le \overline{f(x)} \le B$ for all x in D and accordingly if the sequence converges the limit function f must be bounded.

3.4 Hint: Let C be fixed $(0 < C < 1)$. Given any ϵ $(0 < \epsilon < 1)$ the difference $|x^n - 0| < \epsilon$ for all indices n satisfying: $n \ge (\log \epsilon)/(\log B)$ and for all x in $[0,B]$. The diagram should illustrate the fact that given any ϵ $(0 < \epsilon < 1)$ and any index N there can always be found an index $n \ge N$ and a point x within $[0,1)$ such that $|x^n - 0| > \epsilon$.

3.5 Hint: Develop the proof by considering the results and methods of Theorem 1.2.

3.6 Hint: If $b_n \to 0$ then given any $\epsilon > 0$ there exists an index N sufficiently large so that $b_n < \epsilon$ for all indices $n \ge N$. On the other hand, $b_n = \sup_{x \in D} |f_n(x) - f(x)| \to 0$ as $n \to \infty$ in case of uniform convergence.

3.7 Hint: For $c = a$ the points x must be chosen such that $x > a$; for $c = b$ the points x must be chosen such that $x < b$.

3.8 Hint: f is continuous at $x = c$ iff $\lim_{x \to c} f(x) = f(c)$.

3.10 Hint: For $x \ge 0$ consider the values $n = 1,2,\ldots$ of the positive real variable n.

3.11 Hint: Clearly $x^{n+1} \le x^n$ $(n = 1,2,\ldots)$ for all x in $[0,B]$ $(0 < B < 1)$.

3.12 Hint: See the relationship preceding this Exercise.

3.13 Hint: It has been established that $(1 + z/n)^n \twoheadrightarrow e^z$ for all z in $[0,B]$ $(0 < B < \infty)$.

3.14 Hint: Rewrite the statement $\frac{d}{dx} f_n \twoheadrightarrow \frac{d}{dx} f$ over $[a,b]$ in the form: $\lim_{n\to\infty} \frac{d}{dx} f_n(x) = \frac{d}{dx} \lim_{n\to\infty} f_n(x)$.

3.15 Hint: A simple example would be the sequence $\{f_n\}$ of functions $f_n(x) = x^n$, $0 \le x \le B$ $(0 < B < 1)$ $n = 1,2,...$ For the second part, use Theorem 3.3 with $\frac{d}{dx} f_n$ as integrand.

3.16 Hint: Use the results of Exercise 3.5. Apply them to the sequence of partial sums.

3.17 Hint: (i) $|af_n + bg_n - af - bg| \le |a|\,|f_n - f| + |b|\,|g_n - g|$
(ii) $|hf_n - hf| = |h|\,|f_n - f| \le |B|\,|f_n - f|$.

3.19 Hint: (i) $|\tanh x| \le 1$ for all real x.
(ii) Note that $0 \le e^{nx}/2^n = (e^x/2)^n \le (3/4)^n$ for $x \le \log \frac{3}{2}$.
(iii) Observe that $\sum_1^\infty nx^n = x \sum_1^\infty nx^{n-1}$. Now use the results of Section 3 dealing with uniform convergence and term-by-term differentiation of $\sum_0^\infty x^n$.

3.20 Hint: $\sum_1^\infty f_n(x)$ clearly converges absolutely for every x in D by comparison. Uniform, absolute convergence follows easily. For generalizations consider the results of Theorem 1.14.

3.21 Hint: Apply the results of Theorem 3.4 to the sequence of partial sums.

3.22 Hint: $\frac{d}{dx} S(x) \equiv \frac{d}{dx} \sum_{1}^{\infty} f_n(x) = \sum_{1}^{\infty} \frac{d}{dx} f_n(x)$.

3.23 Hint: For example, try either series (i) or series (ii) of Exercise 3.19.

References to Additional and Related Material: Section 3.

1. Boas, R., "A Primer of Real Functions", Carus Mathematical Monograph 13, Mathematical Association of America (1961).

2. Bromwich, T., "An Introduction to the Theory of Infinite Series", Macmillan and Co. (1926).

3. Brand, L., "Advanced Calculus", John Wiley and Sons, Inc. (1958).

4. Francis, E., "Examples in Infinite Series, with Solutions", Deighton, Bell and Co. (1953).

5. Goffman, C., "Real Functions", Holt, Reinhart and Winston, Inc. (1961).

6. Green, J., "Sequences and Series", Glencoe, Ill. Free Press (1958).

7. Halberstam, H. and K. Roth, "Sequences", Clarendon Press (1966).

8. Hirschman, I., "Infinite Series", Holt, Reinhart and Winston, Inc. (1962).

9. Hobson, E., "The Theory of Functions of a Real Variable", Vol. II, Dover Publications, Inc.

10. Rainville, E., "Infinite Series", Macmillan and Co. (1967).

11. Stanaitis, D., "An Introduction to Sequences, Series, and Improper Integrals", Holden-Day, Inc. (1967).

12. Titchmarsh, E., "The Theory of Functions", Second Edition, Oxford University Press (1960).

4. Real Power Series

We now center attention upon a particular type of infinite series of functions that occurs frequently in Applied Mathematics. This special type of series is known as a power series.

By a <u>power series about $x = a$</u> is meant a series of real functions of the form:

$$\sum_{0}^{\infty} a_n(x-a)^n$$

where the coefficients a_n $(n = 0,1,2,...)$ are constants. While investigating such series and their properties, we shall be specializing many results of previous Sections.

<u>EXAMPLE 4.1</u> Two examples of power series are:

(i) $\sum_{0}^{\infty} x^n/n!$ $\qquad a = 0, \quad a_n = 1/n!$

(ii) $\sum_{0}^{\infty} [\frac{(x+1)}{2}]^n$ $\qquad a = -1, \quad a_n = (\frac{1}{2})^n$.

Clearly every power series about $x = a$ converges (to a_0) at $x = a$. The following result establishes more about convergence properties of power series.

<u>THEOREM 4.1</u> (Radius of Convergence of Power Series) Associated with any power series $\sum_{0}^{\infty} a_n(x-a)^n$ about $x = a$ is a non-negative number R termed its <u>radius of convergence</u> about $x = a$. If $R = 0$ then the series

converges only for $x = a$. If $0 < R < +\infty$ then the series converges absolutely for $|x - a| < R$. If $R = +\infty$ then the series converges absolutely for all real x.

PROOF. Consider a fixed value of x, and apply Cauchy's Root Test (Theorem 1.16) to the infinite series of real numbers $\sum_0^\infty a_n(x-a)^n$. Accordingly, this series <u>converges absolutely</u> if

$$\limsup \sqrt[n]{|x-a|^n |a_n|} =$$

$$\limsup |x-a| \sqrt[n]{|a_n|} =$$

$$|x-a| \limsup \sqrt[n]{|a_n|} =$$

$$|x-a|\,\bar{c} < 1$$

and <u>diverges</u> if

$$|x-a|\,\bar{c} > 1$$

(with no conclusion if $|x-a|\,\bar{c} = 1$).

If $\bar{c} = +\infty$ then $R = 0$ and the series converges only for $x = a$. If $0 < \bar{c} < +\infty$ then the series converges absolutely for $|x-a| < R = 1/\bar{c}$ and diverges for $|x-a| > R = 1/\bar{c}$, with no conclusion if $|x-a| = R$. Finally, if $\bar{c} = 0$ then the series converges absolutely for any real x and $R = +\infty$. This completes the proof.

Although radius of convergence R is defined in terms of the quantity $\bar{c}$ of Cauchy's Root Test (which always exists), there may be simpler means of obtaining R in certain cases.

(i) if $\lim \sqrt[n]{|a_n|} = c$ then $R = 1/c$

(ii) if $\lim |a_{n+1}/a_n| = r$ then $R = 1/r$.[1]

In view of the connection between Cauchy's Root Test and D'Alembert's Ratio Test for convergence, if the limit in (ii) exists, then $r = c = \bar{c}$ and all methods are equivalent.

If the radius of convergence of a power series is non-zero, then it possesses certain uniform convergence properties. These are given in the following Theorem.

<u>THEOREM 4.2</u> (Uniform Convergence of Power Series) If the power series $\sum_0^\infty a_n(x-a)^n$ has a non-zero radius of convergence R, then for any number R', $0 < R' < R$, the series converges uniformly for $|x-a| \le R'$.

PROOF. Since $|a_n(x-a)^n| \le |a_n|(R')^n$ for all $|x-a| \le R'$ and $n = 0,1,2,\ldots$, the proof follows immediately from the Weierstrass M-Test because the series $\sum_0^\infty |a_n|(R')^n$ is clearly convergent. This completes the proof.

<u>EXAMPLE 4.2</u> Consider the power series $\sum_1^\infty x^n/n^2$. Since $|a_{n+1}/a_n| \to 1$ as $n \to \infty$ it follows that $R = 1$ and so the series converges for $|x| < 1$, and diverges for $|x| > 1$. Upon special examination, the series is seen to converge at

[1]The convention "$1/\infty = 0$" and "$1/0 = \infty$" is adopted.

the 'endpoints' $x = \pm 1$. (Of course, the series converges uniformly for $|x| \le R'$ for any R' with $0 < R' < 1$).

EXERCISE 4.1 Examine the convergence properties of the following power series:

(i) $\sum_{1}^{\infty} x^{2n-1}/(2n-1)!$

(ii) $\sum_{1}^{\infty} (-1)^{n+1}(x-1)^n/n$

We are now in a position to examine some of the important and useful properties possessed by power series in general. The first deals with continuity properties of the sum function.

THEOREM 4.3 (Continuity of Sum Function of a Power Series) Let $\sum_{0}^{\infty} a_n(x-a)^n$ be a power series with non-zero radius of convergence R. Then the sum function $S(x)$, defined as $S(x) = \sum_{0}^{\infty} a_n(x-a)^n$ for $|x-a| < R$ is a continuous function of x.[1]

PROOF. EXERCISE 4.2

Accordingly, in view of the preceding Theorem, we may conclude that if c is any point within the interval of convergence of the power series, that is, the interval $a - R < x < a+R$, then:

$$\lim_{x \to c} S(x) = S(c) = \sum_{0}^{\infty} a_n(c-a)^n \quad .$$

The endpoints of the interval of convergence (assuming R finite),

[1]For a given Power Series, it may not be possible to express $S(x)$ in "closed form".

namely $x = a \pm R$, usually need special attention. These will be treated later.

In Example 3.11 it was shown that the series $\sum_0^\infty x^n$ with sum function $S(x) = 1/(1-x)$ could be integrated term-by-term within $[-a,a]$ $(0 < a < 1)$, and it was concluded, for example, as an application, that $\log 1/(1-x) = x + x^2/2 + x^3/3 + \ldots$ uniformly within the same interval. This is a property possessed by all power series with a non-zero radius of convergence, and the result is contained in the following theorem.

<u>THEOREM 4.4</u> (Term-by-Term Integration of Power Series) Let $\sum_0^\infty a_n(x-a)^n$ be a power series with non-zero radius of convergence R and sum function S. Then, the series may be integrated term-by-term within its interval of convergence in the sense that if $[b,d]$ is any closed, bounded interval therein,

$$\int_b^d S(x)\,dx = \int_b^d \sum_0^\infty a_n(x-a)^n\,dx = \sum_0^\infty \int_b^d a_n(x-a)^n dx\ .$$

PROOF. <u>EXERCISE 4.3.</u>

Accordingly, within the interval of convergence of a Power Series, the operations of summation and integration may validly be interchanged.

EXERCISE 4.4 Prove that: $1/(1+x^2) = 1 - x^2 + x^4 - x^6 + \dots$ over $(-1,1)$. Then show that for any x within this interval:

$$\tan^{-1} x = x - x^3/3 + x^5/5 \quad \dots$$

$$\log(1+x) = x - x^2/2 + x^3/3 + \quad \dots \quad .$$

It will not be surprising, now, that a power series can be differentiated term-by-term within its interval of convergence. The precise result is contained in the following Theorem.

THEOREM 4.5 (Term-by-Term Differentiation of Power Series) Let $\sum_0^\infty a_n(x-a)^n$ be a power series with non-zero radius of convergence R and sum function S. Then for any x within its interval of convergence S is differentiable and $S'(x) = \sum_0^\infty na_n(x-a)^{n-1}$.

PROOF. EXERCISE 4.5.

Accordingly, the preceding Theorem states that, within its interval of convergence, the operations of summation and differentiation may validly be interchanged.

Now within the interval of convergence of the power series $S(x) = \sum_0^\infty a_n(x-a)^n$ the operation of differentiation yields the differentiated series $S'(x) = \sum_0^\infty na_n(x-a)^n$ which itself is a power series. Furthermore, it is not difficult to verify that this power series has the same interval of convergence as the original series. This process may be continued, each time yielding a new power series with the same interval of convergence:

$$s^{(k)}(x) = \sum_{0}^{\infty} n(n-1) \ \ldots \ (n-k+1) a_n x^{n-k}$$

$(k = 0,1,2,\ldots)$.

In particular, for $x = a$ we obtain:

$$s^{(k)}(a) = k! \ a_k \quad \text{or} \quad a_k = s^{(k)}(a)/k!$$

$(k = 0,1,2,\ldots)$. Thus, successive coefficients of a convergent power series about $x = a$ $(R > 0)$ can be expressed in terms of successive derivatives of the sum function of the series evaluated at $x = a$. This is what is meant by the statement that a power series is the <u>Taylor series representation of its sum function</u>, that is:

$$S(x) = \sum_{0}^{\infty} \frac{S^{(k)}(a)}{k!} (x-a)^k$$

over the interval of convergence.

<u>EXERCISE 4.7</u> (The Geometric Series) The Geometric Series

$$1 + x + x^2 + x^3 + \ldots\ldots = 1/(1-x) \quad , \quad |x| < 1$$

has already been considered. In view of Theorem 4.5, verify the convergence of the following series to the indicated sum functions:

$$1 + 2x + 3x^2 + 4x^3 \ + \ldots = 1/(1-x)^2 \quad , \quad |x| < 1$$
$$2 + 6x + 12x^2 + 20x^3 + \ldots = 2/(1-x)^3 \quad , \quad |x| < 1,$$

and in general, for any integer $k > 2$, and $|x| < 1$,

$$\frac{d^k}{dx^k} \sum_{n=1}^{\infty} x^{n-1} = \sum_{m=1}^{\infty} m(m+1)\ldots(m+k-1)x^{m-1} = k!/(1-x)^{k+1}$$

Hints and Answers to Exercises: Section 4

4.1 Answer: (i) absolute convergence for all real x. (ii) absolute convergence for $0 < x < 2$, conditional convergence for $x = 0$ and $x = 2$.

4.2 Hint: Apply Theorem 3.10, noting that its assumptions are satisfied for $|x-a| \le R'$ for any $R' < R$.

4.3 Hint: Apply Theorem 3.11.

4.4 Hint: It is known that $1/(1-y) = 1 + y + y^2 + \ldots$ for $|y| < 1$. Accordingly, use the substitutions $y = -x^2$ and $y = -x$ to obtain the series for $1/(1+x^2)$ and $1/(1+x)$ respectively. The two special cases follow by applying Theorem 4.4 to these two series.

4.5 Hint: Apply Theorem 3.12.

4.6 Hint: Clearly the derivative exists for all $x \neq 0$. For $x = 0$ use the basic definition of the derivative together with L'Hospital's Rule.

4.7 Hint: Successive applications of Theorem 4.5 are used. The general formula can be obtained by first finding the general n-th term of the k-th derivative of the series, then using the change-of-index $m = n - k + 1$.

References to Additional and Related Material: Section 4.

1. Brand, L., "Advanced Calculus", John Wiley and Sons, Inc. (1958)

2. Hobson, E., "The Theory of Functions of a Real Variable", Vol. II, Dover Publications, Inc.

3. Kaplan, W., "Advanced Calculus", Addison-Wesley, Inc. (1952).

5. Behavior of a Function Near a Point: Various Types of Limits

As is sometimes the case in Mathematics, the same (or similar) notation may be used for several different (though possibly related) concepts. Such is the case with the 'lim inf', 'lim sup' and 'lim' notation. We have already dealt with this notation (Sections 1-4) when considering <u>sequences</u> of sets, real numbers and real-valued functions. However, the same notation is also used in describing a different concept, related to the behavior of an <u>individual</u> real-valued function f near a single point $x = c$. In this Section, we examine this alternate concept of 'lim inf', 'lim sup' and 'lim', in an attempt to avoid possible confusion. Some (if not all) of the notions considered may be familiar.

Consider a real-valued function f of the real variable x defined (but not necessarily bounded) over a domain D containing the point $x = c$ as an interior point.[1] We consider five quantities (not all of which need exist) related to the behavior of f near the point $x = c$:

(i) Two-sided limit: $\lim_{x \to c} f(x)$

(ii) Left-hand limit: $\lim_{x \uparrow c} f(x)$

(iii) Right-hand limit: $\lim_{x \downarrow c} f(x)$

(iv) Limit infimum: $\liminf_{x \to c} f(x)$ or $\varliminf_{x \to c} f(x)$

(v) Limit supremum: $\limsup_{x \to c} f(x)$ or $\varlimsup_{x \to c} f(x)$

[1]That is, there exists an open interval centered at $x = c$ completely contained in D. Later, this condition will be relaxed.

We now consider the definitions and conditions for existence of the five quantities given above. The first is from Elementary Calculus.

(i) Two-sided limit. The function f is said to possess a two-sided limit at the interior point $x = c$ iff there exists a finite real number L such that, given any $\epsilon > 0$ there can always be found a $\delta > 0$ sufficiently small so that $|f(x)-L| < \epsilon$ for all x in D satisfying: $0 < |x-c| < \delta$. In this case we then write:

$$\lim_{x \to c} f(x) = L \quad .$$

Since L is finite, and $x = c$ is an interior point of D, it is not difficult to prove that when a two-sided limit L exists at $x = c$ it is unique (hence it is the two-sided limit at $x = c$).

EXERCISE 5.1 Prove uniqueness of a two-sided limit when it exists.

However, even though f may possess a two-sided limit L at $x = c$, it is not necessarily true that the value of f at $x = c$, that is $f(c)$, is equal in value to L.

EXERCISE 5.2 Consider the function f defined as follows:

$$f(x) = \begin{cases} |x| & \text{if } x \neq 0 \\ 1 & \text{if } x = 0 \end{cases}$$

Prove that the two-sided limit of f exists and equals 0 at $x = 0$ yet clearly $f(0) = 1$.

However, a two-sided limit need not exist at a given interior point $x = c$, as is illustrated in the next Exercise.

EXERCISE 5.3 Consider the function f defined as follows:

$$f(x) = \begin{cases} 1+|x| & \text{if } x < 0 \\ 0 & \text{if } x = 0 \\ |x|-1 & \text{if } x > 0 \end{cases} .$$

Prove that f does not possess a two-sided limit at $x = 0$. A graph is helpful.

When a two-sided limit of f does exist at $x = c$, it conveys some information about the behavior of f near $x = c$ in the sense that, for any given $\epsilon > 0$, there can be found a value of $\delta > 0$, such that $L - \epsilon < f(x) < L + \epsilon$ for all x satisfying: $0 < |x-c| < \delta$. However, it has been shown in Exercise 5.3 that, in general, nothing can be said about the value of f at $x = c$.

As with many definitions, there are alternate, but equivalent forms. The following Exercise illustrates an alternate and useful definition of a two-sided limit (the so-called 'Sequence' version).

EXERCISE 5.4 Prove that the following definition of a two-sided limit is equivalent to the one given above. The function f is said to possess a two-sided limit L (finite) at an interior point $x = c$ of D iff for every sequence $\{x_n\}$ of points in a deleted [1] neighborhood

[1]Meaning the point $x = c$ itself is excluded.

of $x = c$ that converges to c, the corresponding sequence $\{f(x_n)\}$ of function values converges to L.

It is to be noted that the familiar notion of continuity of a function at a point is defined in terms of a two-sided limit. That is, the function f is said to be continuous at the interior point $x = c$ of its domain iff the two-sided limit L of f exists at $x = c$ and equals $f(c)$ in value. In other words, f is continuous at $x = c$ iff $\lim_{x \to c} f(x) = f(c)$.

(ii) Left-hand limit. The function f is said to possess a left-hand limit at the interior point $x = c$ iff there exists a finite real number L^- such that, given any $\epsilon > 0$ there can always be found a $\delta > 0$ sufficiently small so that $|f(x) - L^-| < \epsilon$ for all x in D satisfying: $c - \delta < x < c$. In this case we write:

$$\lim_{x \uparrow c} f(x) = L^- \quad .$$

As is also the case for a two-sided limit, when a left-hand limit exists at an interior point $x = c$ it is unique, but need not equal $f(c)$ in value. However it does follow that if f possesses the two-sided limit L at $x = c$ then it possesses the left-hand limit L^- at $x = c$ and $L^- = L$.

EXERCISE 5.5 Give an example of a function possessing the left-hand but not the two-sided limit at a point.

EXERCISE 5.6 If f is defined over a domain D such as $[a, c]$ or $[a,c)$ where $a < c$, we can still define what is meant by f possessing a left-hand limit at $x = c$

even though $x = c$ is not an interior point of (in fact, may not even belong to) D. Explain, and illustrate with diagrams.

<u>EXERCISE 5.7</u> (The Left-hand Limit Need Not Exist) Consider the function f defined as follows:

$$f(x) = \begin{cases} 1 & \text{if } x \text{ is rational} \\ 0 & \text{if } x \text{ is irrational}. \end{cases}$$

Prove that at <u>no</u> point does f possess a left-hand limit. A graph may help.

<u>EXERCISE 5.8</u> Formulate and prove a 'Sequence' version of the definition of the left-hand limit of a function at a point (either interior, or of the type described in Exercise 5.6).

When $\lim_{x \uparrow c} f(x) = L^-$ exists, this left-hand limit L^- is also designated by $f(c-)$. Now, it has been pointed out in Exercise 5.6 that $f(c-)$ may exist in cases where f itself is not even defined at $x = c$. If, however, $f(c)$ is defined, and in addition, $f(c-) = f(c)$, i.e. $\lim_{x \uparrow c} f(x) = f(c)$, then the function f is said to be <u>continuous from the left</u> at $x = c$.

<u>EXERCISE 5.9</u> Give an example of a function that is continuous from the left but not continuous at a point. Draw a graph illustrating this.

However, it is clear that if a function f is continuous at a point $x = c$ it is automatically continuous from the left at $x = c$.

EXERCISE 5.10 What information is conveyed about the behavior of f near $x = c$ if the left-hand limit of f exists at $x = c$? Can this be improved if f is known to be continuous from the left at $x = c$?

(iii) Right-hand limit. The function f is said to possess a right-hand limit at the interior point $x = c$ iff there exists a finite real number L^+ such that, given any $\epsilon > 0$ there can always be found a $\delta > 0$ sufficiently small so that $|f(x) - L^+| < \epsilon$ for all x in D satisfying: $c < x < c + \delta$. In this case we write:

$$\lim_{x \downarrow c} f(x) = L^+ \quad .$$

EXERCISE 5.11 As with the case of the left-hand limit, the right-hand limit may be defined at a point $x = c$ even though it is not an interior point of D. Specifically, if f is defined over a domain D such as $[c,b]$ or $(c,b]$ with $c < b$, we may still define what is meant by f possessing a right-hand limit at $x = c$ even though f may not be defined at $x = c$. Explain.

As has been the case with the other types of limits considered so far, when a right-hand limit exists at a point, it is unique. However, the right-hand limit need not exist, as is illustrated in the next Exercise.

EXERCISE 5.12 (The Right-hand Limit Need Not Exist) Use the function in Exercise 5.7 to prove that a function need not possess a right-hand limit at any point. Construct another example.

When the right-hand limit of f exists at a point $x = c$, it is further denoted by $f(c+)$. As may be anticipated, even though $f(c+)$ exists at a point $x = c$ it need not equal $f(c)$ when f is defined at $x = c$. If, however, it occurs that $f(c+) = f(c)$, i.e. $\lim_{x \downarrow c} f(x) = f(c)$, then the function f is said to be <u>continuous from the right</u> at $x = c$.

<u>EXERCISE 5.13</u> Prove that if f possesses the two-sided limit L at the interior point $x = c$, then both left- and right-hand limits of f exist at $x = c$ and $L^- = L^+ = L$, and conversely.

<u>EXERCISE 5.14</u> (Continuation) Prove that f is continuous at the interior point $x = c$ iff it is both continuous from the right and continuous from the left at $x = c$, that is, iff $f(c-) = f(c+) = f(c)$.

<u>EXERCISE 5.15</u> Formulate and prove a 'Sequence' version of the definition of a right-hand limit of a function f at a point $x = c$ (either an interior point, or a point of the type described in Exercise 5.11).

<u>EXAMPLE 5.1</u> If it is known that f possesses the right-hand limit $f(c+)$ at the point $x = c$, then given any $\epsilon > 0$ there can always be found a $\delta > 0$ sufficiently small so that $f(c+) - \epsilon < f(x) < f(c+) + \epsilon$ for all x in D satisfying: $c < x < c + \delta$. Furthermore, if $x = c$ is a point in D at which f is continuous from the right, it may additionally be asserted that for all x in D satisfying: $c \le x < c + \delta$, we have that $f(c) - \epsilon < f(x) < f(c) + \epsilon$. In this sense, existence of

the right-hand limit of a function at a point conveys information about the behavior of the values of the function near the point.

We have considered so far only the case of existence (or non-existence) of <u>finite</u> two-sided, left and right-hand limits of a function f at a point $x = c$. It is not difficult to modify the definitions given above so as to permit consideration of infinite limits (subject to existence) of each of the three types, at a point $x = c$ where f becomes unbounded (either positively or negatively). This extension constitutes the following Exercise.

<u>EXERCISE 5.16</u> Review the definitions of two-sided, left-and right-hand limits (finite). Modify each so as to permit consideration of infinite limits (of each type) at a point $x = c$ in D where f becomes unbounded. Such modified definitions applied to:

$$f(x) = \begin{cases} 1/x, & x \neq 0 \\ 0, & x = 0 \end{cases}$$

should yield

$$\lim_{x \uparrow 0} f(x) = -\infty \quad \text{and} \quad \lim_{x \downarrow 0} f(x) = +\infty$$

with non-existence of the two-sided limit at $x = 0$. In addition, apply the definitions to the function:

$$f(x) = \begin{cases} 1/|x|, & x \neq 0 \\ 0, & x = 0 \end{cases}$$

and verify that

$$\lim_{x \to 0} f(x) = +\infty .$$

(Clearly $\lim_{x \uparrow 0} f(x)$ and $\lim_{x \downarrow 0} f(x)$ have the same value). The concept of continuity is not relevant at a point possessing an infinite limit.

<u>EXAMPLE 5.2</u> (Existence of limit as $x \to +\infty$ or $x \to -\infty$) Familiar from Elementary Calculus is the extension of the notion of existence of a limit as x approaches a point $x = c$ to that of existence of a limit as x approaches $+\infty$ or $-\infty$. We consider the finite case first. Specifically, the real-valued function f, defined over a suitable unbounded domain, is said to possess a finite limit L as $x \to +\infty$ $(x \to -\infty)$ iff for any given $\epsilon > 0$ there can always be found a corresponding positive constant K sufficiently large so that $|f(x) - L| < \epsilon$ for all x satisfying: $x > K$ $(x < -K)$.[1] In such a case we write: $\lim_{x \to +\infty} f(x) = L$ $(\lim_{x \to -\infty} f(x) = L)$. As with the case of a fixed point $x = c$, f need not possess a limit (finite or otherwise) as $x \to +\infty$ or $x \to -\infty$. This is demonstrated by the function $f(x) = \sin x$. However, when such a limit does exist, it is unique. The following are two examples of functions possessing the (finite) limit as $x \to +\infty$ or $x \to -\infty$.

(i) $\lim_{x \to +\infty} (\sin x)/x = 0$ since, given any $\epsilon > 0$ we have $|(\sin x)/x - 0| < \epsilon$ for all $x > K$ with $K = 1/\epsilon$.

[1] A set of this type is called a neighborhood of $+\infty$ $(-\infty)$.

Similarly, $\lim_{x \to -\infty} (\text{Sin } x)/x = 0$ since, given any $\epsilon > 0$ we have $|(\text{Sin } x)/x - 0| < \epsilon$ for all $x < -K$. (Recall, $|\text{Sin } x| \leq 1$ for all x).

(ii) $\lim_{x \to +\infty} \int_0^x e^{-y} dy = \lim_{x \to +\infty} (1-e^{-x}) = 1$ since given any $\epsilon > 0$ we have $|1-e^{-x}-1| = |e^{-x}| < \epsilon$ for all $x > K$ with $K = \ln(1/\epsilon)$, $\epsilon < 1$.

<u>EXERCISE 5.17</u> Formally extend the notion of existence of a finite limit L as $x \to +\infty$ or $x \to -\infty$ so as to permit consideration of existence of an infinite limit as $x \to +\infty$ or $x \to -\infty$. Give several examples of functions possessing such limits.

<u>EXERCISE 5.18</u> (Monotone Functions) For simplicity, assume f is defined for all real x. If f is either non-increasing or non-decreasing for sufficiently large positive (negative) values of x, then $\lim_{x \to +\infty} f(x)$ ($\lim_{x \to -\infty} f(x)$) always exists, though may not be finite. Prove the preceding statements.

<u>EXERCISE 5.19</u> (Relating limits of different types) Provided f is defined over the suitable domain:

(i) $\lim_{x \to +\infty} f(x) = \lim_{x \downarrow 0} f(1/x)$

and

(ii) $\lim_{x \to -\infty} f(x) = \lim_{x \uparrow 0} f(1/x)$

in the sense that if either limit of a pair above exists, then both do, and the two are equal. Prove the preceding statement. Accordingly, since it is known (from Elementary Calculus) that $\lim_{x\downarrow 0} (\text{Sin } x/x) = 1$, it follows that $\lim_{x\to+\infty} x \text{ Sin}(1/x) = 1$.

EXAMPLE 5.3 (Operations With Limits: Algebraic Properties) If f and g are real-valued functions defined over a common domain, and a and b are positive constants, then subject to existence of each limit in a given expression,[1] the three types of limits considered so far [2] obey the following rules.

(1) Addition. $\lim[a\, f(x) \pm bg(x)] = a \lim f(x) \pm b \lim g(x)$.

(a) Exceptions to Rule.[3]

(i) $a \cdot b > 0$ (i.e. a and b same sign)

with $\lim f(x) = +\infty$ and $\lim g(x) = -\infty$

or $\lim f(x) = -\infty$ and $\lim g(x) = +\infty$

(ii) $a \cdot b < 0$ (i.e. a and b opposite sign)

with $\lim f(x) = +\infty$ and $\lim g(x) = +\infty$

or $\lim f(x) = -\infty$ and $\lim g(x) = -\infty$

---

[1]This assumption is necessary. Consider, for example, the case where $f(x) = 1+\sin(1/x)$ and $g(x) = 1-\sin(1/x)$ as $x \to 0$. Clearly $[f(x)+g(x)] \equiv 2$, whence $\lim_{x\to 0} [f(x)+g(x)] = 2$, yet the Rule 1 does not apply since neither $\lim_{x\to 0} f(x)$ nor $\lim_{x\to 0} g(x)$ exists.

[2]The notation "lim" is used in stating the Rules. It indicates (subject to existence) validity of the expression for $\lim_{x\to c}$, $\lim_{x\downarrow c}$, $\lim_{x\uparrow c}$, $\lim_{x\to+\infty}$, or $\lim_{x\to-\infty}$.

[3]Certain important exceptions will be treated in a subsequent Example as a review of L'Hospital's Rule.

(b) Special Computations with Infinite Limits.

(i) $k + (+\infty) = (+\infty) + k = +\infty$ for k finite or $k = +\infty$.

(ii) $k + (-\infty) = (-\infty) + k = -\infty$ for k finite or $k = -\infty$.

(iii) $c \cdot (\pm\infty) = \pm\infty$ for $c > 0$ and $\mp\infty$ for $c < 0$.

(2) Multiplication. $\lim[f(x)g(x)] = \lim f(x) \cdot \lim g(x)$

(a) Exceptions to Rule.

(i) $\lim f(x) = 0$ and $\lim g(x) = \pm\infty$

(ii) $\lim f(x) = \pm\infty$ and $\lim g(x) = 0$

(b) Special Computations with Infinite Limits.

(i) $k \cdot (\pm\infty) = (\pm\infty) \cdot k = \pm\infty$ for $k > 0$ or $k = +\infty$

(ii) $k \cdot (\pm\infty) = (\pm\infty) \cdot k = \mp\infty$ for $k < 0$ or $k = -\infty$

(3) Division. $\lim[f(x)/g(x)] = \lim f(x) / \lim g(x)$

(a) Exceptions to Rule.

(i) $\lim f(x) = \pm\infty$ and $\lim g(x) = \pm\infty$

(ii) $\lim f(x) = 0$ and $\lim g(x) = 0$

(b) Special Computations with Infinite Limits.

(i) $k/(\pm\infty) = 0$ for k finite

(ii) $(\pm\infty)/k = \pm\infty$ for $k > 0$

(iii) $(\pm\infty)/k = \mp\infty$ for $k < 0$ [1]

[1]The case $k = 0$ must be treated carefully, depending upon the manner in which $g(x)$ approaches 0 as its limit.

(4) Composition. If f is continuous and $f(g(x))$ is defined for all appropriate values of x then

$$\lim f(g(x)) = f(\lim g(x)) .$$

(a) Special Convention.

$$f(\pm\infty) = \lim_{y \to \pm\infty} f(y) .$$

Properties (1), (2) and (3) are normally proved in detail for the two-sided limit in Elementary Calculus. A review of such proofs will reveal that little modification is necessary to establish the three Rules for the other four types of limits. Rule (4) may be unfamiliar. Consider this Rule for the case of the two-sided limit, in which case it is asserted that:

$$\lim_{x \to c} f(g(x)) = f(\lim_{x \to c} g(x)) .$$

For simplicity, assume that $\lim_{x \to c} g(x) = L_1$ is finite. It will suffice to establish that:

$$\lim_{x \to c} f(g(x)) = f(L_1) .$$

First, by the assumed continuity of f at L_1, given any $\epsilon > 0$ there can always be found a $\delta_1 > 0$ sufficiently small so that $|f(y) - f(L_1)| < \epsilon$ for all y satisfying: $0 < |y-L_1| < \delta_1$. Next, by definition of $\lim_{x \to c} g(x)$, given δ_1 there can always be found a corresponding $\delta > 0$ so that $|g(x) - L_1| < \delta_1$ for all x satisfying: $0 < |x-c| < \delta$. Upon combining the two preceding results, it follows that $|f(g(x)) - f(L_1)| < \epsilon$ for all x satisfying:

$0 < |x-c| < \delta$, thereby establishing the required result. The cases involving limits of the other types, follow in a similar manner.

<u>EXERCISE 5.20</u> Apply Rule 4 to evaluate the following limits:

(i) $\lim_{x \to \pm\infty} e^{-(\sin x)/x}$ and (ii) $\lim_{x \to 0} e^{-1/x^2}$.

<u>EXAMPLE 5.4</u> (L'Hospital's Rules) <u>Certain</u> exceptions to Rules (1), (2) and (3) may still be treated by using the so-called L'Hospital's Rules, provided the functions involved possess suitable derivatives (this was not required in the general case). We shall illustrate one such rule. Rules applying to other exceptional cases are generally found in any good Calculus book.

Suppose, for example, we wish to evaluate $\lim f(x)/g(x)$ [1] but $\lim f(x) = 0$ and $\lim g(x) = 0$.[3] This, then, is an exception to Rule (3). If both f and g possess finite first derivatives for those values of x required [2], then L'Hospital's Rule for this case may be stated as follows: if $\lim f'(x)/g'(x)$ exists, then $\lim f(x)/g(x)$ exists and the two are equal in value.

---

[1]As before, "lim" denotes any one of the five types of limits, viz. $\lim_{x \to c}$, $\lim_{x \downarrow c}$, $\lim_{x \uparrow c}$, $\lim_{x \to +\infty}$, or $\lim_{x \to -\infty}$.

[2]Depending upon what type of limit is involved, both $f'(x)$ and $g'(x)$ are required to exist in some deleted neighborhood (two-sided, left or right-hand) of a fixed point $x = c$, or neighborhood of $+\infty$ or $-\infty$, depending upon the case.

[3]Without loss of generality, assume $f(c) = g(c) = 0$.

We prove only the special case of the two-sided limit at a fixed point $x = c$. This proof may, in turn, be modified so as to account for various other cases.

Using the hypotheses, and the Mean Value Theorem for Derivatives, for any point x in a deleted neighborhood of $x = c$ we may write:

$$f(x) = f'(\xi_x)(x-c), \quad \text{where} \quad 0 < |c-\xi_x| < x$$

and

$$g(x) = g'(\eta_x)(x-c), \quad \text{where} \quad 0 < |c-\eta_x| < x$$

hence $\xi_x \to 0$ and $\eta_x \to 0$ as $x \to c$.

Accordingly, for any such x we have:

$$f(x)/g(x) = f'(\xi_x)/g'(\eta_x) \quad .$$

It is not difficult to see that the special properties of ξ_x and η_x guarantee that

$$\lim_{x \to c} f'(x)/g'(x) = \lim_{x \to c} f'(\xi_x)/g'(\eta_x)$$

when the former limit exists, in which case

$$\lim_{x \to c} f'(x)/g'(x) = \lim_{x \to c} f(x)/g(x)$$

This completes the proof.

<u>EXERCISE 5.21</u> Apply L'Hospital's Rules to evaluate:

(i) $\lim_{x \downarrow 0} (\sin x)/x$ and (ii) $\lim_{x \to 1} (\log x)/(1-x)$.

It should be pointed out that most of the notions involving limits for functions of a single real variable have natural generalizations to the case of functions of two or more real variables. These generalizations are quite straightforward. In particular, L'Hospital's Rules have analogues for functions of n real variables which involve the so-called 'Directional Derivative', familiar from Vector Calculus.

It has been seen that, in general, the two-sided, left- or right-hand limit of a function <u>need not exist</u>. Accordingly, whatever information they convey about the behavior of the function (when they exist) may not be available. This difficulty is, in part, overcome by dealing with a more general pair of quantities which, in contrast with limits, <u>always exist</u> (though may be infinite). These quantities convey some information about the behavior of the function, and both reduce to the ordinary two-sided limit when the latter exists (see Exercise 5.34). The quantities are the so-called two-sided[1] limit infimum and limit supremum of the function. We now consider these quantities.

As before, let f be a real-valued function defined over a domain D containing $x = c$ as an interior point.[2]

---

[1]More generally, we could consider the left and right-hand limit infimum and limit supremum. These quantities also always exist (though may be infinite) and reduce to the familiar left and right-hand limits when the latter exist. For most purposes, however, the two-sided quantities suffice.

[2]Later this condition will be relaxed.

First we consider the formal definition of the limit infimum of f at $x = c$, then remark about this definition.

(iv) <u>Limit infimum.</u>[3]

$$\liminf_{x \to c} f(x) = \lim_{\delta \downarrow 0} \inf_{0<|x-c|<\delta} f(x) \ .$$

Although apparently quite complicated, the expression on the right-hand side above has a simple interpretation. First note that

$$I(\delta) = \inf_{0<|x-c|<\delta} f(x)$$

is actually just an ordinary real-valued function of the real variable δ for $\delta > 0$, and essentially represents the 'smallest' possible value $f(x)$ can assume within the deleted neighborhood $0 < |x-c| < \delta$ of the point $x = c$. Furthermore, it is not difficult to see that $I(\delta)$ is non-decreasing in δ as $\delta \downarrow 0$ so that $I(\delta)$ <u>must</u> tend to a limit (possibly infinite) as $\delta \downarrow 0$. This limit is, then, by definition the limit infimum of f at $x = c$.

<u>EXERCISE 5.22</u> Prove the following statements. Diagrams may be useful. $\liminf_{x \to c} f(x)$ may assume any value between $-\infty$ and $+\infty$ inclusive, for, if f is unbounded <u>below</u> near the point $x = c$, as is the case with $f(x) = 1/x$ near $x = 0$, then $I(\delta) \equiv -\infty$ for all $\delta > 0$ (why?),

---

[3]Recall that $\liminf_{x \to c} f(x)$ is also denoted by $\underline{\lim}_{x \to c} f(x)$.

so that $\liminf_{x\to 0} (1/x) = -\infty$. Next, if f is bounded above and below in some deleted neighborhood $0 < |x-c| < \delta'$ of $x = c$ then $\liminf_{x\to c} f(x)$ is finite. Such is the case for the function $f(x) = \text{Sin}(1/x)$ near $x = 0$. Finally, if $\lim_{x\to c} f(x) = +\infty$ then $\underline{\lim}_{x\to c} f(x) = +\infty$ and conversely. Such a case is demonstrated by the function $f(x) = 1/|x|$ near $x = 0$.

<u>EXERCISE 5.23</u> Prove that the 'lim inf' of the function f defined in Exercise 5.7 is 0 at any point.

Accordingly, the lim inf of a function at a point essentially conveys information about 'how small' the value of the function can be in a deleted neighborhood about the point. The following Exercise makes this point more precise.

<u>EXERCISE 5.24</u> (A Property of 'lim inf') Suppose that $\underline{L} = \liminf_{x\to c} f(x)$. If $\underline{L}$ is finite, prove that given any $\epsilon > 0$ there can always be found a $\delta > 0$ sufficiently small so that $f(x) \geq \underline{L} - \epsilon$ whenever $0 < |x-c| < \delta$. Modify this result in case $\underline{L} = \pm\infty$. Without knowing more about the function itself, nothing can be said about the value $f(c)$ of f at $x = c$.

In view of the preceding Exercise, the 'lim inf', which always exists, does provide potentially useful information about how 'small' the values of a function can be in a deleted neighborhood of a point. This may be sufficient in many practical applications.

EXERCISE 5.25 Either review some of the 'maveric' functions of Elementary Calculus or construct some examples of your own, of functions not possessing a limit of any type (neither two-sided, left nor right hand) at certain points. Evaluate the 'lim inf' of such functions at these points.

For completeness, we now consider a standard 'Sequence' version of the definition of the lim inf of a function at a point. This alternate definition may, at times, prove useful, and is contained in the following Exercise.

EXERCISE 5.26 ('Sequence' Version of 'lim inf' Definition) For any sequence of points $\{x_n\}$ in a deleted neighborhood of $x = c$ converging to c, the corresponding sequence of function values $\{f(x_n)\}$ need not converge. However, the quantity $\liminf f(x_n)$ (this is the 'lim inf' of a sequence of real numbers now) always exists. As $\{x_n\}$ ranges over all possible such sequences, we obtain the set S consisting of the respective lim inf's of the corresponding sequences $\{f(x_n)\}$ of function values. Prove that:

$$\liminf_{x \to c} f(x) = \inf S \quad .$$

Next we consider the related concept of the limit supremum of a function at a point. The formal definition follows.

(v) Limit supremum.[1]

$$\limsup_{x \to c} f(x) = \lim_{\delta \downarrow 0} \sup_{0<|x-c|<\delta} f(x) .$$

As with the definition of limit infimum, the quantity on the right-hand side above is, at first glance, quite complicated. However, it has a straightforward interpretation. Observe first that

$$J(\delta) = \sup_{0<|x-c|<\delta} f(x)$$

is an ordinary real-valued function of the real variable δ for $\delta > 0$, and essentially represents the 'largest' possible value $f(x)$ may assume within the deleted neighborhood $0 < |x-c| < \delta$ of the point $x = c$. Furthermore, $J(\delta)$ is non-increasing in δ as $\delta \downarrow 0$ so that the function $J(\delta)$ must tend to a limit (possibly infinite) as $\delta \downarrow 0$. This limit is, then, by definition the limit supremum of f at $x = c$.

EXERCISE 5.27 Prove that if f becomes unbounded above near $x = c$ then $J(\delta) \equiv +\infty$ for all $\delta > 0$ so that $\limsup_{x \to c} f(x) = +\infty$. If f is bounded above and below in some deleted neighborhood $0 < |x-c| < \delta$ of $x = c$ then $\limsup_{x \to c} f(x)$ is finite. If $\lim_{x \to c} f(x) = -\infty$, then $\limsup_{x \to c} f(x) = -\infty$ and conversely. Prove each statement, and illustrate each case with a particular function.

EXERCISE 5.28 Prove that the 'lim sup' of the function f defined in Exercise 5.7 is 1 at any point.

[1] Recall that $\limsup_{x \to c} f(x)$ is also denoted by $\overline{\lim}_{x \to c} f(x)$.

The following Exercise makes more precise the statement that the limit supremum conveys information about 'how large' the values of a function can be in a deleted neighborhood of a point.

EXERCISE 5.29 (A property of 'Lim sup') Suppose that $\overline{L} = \limsup_{x \to c} f(x)$. If $\overline{L}$ is finite, prove that given any $\epsilon > 0$ there can always be found a $\delta > 0$ sufficiently small so that $f(x) \le \overline{L} + \epsilon$ for all x satisfying: $0 < |x-c| < \delta$. Modify this result in case $\overline{L} = \pm \infty$. Without knowing more about the function f itself, nothing can be said about $f(c)$, the value of f at $x = c$.

EXERCISE 5.30 Repeat Exercise 5.24 for limit supremum.

EXERCISE 5.31 Using Exercise 5.26 as a pattern, formulate and prove a 'Sequence' version of the definition of the limit supremum of a function at a point.

EXERCISE 5.32 (Combined Property of 'lim inf' and 'lim sup') Prove that if $A \le f(x) \le B$ for all x satisfying: $0 < |x-c| < \delta'$, for some $\delta' > 0$, then:

$$A \le \liminf_{x \to c} f(x) \le \limsup_{x \to c} f(x) \le B .$$

EXERCISE 5.33 Consider the function f defined by:

$$f(x) = \begin{cases} 1/x & \text{for } x > 0 \text{ and rational} \\ 1 & \text{for } x > 0 \text{ and irrational} \\ 0 & \text{for } x = 0 \\ -1 & \text{for } x < 0 \text{ and irrational} \\ 1/x & \text{for } x < 0 \text{ and rational} \end{cases}$$

Verify that neither two-sided, left nor right-hand limits exist for f at $x = 0$ yet:

$$\liminf_{x \to 0} f(x) = -\infty \quad \text{and} \quad \limsup_{x \to 0} f(x) = +\infty .$$

The following Exercises establish the connection between lim inf, lim sup and existence of the two-sided limit of a function at a point, as well as the basic algebraic properties of 'lim inf' and 'lim sup' of a function at a point.

<u>EXERCISE 5.34</u> (Relating '$\underline{\lim}$', '$\overline{\lim}$' and 'lim')

Prove that if $\lim_{x \to c} f(x) = L$ exists, then:

$$\liminf_{x \to c} f(x) = \limsup_{x \to c} f(x) = L$$

and conversely. Demonstrate the result with:

$$f(x) = \begin{cases} 1/|x| & \text{for } x \neq 0 \\ 0 & \text{for } x = 0 \end{cases}$$

at $x = 0$.

<u>EXERCISE 5.35</u> (Some Algebraic Properties of $\underline{\lim}$ and $\overline{\lim}$)

Establish the following algebraic properties of the $\underline{\lim}$ and/or $\overline{\lim}$ at any fixed point common and interior to the domain(s) of the function(s) indicated.

(i) $\underline{\lim}\ -[f(x)] = -\overline{\lim}\ f(x)$

$\overline{\lim}\ -[f(x)] = -\underline{\lim}\ f(x)$

(ii) $\underline{\lim}\ [f(x) + g(x)] \geq \underline{\lim}\ f(x) + \underline{\lim}\ g(x)$

$\overline{\lim}\ [f(x) + g(x)] \leq \overline{\lim}\ f(x) + \overline{\lim}\ g(x)$

(iii) Provided f and g are non-negative:

$$\underline{\lim}\,[f(x)g(x)] \geq \underline{\lim}\, f(x) \cdot \underline{\lim}\, g(x)$$

$$\overline{\lim}\,[f(x)g(x)] \leq \overline{\lim}\, f(x) \cdot \overline{\lim}\, g(x)$$

$$\underline{\lim}\,[1/f(x)] = 1/\overline{\lim}\, f(x)$$

$$\overline{\lim}\,[1/f(x)] = 1/\underline{\lim}\, f(x)$$

These properties may then be combined. For example, if f, g and h are non-negative:

$$\underline{\lim}\,[f(x)g(x) - 1/h(x)] \geq \underline{\lim}\, f(x) \cdot \underline{\lim}\, g(x) - 1/\overline{\lim}\, h(x) .$$

EXAMPLE 5.5 Both $\liminf_{x\to+\infty} f(x) = \underline{L}^{o}$ and $\limsup_{x\to+\infty} f(x) = \overline{L}^{o}$ convey information about the behavior of f in a neighborhood of $+\infty$, i.e. a set of the form: $x > K$ for some fixed (but arbitrarily large) positive constant K. Specifically, if $\underline{L}^{o}$ and $\overline{L}^{o}$ are finite, then given any $\epsilon > 0$ there can always be found a positive constant K sufficiently large so that:

$$\underline{L}^{o} - \epsilon \leq f(x) \leq \overline{L}^{o} + \epsilon$$

for all x satisfying:

$$x > K .$$

Analogously for $\liminf_{x\to-\infty} f(x) = \underline{L}_{o}$ and $\limsup_{x\to-\infty} f(x) = \overline{L}_{o}$. A slight modification is necessary if one or more of the above quantities is not finite.

EXAMPLE 5.6 (Dominated Convergence and Fatou's Lemma Generalized) In Section 3 we dealt with the interchange of the two operations $\lim_{x\to c}$ and $\sum_{n=1}^{\infty}$ for infinite series of functions. In the general case, individual terms $f_n(x)$ as well as the sum function $S(x)$ of a pointwise convergent series need not possess a limit at $x = c$, in which case the results in Section 3 would not apply.

By using $\underline{\lim}_{x \to c}$ and $\overline{\lim}_{x \to c}$ (which always exist) we can obtain the following generalization of both Dominated Convergence and Fatou's Lemma in one chain of inequalities.

Specifically, if $\sum_{n=1}^{\infty} f_n(x)$ converges pointwise to $S(x)$ for all x in D, and in addition,

(i) $|f_n(x)| \le b_n$ for all x in D

and

(ii) $\sum_{n=1}^{\infty} b_n$ converges

then

$$\sum_{1}^{\infty} \underline{\lim}_{x \to c} f_n(x) \le \underline{\lim}_{x \to c} \sum_{1}^{\infty} f_n(x) \le \overline{\lim}_{x \to c} \sum_{1}^{\infty} f_n(x) \le \sum_{1}^{\infty} \overline{\lim}_{x \to c} f_n(x) ,$$

alternately,

$$\sum_{1}^{\infty} \underline{\lim}_{x \to c} f_n(x) \le \underline{\lim}_{x \to c} S(x) \le \overline{\lim}_{x \to c} S(x) \le \sum_{1}^{\infty} \overline{\lim}_{x \to c} f_n(x) .$$

PROOF. We prove the right-hand portion (the center is obvious and the left-hand portion can be proved similarly). It is an instructive lesson in applying many of the results in previous Sections. By Exercise 5.32, $-b_n \le \overline{\lim}_{x \to c} f_n(x) \le b_n$ for $n = 1,2,\ldots,$ and, in view of conditions (i) and (ii) and properties of infinite series of real numbers, $\sum_{1}^{\infty} \overline{\lim}_{x \to c} f_n(x)$ is convergent.

Accordingly, given any $\epsilon > 0$ we can find an index N sufficiently large so that for all $m \geq N$,

$$\left|\sum_1^\infty \overline{\lim_{x\to c}} f_n(x) - \sum_1^m \overline{\lim_{x\to c}} f_n(x)\right| < \epsilon/2 \quad ,$$

and an index M sufficiently large so that for all $m \geq M$,

$$\left|\sum_1^\infty b_n - \sum_1^m b_n\right| < \epsilon/2 \quad ,$$

so that both conditions are satisfied for all indices $m \geq K = \max(N,M)$.

Now let m be any fixed integer $\geq K$. Then,

$$\begin{aligned} S(x) = \sum_1^\infty f_n(x) &= \sum_1^m f_n(x) + \sum_{m+1}^\infty f_n(x) \\ &\leq \sum_1^m f_n(x) + \sum_{m+1}^\infty b_n \\ &\leq \sum_1^m f_n(x) + \epsilon/2 \quad . \end{aligned}$$

By Exercise 5.35, we may then conclude that:

$$\overline{\lim_{x\to c}} S(x) \leq \sum_1^m \overline{\lim_{x\to c}} f_n(x) + \epsilon/2 \quad ,$$

and because $m \geq K$,

$$\overline{\lim_{x\to c}} S(x) \leq \sum_1^\infty \overline{\lim_{x\to c}} f_n(x) + \epsilon \quad .$$

Since this is true for any choice of $\epsilon > 0$, the required result is established.

<u>EXAMPLE 5.7</u> (Simple Illustration) If $\sum_{1}^{\infty} a_n$ is any convergent infinite series of real numbers, then the series of functions $\sum_{1}^{\infty} a_n \operatorname{Sin}(n/x)$ converges pointwise for all $x \neq 0$. If $S(x)$ denotes its sum (function), we can conclude from the previous Example that,

$$-\sum_{1}^{\infty} a_n \leq \underline{\lim}_{x\to 0} S(x) \leq \overline{\lim}_{x\to 0} S(x) \leq \sum_{1}^{\infty} a_n$$

since

$$\underline{\lim}_{x\to 0} \operatorname{Sin}(n/x) = -1 \quad \text{for} \quad n = 1,2,\ldots$$

and

$$\overline{\lim}_{x\to 0} \operatorname{Sin}(n/x) = 1 \quad \text{for} \quad n = 1,2,\ldots \quad .$$

<u>EXERCISE 5.36</u> (Integer-variable Version) Using the notation of Example 3.7, formulate the integer-variable version of the result in Example 5.6.

<u>Hints and Answers to Exercises: Section 5</u>

5.1 Hint: To establish uniqueness, assume that:

$$\lim_{x\to c} f(x) = L_1 \quad \text{and} \quad \lim_{x\to c} f(x) = L_2$$

yet

$$L_1 \neq L_2 \ .$$

Next, choose any $\epsilon > 0$ for which:

$$0 < \epsilon < |L_1 - L_2|/2 \quad .$$

Now establish a contradiction by showing that for any such choice of ϵ there cannot be found a corresponding $\delta > 0$ for which, simultaneously, the following two conditions are satisfied by all x in D for which $0 < |x-c| < \delta$:

(i) $|f(x) - L_1| < \epsilon$

and

(ii) $|f(x) - L_2| < \epsilon$.

Why? A diagram may help.

5.2 Hint: For any $\epsilon > 0$ we have $|f(x) - 0| < \epsilon$ for all x in D satisfying: $0 < |x| < \delta$ with $\delta = \epsilon$. However, it is clear that: $|f(0) - 0| = 1$.

5.3 Hint: First prove that for any given real number L (being a possible value for a two-sided limit of f at $x = 0$) there always exist values of $\epsilon > 0$, as small as desired, for which no corresponding $\delta > 0$ can be found such that $|f(x) - L| < \epsilon$ whenever $0 < |x-c| < \delta$. Next, verify that the preceding is, in fact, sufficient to prove that a two-sided limit does not exist at $x = 0$. A graph of f may be helpful in considering the various cases (e.g. $L = 0$, $0 < L \leq 1$, $-1 \leq L < 0$, $L > 1$ and $L < -1$) .

5.4 Hint: First, suppose that $\lim_{x \to c} f(x) = L$ exists (finite) and that $\{x_n\}$ is any sequence of the type described. Given any $\epsilon > 0$ there can always be found a corresponding $\delta > 0$ such that $|f(x) - L| < \epsilon$ for all x in D

satisfying: $0 < |x-c| < \delta$. Next, corresponding to this δ there can be found an index N sufficiently large so that $0 < |x_n-c| < \delta$ for all indices $n \geq N$. Accordingly, $|f(x_n) - L| < \epsilon$ for all indices $n \geq N$, thereby establishing that $f(x_n) \to L$ as $n \to \infty$. For the remaining part of the equivalence of the two definitions, suppose that $f(x_n) \to L$ as $n \to \infty$ for any sequence of the type described, yet $\lim_{x \to c} f(x)$ either does not exist, or else exists but is unequal to L in value. From this assumption deduce a contradiction and accordingly prove that $\lim_{x \to c} f(x) = L$. The Hint for Exercise 5.3 may be helpful.

5.5 Hint: Concerning the function in Exercise 5.3, for example, it is known that the two-sided limit does not exist at $x = 0$ yet it can be shown that $\lim_{x \uparrow 0} f(x) = 1$. Note that $f(0) = 0$. Construct a different example.

5.6 Hint: Either criterion for existence of the left-hand limit of f at $x = c$ depends only upon values of $f(x)$ at points $x < c$ (regardless of how f is defined, if at all, for $x \geq c$). For example, if $f(x) = x$ for $x < 0$ (or $x \leq 0$) then $\lim_{x \uparrow 0} f(x) = 0$, regardless of how f is defined, if at all, for values of $x \geq 0$ (or $x > 0$).

5.7 Hint: Let $x = c$ be any fixed real number (either rational like $\frac{1}{2}$ or irrational like π or $\sqrt{2}$). Within <u>any</u> left-hand neighborhood $c - \delta < x < c$ of $x = c$ there <u>always</u> exist points x for which $f(x) = 0$ and for which $f(x) = 1$.

5.8 Hint: Refer to the Hint for Exercise 5.4.

5.9 Hint: A simple example is the function:

$$f(x) = \begin{cases} x & \text{for } x \le 0 \\ 1 & \text{for } x > 0 \end{cases} ,$$

since at the point $x = 0$ we have:

$$f(0-) = \lim_{x \uparrow 0} f(x) = 0 = f(0) .$$

However, the function defined by:

$$f(x) = \begin{cases} x & \text{for } x < 0 \\ 1 & \text{for } x \ge 0 \end{cases}$$

is <u>not</u> continuous from the left at $x = 0$. Explain why. A graph may be helpful. Construct another example.

5.10 Hint: If $\lim_{x \uparrow c} f(x) = L^-$ exists (finite), then given any $\epsilon > 0$ there can always be found a corresponding $\delta > 0$ such that:

$$L^- - \epsilon < f(x) < L^- + \epsilon \quad \text{whenever} \quad c - \delta < x < c \quad .$$

If f is continuous from the left at $x = c$ then

$$f(c) - \epsilon < f(x) < f(c) + \epsilon \quad \text{whenever} \quad c - \delta < x \le c \quad .$$

5.11 Hint: Adapt the Hint for Exercise 5.6 to this case of the right-hand limit. Construct another example.

5.13 Hint: For the first part, note that the statement: "$|f(x) - L| < \epsilon$ for all x in D satisfying $0 < |x-c| < \delta$", implies that $|f(x) - L| < \epsilon$ for all x in D satisfying $c - \delta < x < c$ or $c < x < c + \delta$. Accordingly, L^- and L^+ exist with $L^- = L$ and $L^+ = L$. For the converse, note that if $|f(x) - L| < \epsilon$ for all x in D such that $c - \delta < x < c$ <u>and</u> $c < x < c + \delta$ then $|f(x) - L| < \epsilon$ for all x in D satisfying $0 < |x-c| < \delta$. Therefore L exists and $L^- = L^+ = L$.

5.14 Hint: Use the results of Exercise 5.13, noting that continuity of f at $x = c$ means that $L = f(c)$. Recall, the alternate notation for L^- is $f(c-)$ and for L^+ is $f(c+)$.

5.15 Hint: Refer to the Hint for Exercise 5.4.

5.16 Hint: Consider the case of the two-sided limit first, and let $x = c$ be any interior point of D. If, given any fixed (but arbitrarily large) positive constant M, there can always be found a corresponding $\delta > 0$ sufficiently small so that $f(x) > M$ for all x in D satisfying: $0 < |x-c| < \delta$, then $\lim_{x \to c} f(x) = +\infty$. Analogously, if $f(x) < -M$ for all x in D satisfying: $0 < |x-c| < \delta$, then $\lim_{x \to c} f(x) = -\infty$. For the case of the left-hand limit, if, given any fixed (but arbitrarily large) positive constant M there can always be found a $\delta > 0$ sufficiently small so that $f(x) > M$ for all x in D satisfying: $c - \delta < x < c$, then $\lim_{x \uparrow c} f(x) = +\infty$. Similarly for $\lim_{x \uparrow c} f(x) = -\infty$, where $-M$ replaces M and $f(x) < -M$

replaces $f(x) > M$. The case of infinite right-hand limits follows immediately.

5.17 Hint: For instance, we say that $\lim_{x \to +\infty} f(x) = +\infty$ if, given any fixed (but arbitrarily large) positive constant M there can always be found a corresponding positive constant K sufficiently large so that $f(x) > M$ for all x in D satisfying: $x > K$. As a simple example, consider $f(x) = x^2$. Clearly $x^2 > M$ for all $x > \sqrt{M}$ so that $\lim_{x \to +\infty} (x^2) = +\infty$.

5.18 Hint: For simplicity suppose that $f(x)$ is non-decreasing over $D = (-\infty, \infty)$ and let $L^* = \sup_{x \in D} f(x)$. First prove that if f is bounded above then L^* is finite. Then use properties of the "sup" of a function over a set (and the fact that f is non-decreasing over D) to establish that, given any $\epsilon > 0$, there can always be found a positive constant K sufficiently large so that $|f(x) - L^*| < \epsilon$ for <u>all</u> x satisfying: $x > K$. Accordingly, $\lim_{x \to +\infty} f(x) = L^*$ (finite case). A slight modification is needed when L^* equals $+\infty$, and a similar argument establishes the result in case f is non-increasing over D.

5.19 Hint: In (i), assume for simplicity that f is defined over $D = (0, +\infty)$ and that $\lim_{x \to +\infty} f(x) = L^*$ is finite. Thus, given any $\epsilon > 0$ there can always be found a corresponding positive constant K sufficiently large so that: $|f(y) - L^*| < \epsilon$ for all $y > K$. Whence, upon setting $y = 1/x$ $(y > 0)$ it follows that

$|f(1/x) - L^*| < \epsilon$ for all x satisfying: $0 < x < \delta$ with $\delta = 1/K$. Accordingly, $\lim_{x \downarrow 0} f(1/x) = L^*$. Analogously for the other cases (including infinite limits).

5.20 Hint: (i) $f(g(x)) = f(z) = e^{-z}$ where $z = g(x) = (\text{Sin } x)/$

(ii) $f(g(x)) = f(z) = e^{-z}$ where $z = g(x) = 1/x^2$.

5.21 Hint: (i) $\lim_{x \downarrow 0} (\text{Sin } x)/x = \lim_{x \downarrow 0} (\text{Cos } x)/1 = 1$

(ii) $\lim_{x \to 1} (\log x)/(1-x) = \lim_{x \to 1} (1/x)/(-1) = -1$.

5.22 Hint: If f is unbounded below near $x = c$ then $I(\delta) \equiv -\infty$ for all $\delta > 0$ since <u>any</u> deleted neighborhood of $x = c$ contains points x for which $f(x)$ is arbitrarily large negatively. Next, if f is bounded above and below for $0 < |x-c| < \delta'$, then prove that $A \le I(\delta) \le B$ (A, B fixed constants) for <u>all</u> $0 < \delta < \delta'$ so that $\underline{\lim}_{x \to c} f(x) = \lim_{\delta \downarrow 0} I(\delta)$ must be finite. Finally prove that $I(\delta) \to +\infty$ as $\delta \downarrow 0$ if and only if $f(x) \to +\infty$ as $x \to c$.

5.23 Hint: Since the value of $f(x)$ is either 0 or 1, and every neighborhood $0 < |x-c| < \delta$ of a fixed point $x = c$ contains points x for which $f(x) = 0$ and for which $f(x) = 1$, it follows that for <u>every</u> $\delta > 0$,

$$I(\delta) = \inf_{0<|x-c|<\delta} f(x) = 0 \quad .$$

Accordingly,

$$\underline{\lim}_{x \to c} f(x) = \lim_{\delta \downarrow 0} I(\delta) = 0 \quad .$$

5.24 Hint: First suppose that $\underline{L} = \lim_{\delta \downarrow 0} \inf_{0<|x-c|<\delta} f(x)$ is finite. Since $I(\delta) = \inf_{0<|x-c|<\delta} f(x)$ is an ordinary function of δ for $\delta > 0$, it follows from the very definition of $\lim_{\delta \downarrow 0}$ that, given any $\epsilon > 0$, $I(\delta) \geq \underline{L} - \epsilon$ for (all) δ sufficiently small. Accordingly, by properties of the "inf" of a function over a set, it follows that: $f(x) \geq \underline{L} - \epsilon$ for all x such that $0 < |x-c| < \delta$. If $\underline{\lim}_{x \to c} f(x) = +\infty$, it can be shown that given any fixed (but arbitrarily large) positive constant M there can be found a corresponding $\delta > 0$ such that $f(x) > M$ for all x for which $0 < |x-c| < \delta$. (See Exercise 5.22). Finally, if $\underline{\lim}_{x \to c} f(x) = -\infty$, no corresponding inequalities can be established. Why?

5.25 Hint: The function in Exercise 5.6 is such a function. At <u>any</u> point $x = c$, $\underline{\lim}_{x \to c} f(x) = 0$. Another example is the function $f(x) = \mathrm{Sin}(1/x)$ for $x \neq 0$ (it may be defined arbitrarily at $x = 0$). It is not difficult to show that within <u>any</u> arbitrarily small deleted neighborhood of $x = 0$, the function assumes <u>all</u> values between -1 and 1 inclusive. Accordingly, neither two-sided, left nor right-hand limits exist at $x = 0$, yet $\underline{\lim}_{x \to 0} \mathrm{Sin}(1/x) = -1$.

5.26 Hint: First assume that $\underline{L}$ is finite and establish that there exists at least one sequence of the type described for which $f(x_n) \to \underline{L}$. This may be done as follows: let $\{\epsilon_n\}$ be a sequence of positive real numbers converging to zero. For each index n $(n = 1,2,\ldots)$ let δ_n denote a corresponding positive real number chosen so that:

$$|I(\delta_n) - \underline{L}| < \epsilon_n/2 ,$$

where :

$$I(\delta_n) = \inf_{0<|x-c|<\delta_n} f(x) .$$

This can always be done. The sequence $\{\delta_n\}$ may clearly be chosen so as also to converge to zero. Accordingly, for each index n $(n = 1,2,\ldots)$ we may choose a point x_n so that $0 < |x_n - c| < \delta_n$, and for which $|f(x_n) - I(\delta_n)| < \epsilon_n/2$. Hence $|f(x_n) - \underline{L}| < \epsilon_n$ so that $\{x_n\}$ is a sequence of the type required.

Next, show that there cannot be a sequence of the type described in the Exercise for which $f(x_n) \to L' < \underline{L}$. Finally, prove that the two preceding results are, in fact, sufficient to establish the required result. The case where $\underline{L} = \pm \infty$ follows similarly.

5.27 Hint: Adapt the Hint for Exercise 5.22.

5.28 Hint: Refer to the Hint for Exercise 5.23

5.29 Hint: Apply the information given in the Hint for Exercise 5.24 to the present Exercise.

5.31 Hint: Using the terminology of Exercise 5.26, if S' is the set consisting of the respective 'lim sup's' of the sequences $\{f(x_n)\}$ as $\{x_n\}$ varies over all possible sequences of points in a deleted neighborhood of $x = c$ converging to c, then $\limsup_{x \to c} f(x) = \sup S'$. The technique of proof suggested in the Hint for Exercise 5.26 can also be used here.

5.32 Hint: Prove that if for fixed constants A and B we have $A \leq f(x) \leq B$ for all x satisfying $0 < |x-c| < \delta'$ (for some $\delta' > 0$), then $A \leq I(\delta) \leq J(\delta) \leq B$ for all $0 < \delta < \delta'$. The result then follows immediately.

5.34 Hint: First, suppose that $L = \lim_{x \to c} f(x)$ exists (finite). Accordingly, given any $\epsilon > 0$ there can always be found a corresponding $\delta > 0$ so that:

$$|f(x) - L| < \epsilon \quad \text{whenever} \quad 0 < |x-c| < \delta ,$$

or equivalently,

$$L - \epsilon < f(x) < L + \epsilon \quad \text{whenever} \quad 0 < |x-c| < \delta \quad .$$

Therefore, by properties of the "inf" and "sup" of a function over a set it follows that:

$$L - \epsilon \leq \inf_{0<|x-c|<\delta} f(x) \leq \sup_{0<|x-c|<\delta} f(x) \leq L + \epsilon .$$

Accordingly,

$$L - \epsilon \leq \underline{\lim}_{x \to c} f(x) \leq \overline{\lim}_{x \to c} f(x) \leq L + \epsilon . \quad \text{(Why?)} .$$

Since ϵ is arbitrary, it then follows that $L = \underline{L} = \overline{L}$.

Next, suppose that $L = \underline{L} = \overline{L}$ (finite); that is,

$$\underline{\lim}_{x \to c} f(x) = \overline{\lim}_{x \to c} f(x) = L \quad .$$

By properties of $\underline{\lim}$ and $\overline{\lim}$ it then follows that, given any $\epsilon > 0$ there can always be found a $\delta > 0$ sufficiently small so that:

$$L - \epsilon \le \inf_{0<|x-c|<\delta} f(x) \le \sup_{0<|x-c|<\delta} f(x) \le L + \epsilon .$$

Therefore, by properties of the "inf" and "sup" of a function over a set, we have that:

$$L - \epsilon \le f(x) \le L + \epsilon \quad \text{for} \quad 0 < |x-c| < \delta ,$$

or equivalently,

$$|f(x) - L| < \epsilon \quad \text{for} \quad 0 < |x-c| < \delta \quad ,$$

and accordingly,

$$\lim_{x\to c} f(x) = L \quad .$$

The preceding proof may be modified to establish the result in case of infinite values.

5.35 Hint: Modify the technique of Exercise 1.41 (Section 1) and use the definition of $\underline{\lim}$ and $\overline{\lim}$ as the limit of $I(\delta)$ and $J(\delta)$, respectively, as $\delta \downarrow 0$.

5.36 Answer: If $\sum_1^\infty f(m,n)$ converges to the sum $g(m)$ for $m = 1,2,\ldots$, and in addition,

(i) $|f(m,n)| \le b_n$ for $n = 1,2,\ldots$

and

(ii) $\sum_1^\infty b_n$ converges ,

then

$$\sum_1^\infty \underline{\lim}_{m\to\infty} f(m,n) \le \underline{\lim}_{m\to\infty} g(m) \le \overline{\lim}_{m\to\infty} g(m) \le \sum_1^\infty \overline{\lim}_{m\to\infty} f(m,n) \quad .$$

References to Additional and Related Material: Section 5

1. Apostol, T., "Mathematical Analysis", Addison-Wesley, Inc., Reading, Mass. (1960).

2. Buck. R., "Advanced Calculus", (Second Ed.), McGraw-Hill, Inc. New York (1965).

3. Devinatz, A., "Advanced Calculus", Holt, Reinhart and Winston, Inc., New York (1968).

4. Hight, D., "A Concept of Limits", Prentice-Hall, Inc., Engelwood Cliffs, N.J. (1966).

5. Widder, D., "Advanced Calculus", (Second Ed.), Prentice-Hall, Inc., Engelwood Cliffs, N.J. (1963).

6. Orders of Magnitude: The 0, o, ~ Notation

In many practical applications of Mathematics it is necessary to consider the behavior of some function $f(x)$ of x as x tends to some limit (finite or infinite). However, many times the function $f(x)$ is extremely complicated in nature, or incompletely known, and it is preferable (in fact, perhaps only possible) to describe the asymptotic behavior of $f(x)$ relative to (or compared with) some other function $g(x)$ of x as x tends to the same limit. In practice, the comparison function g is often chosen as a "simpler" function, such as a power or exponential function.

The above considerations indicate the background that gives rise to the so-called Order-of-Magnitude notation: 0, o, $\sim$. We now consider this notation and some applications.

DEFINITION 6.1 (The $\sim$ Notation: Same Order of Magnitude) If $f(x)$ and $g(x)$ are two real-valued functions of the real variable x [1] defined over a common domain D, then we say that $f(x)$ has the same order of magnitude as $g(x)$ as x tends to some limit, written $f(x) \sim g(x)$, provided that $f(x)/g(x) \to 1$ as x tends to its limit.

EXAMPLE 6.1 $\frac{x^2}{x+\log x} \sim x$ as $x \to \infty$ and $\operatorname{Tan} x \sim x$ as $x \to 0$.

[1] The 0, o, $\sim$ concepts apply to complex functions as well, provided only modulus of the function and the variable are involved. See References.

EXERCISE 6.1 Construct various pairs (three will do) of functions $f(x)$ and $g(x)$ that have the same order of magnitude as x tends to certain limits.

EXERCISE 6.2 Clearly the $\sim$ relationship is symmetric in the sense that $f(x) \sim g(x)$ is equivalent to $g(x) \sim f(x)$. Establish the following result, that gives some insight into the relationship between $f(x)$ and $g(x)$ in case they have the same order of magnitude: if $f(x) \sim g(x)$ as x tends to some limit, then given any $\epsilon > 0$, it can be said that $g(x)(1-\epsilon) < f(x) < g(x)(1+\epsilon)$ for all x sufficiently near the given limit. Interpret this result using the example of $\text{Tan}x \sim x$ as $x \to 0$. Illustrate with a diagram.

EXERCISE 6.3 (The Gamma Function: a Preliminary Result) For any $x > 0$ the function $\Gamma(x)$, termed the Gamma Function, is defined as follows:

$$\Gamma(x) = \int_0^\infty z^{x-1}e^{-z}dz \quad .$$

Using integration-by-parts, prove that $\Gamma(x) = (x-1)\,\Gamma(x-1)$, whence, because $\Gamma(1) = 1$, it follows that for any positive integer n, $\Gamma(n) = (n-1)!$ [1]

[1]For an extensive treatment of the Gamma Function, the closely related Incomplete Gamma Function, and their relationships with the Beta Function, see the References. These functions are tabled (also see the References).

EXERCISE 6.4 (Stirling's Formulas) Prove that $\Gamma(x) \sim \sqrt{2\pi}\, e^{-x} x^{x-\frac{1}{2}}$ as $x \to \infty$, then establish the two following forms of Sterling's Formula: (i) $n! \sim \sqrt{2\pi}\,(n+\frac{1}{2})^{n+\frac{1}{2}} e^{-n}$ and (ii) $n! \sim \sqrt{2\pi}\, n^{n+\frac{1}{2}} e^{-n}$ as $n \to \infty$. (This is not a simple Exercise).

EXERCISE 6.5 (Applications) Prove that for any positive integers a and b,

$$\frac{(a+1)(a+2)\ldots(a+n)}{(b+1)(b+2)\ldots(b+n)} \sim \frac{b!}{a!}\, n^{a-b}$$

as $n \to \infty$. Hence, prove that as $n \to \infty$,

$$\binom{2n}{n} \sim \frac{4^n}{\sqrt{\pi n}}, \quad \text{and accordingly} \quad \frac{\binom{2n}{n}}{2^{2n}} \to 0 .$$

EXERCISE 6.6 (Slow Variation) A positive, monotone function $L(x)$ is said to vary slowly at infinity provided that for every (any) fixed positive number c, $L(cx)/L(x) \to 1$, i.e. $L(cx) \sim L(x)$, as $x \to \infty$. For example, the function $L(x) = \log(1+x)$ varies slowly at infinity. Prove that for any such function $L(x)$, $x^t L(x) \to \infty$ and $x^{-t} L(x) \to 0$ as $x \to \infty$ for any fixed $t > 0$. (See also Section 6).

EXAMPLE 6.2 (Cumulative Normal Distribution Function)[1] The standard Normal cumulative distribution function $\Phi(x)$ is defined for all real x by the relation:

$$\Phi(x) = \int_{-\infty}^{x} \frac{1}{\sqrt{2\pi}}\, e^{-\frac{1}{2}z^2}\, dz .$$

[1]Also called Gaussian.

Since $\Phi(x) \to 1$ as $x \to \infty$, we also have:

$$1 - \Phi(x) = \int_x^\infty \frac{1}{\sqrt{2\pi}} e^{-\frac{1}{2}z^2} dz .$$

Now for every $x > 0$,

$$\frac{1}{\sqrt{2\pi}} e^{-\frac{1}{2}x^2} \{\frac{1}{x} - \frac{1}{x^3}\} < 1 - \Phi(x) < \frac{1}{\sqrt{2\pi}} e^{-\frac{1}{2}x^2} \{\frac{1}{x}\}$$

whence,

$$1 - \Phi(x) \sim \frac{1}{\sqrt{2\pi x}} e^{-\frac{1}{2}x^2} \quad \text{as} \quad x \to \infty .$$

To verify this note that:

$$\frac{1}{\sqrt{2\pi}} e^{-\frac{1}{2}x^2} \{\frac{1}{x}\} = \int_x^\infty \frac{1}{\sqrt{2\pi}} e^{-\frac{1}{2}z^2} \{1 + \frac{1}{z^2}\}dz > 1 - \Phi(x) ,$$

and

$$\frac{1}{\sqrt{2\pi}} e^{-\frac{1}{2}x^2} \{\frac{1}{x} - \frac{1}{x^3}\} = \int_x^\infty \frac{1}{\sqrt{2\pi}} e^{-\frac{1}{2}z^2} \{1 - \frac{3}{z^4}\}dz < 1 - \Phi(x),$$

so the stated result follows immediately. (Why?).

EXERCISE 6.7 Assuming all functions are positive, prove that if $f(x) \sim g(x)$ and $h(x) \sim k(x)$ as x tends to its limit, then $[f(x)]^a[h(x)]^b \sim [g(x)]^a[k(x)]^b$ as x tends to its limit. Is it necessarily true that $f(x) \pm h(x) \sim g(x) \pm k(x)$ as x tends to its limit? Is it necessarily true that $f(x) - g(x) \to 0$ as x tends to its limit? The answers to the latter questions should suggest caution in dealing with the $\sim$ relation, lest more properties be attributed to it than are, in general, justified.

Two other means of comparing the asymptotic behavior of a pair of functions $f(x)$ and $g(x)$ as x tends to its limit employ the 0 and o notation. Their definitions follow.

DEFINITION 6.2 (The 0 Notation: at Most the Order of Magnitude) If $f(x)$ and $g(x)$ are two real-valued functions of the real variable x [1] defined over a common domain D, then we say that $f(x)$ has at most the order of magnitude of $g(x)$ as x tends to some limit, written $f(x) = 0(g(x))$, provided that $f(x)/g(x)$ remains bounded as x tends to its limit.

This weaker relationship between $f(x)$ and $g(x)$ is often sufficient when the stronger $\sim$ does not hold. Its uses are the same.

EXAMPLE 6.3 $\frac{x(1 \pm \mathrm{Sin}x)}{1+x^3} = 0(x^{-2})$ as $x \to \infty$; the bound referred to in the Definition above can be taken as 2. If $P_n(x)$ is a polynomial of degree n, then $P_n(x) = 0(x^n)$ as $n \to \infty$; $x\mathrm{Cos}(1/x) = 0(x)$ as $x \to 0$.

EXERCISE 6.8 Rework Exercise 6.7 with 0 replacing $\sim$.

EXERCISE 6.9 Prove that if $f(x) = 0(g(x))$ as x tends to its limit, then there exists a positive constant B such that

$$-Bg(x) \leq f(x) \leq Bg(x)$$

for all x sufficiently near the limit.

---

[1]See the Footnote on page 6-1.

EXERCISE 6.10 (Remainder Term in Taylor Expansion) Prove that if the function $f(x)$ possesses $n+1$ continuous derivatives at $x = 0$, then

$$f(x) = \sum_0^n \frac{f^{(j)}(0)}{j!} x^j + 0(x^{n+1}) \quad \text{as} \quad x \to 0.$$

Illustrate the result with a specific example.

DEFINITION 6.3 (The o Notation: of smaller order of magnitude) If $f(x)$ and $g(x)$ are two real-valued functions of the real variable x [1] defined over a common domain D, then we say that $f(x)$ has a Smaller Order of Magnitude than $g(x)$ as x tends to some limit, written $f(x) = o(g(x))$, provided that $f(x)/g(x) \to 0$ as x tends to its limit.

EXAMPLE 6.4 $\mathrm{Tan}(x^3) = o(x^2)$ as $x \to 0$; $x = o(\sqrt{x})$ as $x \downarrow 0$; for $n > 0$, $x^n = o(e^x)$ as $x \to \infty$.

EXERCISE 6.11 Construct several pairs of functions $f(x)$ and $g(x)$ for which (a) $f(x) = o(g(x))$ as x tends to its limit, and (b) $f(x) = 0(g(x))$ as x tends to its limit.

EXERCISE 6.12 Rework Exercise 6.7 with o replacing $\sim$.

EXAMPLE 6.5 By applying the definitions of the 0, o symbols, note that $f(x) = 0(1)$ means that the function $f(x)$ remains bounded as x tends to its limit, and $f(x) = o(1)$ means that $f(x)$ tends to zero as x tends to its limit. Clearly $f(x) = a\mathrm{Cos}x + b\mathrm{Sin}x = 0(1)$ as $x \to \infty$, and $f(x) = x^{-n} = o(1)$ as $x \to \infty$ for any $n > 0$.

[1] See the footnote on page 6-1.

EXAMPLE 6.6 (A Useful Exponential-type Limit) If $f(n) = o(1)$ as $n \to \infty$, then $[1 + \frac{c}{n} + \frac{f(n)}{n}]^n \to e^c$ as $n \to \infty$. To prove this note first that by Taylor's Theorem,
$\log(1+x) = x + \frac{K}{(1+a)^2} x^2$, where K is a constant and $0 < a < x$. Thus,

$$n \log [1 + \frac{c}{n} + \frac{f(n)}{n}] = n [\frac{c}{n} + \frac{f(n)}{n} + \frac{K}{(1+a_n)^2} \{\frac{c}{n} + \frac{f(n)}{n}\}^2] ,$$

where $0 < a_n < \frac{c+f(n)}{n}$; since $f(n) \to 0$ as $n \to \infty$ the quantity above on the right tends to c as $n \to \infty$, and the desired result then follows upon taking anti-logarithms of both sides. This is a generalization of the well-known result from Elementary Calculus that, for any constant c, $\lim_{n\to\infty} (1+c/n)^n = e^c$.

Useful relationships exist amongst the order relations 0, o, and $\sim$. For example, if $f(x) = o(g(x))$, then immediately $f(x) = 0(g(x))$. As another example, if $f(x) = o(g(x))$ and $h(x) = o(k(x))$ as x tends to its limit, then $f(x)h(x) = o(g(x)k(x))$ as x tends to its limit.

EXERCISE 6.13 Prove the preceding statements. Also prove that if $f(x) \sim g(x)$ as x tends to its limit, and $h(x)$ is at most of order (of smaller order) of magnitude as $k(x)$, then $f(x)h(x)$ is at most of order (of smaller order) of magnitude as $g(x)k(x)$.

A convenient "Algebra" naturally develops in working with the order of magnitude concept, and in the process of problem-solving, functions themselves are frequently replaced by their magnitude comparisons, e.g. by symbols such as $o(x^2)$, $O(1)$, $o(e^{-x})$, etc., and use is then made of the relationships amongst the three magnitude relations. The following Example illustrates this point.

<u>EXAMPLE 6.7</u> We shall show that if $f(x) = x^2 + O(x)$ and $g(x) \sim 1/x$ as $x \to \infty$, then $f(x)g(x) = x + o(x)$ as $x \to \infty$. To prove this note that $g(x) \sim x^{-1}$ implies $g(x) = x^{-1} + o(x^{-1})$. Thus $f(x)g(x) = \{x^2 + O(x)\} \cdot \{x^{-1} + o(x^{-1})\} = x + x^2\, o(x^{-1}) + x^{-1}\, O(x) + O(x)\, o(x^{-1}) = x + o(x) + O(1) + O(x)\, o(x^{-1})$. But $O(1)$ and $O(x)\, o(x^{-1}) = o(1)$ are both $o(x)$ as $x \to \infty$, hence $f(x)g(x) = x + o(x)$ as $x \to \infty$.

The situation illustrated in the preceding Example occurs often in Applied Mathematics; the "order" terms might correspond to "remainder" terms in an approximation or asymptotic result where interest centers upon the "size" (in the present terminology, the order of magnitude) of the remainder term.

<u>Hints and Answers to Exercises: Section 6</u>

6.2 Hint: Using the basic definitions of the limit, if $h(x) \to 1$ as x tends to its limit, then for all x sufficiently close to the limit, $1 - \epsilon < h(x) < 1 + \epsilon$. Now choose $h(x) = f(x)/g(x)$. Use a table of Tangent, and compare Tanx and x for x close to zero.

6.3 Hint: In the integration-by-parts, choose $u = z^{x-1}$ and $dv = e^{-z}$.

6.4 Hint: For a derivation of (ii) by geometric arguments, using approximating sums for the Riemann Integral, see Feller, W., "An Introduction to Probability Theory and its Applications", Vol. I, (Second Ed.), Wiley and Sons, Inc., New York (1960), or refer to the book by N. Lebedev referred to in the Reference Section.

6.5 Hint: Apply the results of Exercise 6.4, and the necessary algebraic simplifications.

6.6 Hint: Consider, for concreteness, the case where $L(x)$ is monotone decreasing. Assume, to the contrary, that for some $t > 0$, $x^t L(x)$ is <u>bounded</u>, i.e. $x^t L(x) \le B$ for some B finite, as $x \to \infty$. Now deduce a contradiction to the assumed property possessed by L, namely, that for any positive constant c, $L(cx)/L(x) \to 1$ as $x \to \infty$. The other cases can be handled analogously.

6.7 Hint: For the second parts, note that $f(x) = \text{Sin} x + x \sim x$ as $x \to \infty$; the same is true for $g(x) = \text{Cos} x + x$. However, $f(x) - g(x) \not\sim x$ as $x \to \infty$. Note also, $f(x) - g(x)$ has no limit as $x \to \infty$. Examples exist where $f(x) - g(x)$ becomes unbounded. Construct one.

6.8 Hint: Use the ideas presented in the Hint for Exercise 6.7.

6.10 Hint: Refer to the standard "Taylor Theorem with Remainder Term" result found in the Reference Section in the book by Brand, for example.

6.12 Hint: Use the ideas presented in the Hint for Exercise 6.7.

References to Additional and Related Material: Section 6

1. Abramowitz, M. and I. Stegun (Editors), "Handbook of Mathematical Functions with Formulas, Graphs, and Mathematical Tables", National Bureau of Standards, Applied Mathematics Series No. 55, Washington, D. C. (1964).

2. Brand, L., "Advanced Calculus", Wiley and Sons, Inc., New York (1958).

3. Courant, R., "Differential and Integral Calculus" Vol. I, (Second Ed.), Blackie and Son, Ltd., London (1963).

4. Cramér, H., "Mathematical Methods of Statistics", Princeton University Press (1958).

5. Hobson, E., "Theory of Functions of a Real Variable", Vol. II, Dover Publications, New York.

6. Lebedev, N., "Special Functions and Their Applications", Prentice-Hall, Inc. Englewood Cliffs, N.J. (1965).

7. Pearson, K., (Editor), "Tables of the Incomplete Gamma Function", London (1922).

8. Richardson, C., "An Introduction to the Calculus of Finite Differences", Van Nostrand, Inc., New York (1954).

9. Spiegel, M., "Mathematical Handbook of Formulas and Tables", Schaum's Outline Series, Schaum Publishing Co., New York (1968).

10. Whittaker, E. and G. Watson, "A Course in Modern Analysis", Cambridge University Press (1952).

7. Some Abelian and Tauberian Theorems

As an immediate application of the results of the preceding Sections we now consider briefly some Examples of what are known as Abelian and Tauberian-type Theorems. These results have found use in a variety of Applied fields.

In Applied Mathematics we often deal with the Laplace Transform of a function, either out of necessity, or else as a useful alternative to working directly with the function itself. Abelian and Tauberian Theorems connect the behavior of the function and that of its transform in a manner which we shall presently demonstrate.

For our purposes here, we shall deal exclusively with Riemann-Integrable functions $f(x)$ whose Laplace Transforms $L(s)$ exist for all $s > 0$ (though not necessarily for $s = 0$). For such functions, the Laplace Transform $L(s)$ is defined as follows:

$$L(s) = \int_0^\infty e^{-sx} f(x)\,dx \quad \text{(Riemann)} \quad s > 0 .$$

Although $f(x)$ itself may be Riemann-Integrable over $[0, \infty)$, this is not necessary in order for its Laplace Transform to exist for all $s > 0$.

<u>EXERCISE 7.1</u> Find the Laplace Transform of $f(x) = x$, and, accordingly, demonstrate the truth of the preceding statement.

<u>EXERCISE 7.2</u> Prove that if the Riemann-Integrable function $f(x)$ is bounded over $[0, \infty)$, then its Laplace Transform exists for all $s > 0$. Accordingly, the Laplace Transforms

of F and $1-F$ exist for all $s > 0$ for any c.d.f. F. Can this be generalized to function of bounded variation?

Now we can state more precisely the nature of Abelian and Tauberian-type Theorems. Such Theorems quantify the close connection between the asymptotic behavior of

$$L(s) = \int_0^\infty e^{-sx} f(x)\,dx \quad \text{as} \quad s \downarrow 0$$

and that of

$$I(x) = \int_0^x f(z)\,dz \qquad \text{as} \quad x \to +\infty .$$

The two types of Theorems are complimentary in the sense that Abelian-type Theorems provide information about the behavior of $L(s)$ as $s \downarrow 0$ based upon knowledge about the behavior of $I(x)$ as $x \to +\infty$, whereas Tauberian-type Theorems provide information about the behavior of $I(x)$ as $x \to \infty$ based upon the behavior of $L(s)$ as $s \downarrow 0$.

Perhaps the earliest Classical results of these types dealt with power series. As will be seen, when these Theorems are in their original form, the connection with the preceding description of an Abelian or Tauberian-type Theorem might not be obvious. However, we shall subsequently restate them so that the connection does become clearer.

<u>THEOREM 7.1</u> (Abel) If the interval of convergence of the power series $S(x) = \sum_{1}^{\infty} a_n x^n$ is $(-1,1)$, and if the series $S(1) = \sum_{1}^{\infty} a_n$ converges, then,

$$\lim_{x \uparrow 1} S(x) = S(1) \quad ,$$

that is, the sum function is continuous from the left at $x = 1$.

<u>THEOREM 7.2</u> (Tauber) If the interval of convergence of the power series $S(x) = \sum_{1}^{\infty} a_n x^n$ is $(-1,1)$, and if $\lim_{x \uparrow 1} S(x)$ exists and is finite, whereas $a_n = 0(1/n)$ as $n \to \infty$, then the series $S(1) = \sum_{0}^{\infty} a_n$ converges.[1]

We shall now reformulate these two Classical results in the general form of Abelian and Tauberian-type Theorems.

We begin by defining a function $f(x)$ as follows:

$$f(x) = a_n \quad \text{for} \quad n-1 \leq x < n \quad (n = 1,2,\ldots) .$$

Thus, for any positive integer n, we have

$$\int_0^n f(x)\,dx = S_n = a_1 + a_2 + \ldots + a_n .$$

Next note that (subject to existence) the Laplace transform $L(s)$ of the function f defined above is given by:

[1] For proofs see Hobson, E., Theory of Functions of a Real Variable, Vol. II, pages 175 and 182, Dover Publications, N.Y.

$$L(s) = \int_0^\infty e^{-sx} f(x)\,dx = \sum_1^\infty \int_{n-1}^{n} a_n e^{-sx} dx$$

$$= [\frac{e^s - 1}{s}] \sum_1^\infty a_n e^{-ns}$$

$$= [\frac{e^s - 1}{s}] S(e^{-s}) ,$$

where $S(\cdot)$ is the sum function of the power series $\sum_1^\infty a_n x^n$.

Using the notation just established, we may now restate Abel's and Tauber's original Theorems as follows:

THEOREM 7.3 (Abel: Reformulation) If the Laplace transform $L(s)$ of $f(x)$ exists for some $s > 0$ and $\int_0^n f(x)\,dx$ converges to a finite limit as $n \to +\infty$ then $L(s) \to L(0)$ as $s \downarrow 0$, that is, the Laplace transform of f is continuous from the right at $s = 0$.

THEOREM 7.4 (Tauber: Reformulation) If the Laplace transform $L(s)$ of $f(x)$ exists for some $s > 0$ and $L(s)$ has a finite limit as $s \downarrow 0$, while $f(x) = 0(1/x)$ as $x \to +\infty$, then $\int_0^n f(x)\,dx$ converges as $n \to \infty$.

It will be noted that the Tauberian Theorem requires a little more in the form of assumptions than does the corresponding Abelian-type Theorem (e.g. in Theorem 7.4 an assumption must be made on the order of $f(x)$ as $x \to +\infty$). This is typically the case.

In essence, however, Abelian and Tauberian-type Theorems connect the behaviors of the transform of a function near $s = 0$ with that of the integral of the function over $[0,x]$ as $x \to +\infty$.

Such information often proves useful in Application of transform Mathematics.

We shall conclude this Section with a pair of Abelian and Tauberian Theorems that have found wide Application. (Only a Reference will be given for their proofs.) They are useful for the reason that the results apply to a useful class of functions termed functions of slow and regular variation.

Let $V(x)$ be a non-negative function that is either monotone increasing or monotone decreasing (see footnote, page 3-9). The function $V(x)$ is termed a function of <u>Regular Variation</u> at infinity (with variation constant p) if, for any (every) fixed positive constant c,

$$\lim_{x\to\infty} \frac{V(cx)}{V(x)} = c^p \quad .$$

In the special case where the variation constant is $p = 0$, the function is termed a function of <u>Slow Variation</u> at infinity.

<u>EXERCISE 7.3</u> Verify the following:

(a) All powers of $|\log x|$ are of Slow Variation.

(b) $V(x) = (1+x^2)^s$ is of Regular Variation with $p = 2s$.

(c) $V(x) = e^x$ is of Regular Variation with $p = +\infty$.

(d) Functions involving Sines and Cosines are not, in general, monotone, whence are not included in the class being considered.

(e) $V(x) = \log\log(1+x^2)$ is of Regular Variation. Find p.

(f) Provide several examples of functions of Slow Variation.

<u>EXERCISE 7.4</u> Prove that the Improper Riemann Integral, namely:

$$\int_0^\infty V(x)\,dx = \lim_{z\to\infty} \int_0^z V(x)\,dx ,$$

of a Slowly Varying function $V(x)$, always diverges. Can some statement be made for the case of Regular Variation? Consider p negative, for example.

We may now state a general result applying to the entire class of functions that vary slowly at infinity. The potential value of such a result was pointed out at the beginning of this Section.

<u>THEOREM 7.5</u> (Abelian-Tauberian Theorem and Slowly Varying Functions) Suppose $V(x)$ varies slowly at infinity and that t is non-negative. Then,

$$\int_0^x f(z)\,dz \sim x^t V(x) \quad \text{as} \quad x \to +\infty$$

if and only if the Laplace Transform $L(s)$ of $f(x)$ satisfies:

$$L(s) \sim \frac{V(1/s)\,\Gamma(t+1)}{s^t} \quad \text{as} \quad s \downarrow 0 .$$

PROOF. See Feller, W., "An Introduction to Probability Theory and its Applications", Vol. II, pp. 418-422, Wiley and Sons, N. Y. (1966).

Accordingly, the asymptotic behavior of a (not necessarily convergent) Integral is related to the asymptotic behavior of the Laplace Transform of its integrand.

EXAMPLE 7.1 From the preceding Theorem we may conclude that: $\int_0^x f(z)dz \sim \log\log x$ as $x \to \infty$ iff $L(s) \sim \log\log(1/s)$ as $s \downarrow 0$. Furthermore, since $V(x) \equiv 1$ varies slowly at infinity, $\int_0^x f(z)dz \sim \sqrt{x}$ as $x \to \infty$ iff $L(s) \sim \sqrt{\pi}/2s$ as $s \downarrow 0$.

To conclude this brief Section, we shall reformulate the preceding Theorem so as to apply to infinite series. The result deals with the asymptotic behavior of the n-th partial sum $a_1 + a_2 + \ldots + a_n$ of a possibly divergent infinite series. Such information often proves useful in Applied Mathematics.

THEOREM 7.6 (Abelian-Tauberian Theorem: Series Version) Suppose $V(x)$ varies slowly at infinity and $t \geq 0$. Then,

$$a_1 + a_2 + \ldots + a_n \sim n^t V(n) \quad \text{as} \quad n \to \infty$$

if and only if

$$S(s) \sim \Gamma(t+1)\, V(\tfrac{1}{s})/s^t \quad \text{as} \quad s \downarrow 0,$$

where $S(s) = \sum_1^\infty a_n s^n$.

EXERCISE 7.5 Verify that the above result follows from Theorem 7.5.

Hints and Answers to Exercises: Section 7

7.1 Answer: $L(s) = \int_0^\infty xe^{-sx}dx = 1/s^2$, which clearly exists for all $s > 0$.

7.2 Answer: If $|f(x)| \le B$ for all $x \ge 0$ then

$$\left|\int_0^\infty f(x)e^{-sx}dx\right| \le \int_0^\infty |f(x)|e^{-sx}dx \le B\int_0^\infty e^{-sx}dx = B/s$$

which clearly exists for all $s > 0$.

7.3 Answer: (Illustrations) $V(x) = \{\log\log\cosh(2+x)\}^{\frac{1}{2}}$, $V(x) = 1+1/x$, $V(x) = F(x)$, where F is any 1-dimensional c.d.f.

7.4 Answer: Suppose, to the contrary, that $\int_0^\infty V(x)dx < \infty$. Then, by definition of a function of slow variation, for all x sufficiently large, say $x \ge B$, we have

$$\frac{4}{5} \le \frac{V(2x)}{V(x)} .$$

Accordingly,

$$\int_B^\infty V(x)dx \le \frac{5}{4}\int_B^\infty V(2x)dx = \frac{5}{8}\int_{2B}^\infty V(x)dx ,$$

which is impossible.

7.5 Hint: Consider the function $f(x)$ defined as follows:

$$f(x) = \sum_1^{[x]} a_n,$$ [1]

---

[1] $[x]$ denotes the largest integer not exceeding x.

whose Laplace Transform is given by:

$$L(s) = \sum_{1}^{\infty} a_n e^{-ns} = s(e^{-s}) \quad .$$

Now then apply the results of Theorem 7.5.

References to Additional and Related Material: Section 7

1. Feller, W., "An Introduction to Probability Theory and its Applications", Vol. II, John Wiley and Sons, Inc. (1966).

2. Ford, W., "Divergent Series", Chelsea Publishing Co., Inc. (1960).

3. Grimm, C., "A Unified Method of Finding Laplace Transforms, Fourier Transforms, and Fourier Series", U.M.A.P. Unit 324, Educational Development Center, Newton, Massachusetts (1978).

4. Hobson, E., "Theory of Functions of a Real Variable", Vol. II, Dover Publications, Inc.

5. Pitt, H., "Tauberian Theorems", Oxford University Press. (1958).

6. Smith, W., Unpublished Lecture Notes, University of North Carolina (1964).

7. Spiegel, M., "Schaum's Outline of Theory and Problems of Laplace Transforms", Schaum Publishing Co. (1965).

8. Widder, D., "An Introduction to Transform Theory", Academic Press (1971).

9. Widder, D., "The Laplace Transform", Princeton University Press (1941).

8. 1-Dimensional Cumulative Distribution Functions and Bounded Variation Functions

A real-valued function $F(x)$, defined for all real x, is termed a 1-dimensional cumulative distribution function (c.d.f.) iff it satisfies the following four basic properties:

(i) $x_1 \le x_2 \Rightarrow F(x_1) \le F(x_2)$ [non-decreasing]

(ii) $F(x) \to 1$ as $x \to +\infty$ [$F(+\infty) = 1$]

(iii) $F(x) \to 0$ as $x \to -\infty$ [$F(-\infty) = 0$] [1]

(iv) $F(x-\epsilon) \to F(x)$ as $\epsilon \downarrow 0$ [left-continuity]

Cumulative distribution functions are fundamental, and our concern here is with certain Mathematical properties of c.d.f.'s that will be useful later when studying Bounded Variation Functions and the Riemann-Stieltjes Integral.

<u>EXERCISE 8.1</u> (Mixtures of C.D.F.'s) Prove that if $F_1, F_2, \ldots$ is a sequence of c.d.f.'s and $a_1, a_2, \ldots$ is a sequence of non-negative real numbers with $\Sigma a_i = 1$, then the function F defined by: $F(x) = \Sigma a_i F_i(x)$ is also a c.d.f. (termed a 'mixture' of the F_i's).

<u>EXERCISE 8.2</u> (Discrete Distributions) For any real number a the function F_a defined by: $F_a(x) = 0$ if $x \le a$, $F_a(x) = 1$ if $x > a$, is clearly a c.d.f. Prove that the function F defined by $F(x) = \Sigma p_i F_{a_i}(x)$ (where $p_i \ge 0$, $\Sigma p_i = 1$, and $\{a_i\}$ distinct reals) is a bona-fide c.d.f.,

---

[1] F may or may not actually achieve the values of 0 and/or 1 for some finite value of x.

being actually a 'step' function with a 'step' of height p_i at a_i $(i = 1,2,\ldots)$.

The defining properties of a 1-dimensional c.d.f. do not require that it be an everywhere continuous function. However, since a 1-dimensional c.d.f. is non-decreasing and bounded between zero and one, there can be at most n distinct points of discontinuity with saltus (step) $\geq 1/n$ $(n = 1,2,\ldots)$. Thus there can be at most a countable number of points of discontinuity of any 1-dimensional c.d.f.

EXERCISE 8.3 Construct a 1-dimensional c.d.f. that is discontinuous at an infinite number of points.

EXERCISE 8.4 (Riemann Continuous Case) Prove that if $F' = f$ exists and is continuous for $a \leq x \leq b$, then $f(x) \geq 0$ and $\int_a^b f(x)dx = F(b) - F(a)$ (Riemann Integral). The function f is termed a density function, and F is termed Riemann Continuous over the interval $[a,b]$ in this case. (Generalizations of this result require measure theory).

EXERCISE 8.5 (Related C.D.F.'s) Let F be any continuous[1] 1-dimensional c.d.f. and h a positive number. Prove that the functions Λ and Δ defined by:

$$\Lambda(x) = \frac{1}{h} \int_x^{x+h} F(z)dz$$

and

$$\Delta(x) = \frac{1}{2h} \int_{x-h}^{x+h} F(z)dz$$

are bona-fide 1-dimensional c.d.f.'s.

[1]The result is actually true for any 1-dimensional c.d.f., but a general proof requires properties of the Riemann Integral that can only be developed using measure theory.

EXERCISE 8.6 Let $F_1, \ldots, F_n$ be 1-dimensional c.d.f.'s. Prove that the functions F_* and F^* defined by:
$F_*(x) = \min(F_1(x), \ldots, F_n(x))$ and $F^*(x) = \max(F_1(x), \ldots, F_n(x))$
are bona-fide 1-dimensional c.d.f.'s. Formulate generalizations of this result; e.g. $\prod_1^n F_i^{k_i}$ $(k_i > 0)$ is a 1-dimensional c.d.f. See also the next Exercise.

EXERCISE 8.7 If $\varphi(x) = \varphi$ is continuous and non-decreasing for $0 \le x \le 1$, with $\varphi(0) = 0$ and $\varphi(1) = 1$, then for any 1-dimensional c.d.f. F, the composite function $\varphi \cdot F$ defined by: $\varphi \cdot F(x) = \varphi[F(x)]$ is a bona-fide 1-dimensional c.d.f.

A sequence of 1-dimensional c.d.f.'s is said to converge completely, written $F_n \overset{c}{\rightarrow}$, iff there exists a bona-fide 1-dimensional c.d.f. F such that $F_n(x) \rightarrow F(x)$ for every real x at which the limit function F is continuous. If, however, F is a possibly defective 1-dimensional c.d.f. (meaning that it is non-decreasing, left-continuous and only bounded between zero and one) then the sequence is said to converge weakly, written $F_n \overset{w}{\rightarrow}$.

EXERCISE 8.8 (Complete vs. Pointwise Convergence) Let $\{a_n\}$ be a bounded sequence of real numbers such that $a_n \uparrow a$. Furthermore, define 1-dimensional c.d.f.'s $F_n (n = 1, 2, \ldots)$ and F as follows:

$$F_n(x) = \begin{cases} 0, & x \le a_n \\ 1, & x > a_n \end{cases} \qquad F(x) = \begin{cases} 0, & x \le a \\ 1, & x > a \end{cases} .$$

Prove that $F_n \overset{c}{\rightarrow} F$, yet $F_n(a) \not\rightarrow F(a)$. However, show that $F_n(a+) \rightarrow F(a+)$. What does this mean?

<u>EXERCISE 8.9</u> (Illustration of Weak Convergence) For each positive integer n define a 1-dimensional c.d.f. F_n as follows:

$$F_n(x) = \begin{cases} 0, & x \le -n \\ \frac{x+n}{2n}, & -n < x \le n \\ 1, & x > n \end{cases} .$$

Prove that this sequence of 1-dimensional c.d.f.'s converges to the function that is identically one-half (which is not a c.d.f.).

<u>EXERCISE 8.10</u> (Relating Complete and Weak Convergence) Prove that $F_n \overset{c}{\to} F$ implies $F_n \overset{w}{\to} F$. Then show that if $F_n \overset{w}{\to} F$ and $F_n(\pm\infty) \to F(\pm\infty)$, then F is a bona-fide 1-dimensional c.d.f. and $F_n \overset{c}{\to} F$.

<u>EXERCISE 8.11</u> (Convergence of Limits From the Right) Prove that if either $F_n \overset{w}{\to} F$ or $F_n \overset{c}{\to} F$ then for every real x, $F_n(x+) \to F(x+)$. Exercise 8.8 established that this convergence cannot be replaced by pointwise convergence.

Many of the results that are valid for the class of 1-dimensional c.d.f.'s remain valid for a broader class of functions, namely the class of so-called Bounded Variation Functions. A real-valued function U of the real variable x is termed a <u>Bounded Variation Function</u> (b.v.f.) over [a,b] iff it can be represented as a difference $U = H - G$ of two bounded, non-decreasing, functions H,G over [a,b].

<u>EXAMPLE 8.1</u> (C.D.F.'s and B.V.F.'s) The reason why c.d.f.'s and b.v.f.'s possess many of the same Mathematical properties stems from the close relationship between a c.d.f. and any bounded, non-decreasing function (appearing in the definition of a b.v.f.). First, it is easy to see that if F is any c.d.f., and k a positive constant, then F and kF possess similar properties, even though kF is not a c.d.f. unless $k = 1$. Next, if c is an arbitrary constant, similar reasoning leads us to conclude that F and $kF + c$ possess similar properties. A function of the latter form differs, in general, from a bounded, non-decreasing function only by the fact that it is always left-continuous; such need not be the case at a point of discontinuity for an arbitrarily bounded, non-decreasing function. For instance, the functions:

$$U_1(x) = \begin{cases} 1, & x < 0 \\ 2, & x = 0 \\ 3, & x > 0 \end{cases} \quad \text{and} \quad U_2(x) = \begin{cases} 1, & x < 0 \\ 3, & x \geq 0 \end{cases}$$

are both bounded and non-decreasing, but both fail to be left-continuous at $x = 0$. This one difference (possibly occurring at points of discontinuity) is the only real difference between bounded, non-decreasing functions and functions of the form $kF + c$, where F is a c.d.f.

Finally, since a b.v.f. is defined as linear combination (namely a difference) of bounded, non-decreasing functions, it is not surprising to find that there is little essential difference between c.d.f.'s and b.v.f.'s.

In view of Example 8.1, we shall deal exclusively with b.v.f.'s that are left-continuous, whence, are linear combinations of the form $aF + bG + c$, where F and G are c.d.f.'s. This represents little loss of generality, and simplifies the development considerably. Modifications necessary in results obtained, so that they remain valid for b.v.f.'s of all types, are minor, if any.

EXERCISE 8.12 Verify that a left-continuous b.v.f. can always be represented in the form: $aF + bG + c$, where F and G are c.d.f.'s and a, b, c are constants.

EXERCISE 8.13 (Alternate Characterization of B.V.F.'s) The terminology "Bounded Variation Function" stems from the following characterization. Prove that a left-continuous function U is a b.v.f. over $[a,b]$ iff for all partitions $P_n = \{a = x_o < x_1 <\ldots< x_n = b\}$ of $[a,b]$, the total variation sums defined by: $\sum_1^n |U(x_i) - U(x_{i-1})|$ are bounded above by some fixed positive constant B. (This is not a simple Exercise. See, for example, Royden or Hobson). Interpret a variation sum graphically.

EXERCISE 8.14 (Differentiability and B.V.F.'s) Prove that if U' exists and is bounded over $[a,b]$, then U is a (continuous) b.v.f. over $[a,b]$.

EXERCISE 8.15 (Combinations of B.V.F.'s) Prove that if U and V are b.v.f.'s over $[a,b]$, then so are the functions $\alpha U + \beta V$, $U \cdot V$, $\max(U,V)$, $\min(U,V)$, and $|U|$. Generalize to the case of $k > 2$ b.v.f.'s. Under what condition(s) would the composite function $U(V)$ be a b.v.f.? Prove also that a finite linear combination of b.v.f.'s is again a b.v.f.

<u>EXAMPLE 8.2</u> (Illustrations) In view of Exercise 8.15, finite linear combinations of the form: $\sum_1^n a_i \text{Sin} b_i x + \sum_1^n c_i \text{Cos} d_i x$ are b.v.f.'s over any bounded interval. The function $V(x) = \text{Sin}(1/x)$ is not a b.v.f. over any bounded interval containing the origin; also its derivative does not remain bounded. A graph of V illustrates this point well. The function $U(x) = x^2\text{Sin}(1/x)$ <u>is</u> a b.v.f. over any bounded interval including the origin. Exercise 8.14 establishes this.

<u>EXAMPLE 8.3</u> The so-called Weierstrass Approximation Theorem (see, for example, Simmons, pg. 154) asserts that <u>any</u> continuous function f defined over a closed, bounded interval $[a,b]$ can be approximated uniformly by some polynomial to within any prescribed $\epsilon > 0$, in the sense that, for any given $\epsilon > 0$, there can always be found a polynomial P, for which $|f(x) - P(x)| < \epsilon$ for <u>all</u> x in $[a,b]$. Clearly any polynomial is a b.v.f. (see Exercise 8.14), but not every continuous function over $[a,b]$ is a b.v.f. Such is the case, for example, with the function $U(x) = x\text{Sin}(1/x)$ over the interval $[-1,1]$, say.

The notion of Bounded Variation[1] can be extended to unbounded intervals in a natural manner. Thus, we term a function U a b.v.f. over the closed, unbounded interval I iff U is a b.v.f.

[1]Recall, we restrict ourselves to the case of left-continuous b.v.f.'s. This represents little loss of generality.

over every closed, bounded subinterval of I. Accordingly, such functions as polynomials, finite linear combinations of exponentials and/or Sines and Cosines, etc. are b.v.f.'s over intervals such as $[0, +\infty)$ or $(-\infty, +\infty)$. The derivative criterion of Exercise 8.14 is often useful in establishing b.v. over a closed, unbounded interval.

It should be observed that there is no essential loss of generality in assuming that a left-continuous b.v.f. is defined for <u>all</u> real x. For, if this is not the case, we can always construct another b.v.f. that agrees with the given b.v.f. wherever it was defined,[1] and is zero elsewhere.

> <u>EXERCISE 8.16</u> Prove the preceding statement. Apply the definition of a b.v.f. over an unbounded interval given above.

In view of the preceding Exercise, we may now consider sequences $\{U_n\}$ of b.v.f.'s defined over a common domain (namely the real line), and their convergence properties. Accordingly, we say that the sequence $\{U_n\}$ of b.v.f.'s converges weakly to the (limit) b.v.f. U, written $U_n \overset{w}{\to} U$, iff $U_n(x) \to U(x)$ for all real x at which U is continuous.[2] If, in addition, $U_n(\pm\infty) \to U(\pm\infty)$, then we say that the sequence $\{U_n\}$ of b.v.f.'s converges completely to U, written $U_n \overset{c}{\to} U$.

---

[1] Except, perhaps, at one point.

[2] A b.v.f. is necessarily continuous at all but perhaps a countable number of points. This follows since any b.v.f. U has a representation of the form $U = aF + bG + c$ (where F and G are c.d.f.'s) over any bounded interval $[a,b]$, and c.d.f.'s possess such a property.

EXERCISE 8.17 Prove that if the sequence $\{U_n\}$ of b.v.f.'s converges either weakly or completely to the b.v.f. U, then $U_n(x+) \to U(x+)$ for all real x. In general, this result cannot be strengthened to pointwise convergence. Illustrate the last comment by a specific example.

Finally, it should be noted that a sequence $\{U_n\}$ of b.v.f.'s may converge to a limit function U that is not a b.v.f. Hence, the requirement that the limit function U be a b.v.f. in the definitions of weak and complete convergence is necessary. The following Example illustrates this point.

EXAMPLE 8.4 For $n = 1,2,\ldots$ define functions $f_n(x)$ over $[-1,1]$ as follows:

$$f_n(x) = \begin{cases} 0, & -1 \le x \le 1/n \\ \mathrm{Sin}(\frac{1}{x}), & 1/n < x \le 1 \ . \end{cases}$$

Each f_n is a b.v.f., yet the limit function f of the sequence $\{f_n\}$, namely,

$$f(x) = \begin{cases} 0, & -1 \le x \le 0 \\ \mathrm{Sin}(\frac{1}{x}), & 0 < x \le 1 \ , \end{cases}$$

is not a b.v.f.

Hints and Answers to Exercises: Section 8

8.1 Hint: Property (1) defining a 1-dimensional c.d.f. is easily verified. Properties (ii), (iii) and (iv) require proof and use of the fact that the (possibly infinite) series defining F is uniformly convergent over $(-\infty, \infty)$.

8.2 Hint: For the first part, refer to the Hint for Exercise 7.1. For the second part, use the fact that the a_i's are distinct and consider (using uniform convergence) the difference between the right and left hand limits of F at $x = a_i$.

8.3 Answer: Let $\{r_n\}$ be any enumeration of the rational numbers and let $\{p_n\}$ be any sequence of positive real numbers for which $\sum_1^\infty p_n = 1$. Define a function F as follows: $F(x) = \sum_1^\infty p_n F_{r_n}(x)$ for all real x. This function can be shown to possess the required properties.

8.4 Hint: The fact that $f(x) \geq 0$ follows from the non-decreasing property of F and the definition of the derivative as a limit of a difference quotient. The second part follows from the Fundamental Theorem of Calculus together with the fact that F must be continuous at $x = a$ and $x = b$.

8.5 Hint: It is not difficult to prove that $\Lambda(x)$ is non-decreasing. The Mean Value Theorem for the Riemann Integral together with continuity of F enables proof of the fact that $\Lambda(x) \to 0$ as $x \to -\infty$ and $\Lambda(x) \to 1$ as $x \to +\infty$. Since F is assumed continuous, actual continuity (not only left-continuity) of Λ follows from properties of the Riemann Integral. Analogously for Δ.

8.6 Hint: For either F_* or F^* property (i) defining a 1-dimensional c.d.f. is easily verified and, since the functions $f_*(x_1,\ldots,x_n) = \min(x_1,\ldots,x_n)$ and $f^*(x_1,\ldots,x_n) = \max(x_1,\ldots,x_n)$ are continuous in the n

real variables $\underline{x}' = (x_1,\ldots,x_n)$, the remaining three defining properties can be shown to be satisfied. As for the generalization, note also that $g(x_1,\ldots,x_n) = \prod_1^n x_i^{k_i}$ $(k_i > 0)$ is a continuous function of the n real variables $\underline{x}' = (x_1,\ldots,x_n)$.

8.8 Hint: Clearly F is a c.d.f. and $F_n(x) \to F(x)$ for all $x \neq a$. Also, $F_n(a) = 1$ $(n = 1,2,\ldots)$ whereas $F(a) = 0$, while $F_n(a+) = 0$ $(n = 1,2,\ldots)$ and $F(a+) = 0$.

8.9 Hint: Given any fixed real number x there exists an index N sufficiently large so that $F_n(x) = (x+n)2n$ for all indices $n \geq N$. Clearly $(x+n)/2n \to \frac{1}{2}$ as $n \to \infty$.

8.11 Hint: Observe it must be established that:

$$\lim_{n\to\infty} \lim_{\epsilon\downarrow 0} F_n(x+\epsilon) = \lim_{\epsilon\downarrow 0} F(x+\epsilon) .$$

8.12 Hint: A left-continuous b.v.f. is the difference between two functions $a_1F + a_2$ and $b_1G + b_2$, where the a's and b's are constants, and F and G are c.d.f.'s.

8.13 Hint: Refer to the References cited.

8.14 Hint: Use Exercise 8.13 and the Mean Value Theorem for derivatives to represent $\sum_1^n |U(x_i) - U(x_{i-1})|$ in the form $\sum_1^n |U'(x_i^*)(x_i - x_{i-1})|$ where $x_{i-1} \leq x_i^* \leq x_i$ $(i = 1,2,\ldots,n)$, then use boundedness of U'.

8.15 Hint: Refer to the Hint for Exercises 8.1, 8.6 and 8.7, and use the connection between (left-continuous) b.v.f.'s, and c.d.f.'s.

8.16 Hint: Suppose U is a (left-continuous) b.v.f. defined over the closed, bounded interval $[a,b]$. Define U^* as follows:

$$U^* = \begin{cases} U(x) & \text{for } a < x \le b \\ 0 & \text{elsewhere} \end{cases}$$

Now apply the extended definition of a left-continuous b.v.f. over $(-\infty, +\infty)$. U and U^* may not agree at the single point $x = a$.

8.17 Hint: Use the representation of Exercise 8.12 and the results of Exercise 8.11.

References to Additional and Related Material: Section 8

1. Feller, W., "An Introduction to Probability Theory and its Applications", Vol. II, Wiley and Sons, New York (1966).

2. Hobson, E., "Theory of Functions of a Real Variable", Vol. I, Dover Publications, Inc., New York.

3. McShane, E. and T. Botts, "Real Analysis", VanNostrand, Inc., New York (1959).

4. Royden, H., "Real Analysis", Macmillan Co., New York (1964).

5. Simmons, G., "Topology and Modern Analysis", McGraw-Hill, Inc., New York (1963).

6. Titchmarsh, E., "The Theory of Functions", (Second Ed.), Oxford University Press, (1939).

9. 1-Dimensional Riemann-Stieltjes Integral

Although the familiar Riemann Integral is sufficient for a wide variety of problem solving-purposes in Applied Mathematics, a generalization of it, known as the Riemann-Stieltjes Integral, must be called upon in many situations. In the present Section we shall develop what is termed a 1-dimensional Riemann-Stieltjes Integral, first with respect to a 1-dimensional c.d.f., then, more generally, with respect to b.v.f.

Accordingly, let F be a 1-dimensional c.d.f. and g a continuous, real-valued function, defined over a bounded interval of the form $[a,b)$.[1] For any partition

$$P_n = \{a = x_{0,n} < x_{1,n} < x_{2,n} <\dots x_{n-1,n} < x_{n,n} = b\}$$

of the interval $[a,b)$, we denote by $\|P_n\|$ the maximum gap between neighboring points of the partition, that is

$$\|P_n\| = \max_{i=1,\dots,n} |x_{i,n} - x_{i-1,n}| \ .$$

We now define what is termed an approximating sum corresponding to the partition P_n. Within each segment $[x_{i-1,n}, x_{i,n})$ of the partition, select a point $x_{i,n}^*$ and, upon evaluation of g at each such point, form the following sum:

$$I(P_n) = \sum_1^n g(x_{i,n}^*)[F(x_{i,n}) - F(x_{i-1,n})] \ .$$

---

[1]Later on it will be shown how these conditions can be relaxed. For example, we shall consider intervals of other types, and functions possibly possessing discontinuities. However, the present, simpler assumptions make an introduction clearer.

This is termed an approximating sum corresponding to the partition.

The following diagram may help in visualizing the quantities that constitute an approximating sum.

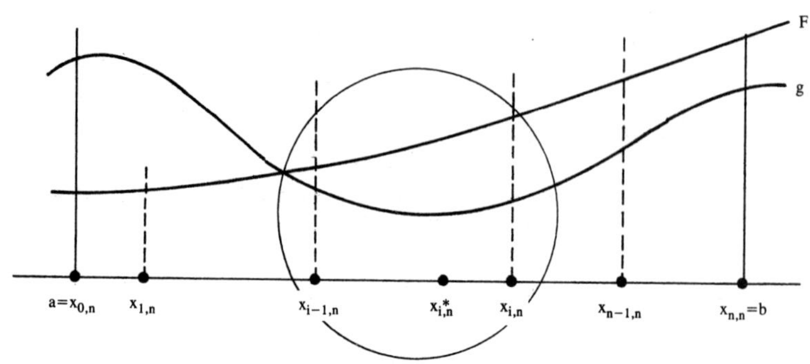

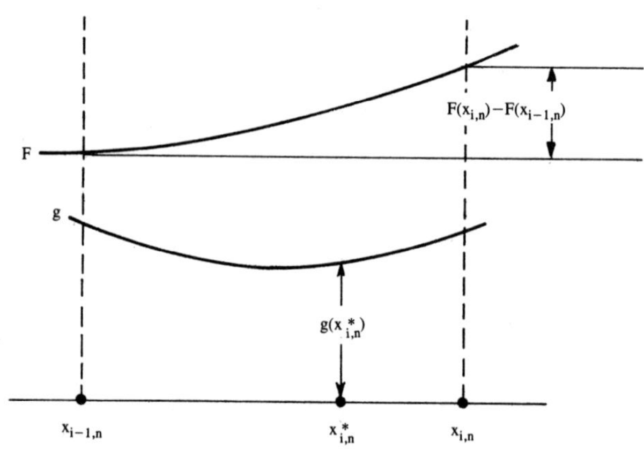

Figure 9-1

Consider now any sequence $\{P_n\}$ of partitions of $[a,b)$ for which each partition in the sequence is a refinement of its predecessor, and for which $\|P_n\| \to 0$ as $n \to \infty$. If, independently of any particular such sequences, all approximating sums $I(P_n)$ tend to the same finite limit as $n \to \infty$, then this limit is called the Riemann-Stieltjes Integral of g with respect to F over $[a,b)$, and is denoted:

$$\int_{a^-}^{b^-} gdF \quad \text{or} \quad \int_{a^-}^{b^-} g(x)dF(x) \ ,$$

where the limits of integration written: a^- and b^-, are intended to indicate that the left-hand point $x = a$ <u>is</u> included within the interval of integration whereas the right-hand point $x = b$ <u>is</u> <u>not</u>. (Later on we shall extend the definition of the Integral to other types of intervals and this distinction will become important. Such a distinction is not necessary for the ordinary Riemann Integral).

Although the definition of the Riemann-Stieltjes Integral given above is straightforward, it does not actually specify conditions under which the Integral itself exists. The following Theorem gives simple sufficient conditions under which the Integral exists.

<u>THEOREM 9.1</u> (Sufficient Conditions for Existence of the Riemann-Stieltjes Integral) Let F be a c.d.f. and g any continuous function over the closed, bounded interval $[a,b]$. Then the Riemann-Stieltjes Integral of g with respect to F exists over $[a,b)$.

PROOF. Let $\{P_n\}$ be any sequence of refinements partitioning $[a,b)$ such that $\|P_n\| \to 0$ as $n \to \infty$, and let $I(P_n)$ and $J(P_n)$, defined as:

$$I(P_n) = \sum_1^n g(x_{i,n}^{*}) [F(x_{i,n}) - F(x_{i-1,n})]$$

$$J(P_n) = \sum_1^n g(x_{i,n}^{**}) [F(x_{i,n}) - F(x_{i-1,n})]$$

for $n = 1,2,\ldots,$ be any pair of corresponding sequences of approximating sums. We must first show that $I(P_n)$ and $J(P_n)$ tend to the same finite limit as $n \to \infty$.

Since g is continuous over the closed, bounded interval $[a,b]$, it is uniformly continuous over this interval.[1] Thus, given any $\epsilon > 0$ there can always be found a $\delta > 0$ such that $|g(x) - g(x')| < \epsilon$ whenever any pair of points x,x' in $[a,b]$ satisfies: $|x-x'| < \delta$.

Now choose an index N sufficiently large so that $\|P_n\| < \delta$ for all indices $n \geq N$. Then,

$$|I(P_n) - J(P_n)| = \left| \sum_1^n [g(x_{i,n}^{*})-g(x_{i,n}^{**})][F(x_{i,n}) - F(x_{i-1,n})]\right|$$
$$\leq \epsilon \sum_1^n [F(x_{i,n}) - F(x_{i-1,n})] \leq \epsilon .$$

[1] Meaning, given any $\epsilon > 0$ there can be found a single $\delta > 0$ such that for <u>any</u> pair x,x' of points within $[a,b]$ for which $|x-x'| < \delta$, we have $|g(x) - g(x')| < \epsilon$. The essential feature is that δ does not depend upon the particular choice for x,x'. A proof of this result can be found in Hobson's book, for example.

Therefore, $I(P_n)$ and $J(P_n)$ must approach the same limit L, say, as $n \to \infty$. This limit is finite, since g is continuous over a closed, bounded interval, and therefore is bounded over that interval.

EXERCISE 9.1 Complete the proof of Theorem 9.1 by showing that if $\{Q_n\}$ is any other sequence of refinements partitioning [a,b) and $I(Q_n)$ $(n = 1,2,\ldots)$ any corresponding sequence of approximating sums, then $I(Q_n)$ also tends to L as $n \to \infty$.

EXAMPLE 9.1 (An Interpretation of the Riemann-Stietjes Integral) Consider the problem of finding the "weight" of the "plate" R bounded by the graph of the continuous function $y = g(x)$ between $x = a$ and $x = b$ (pictured below).

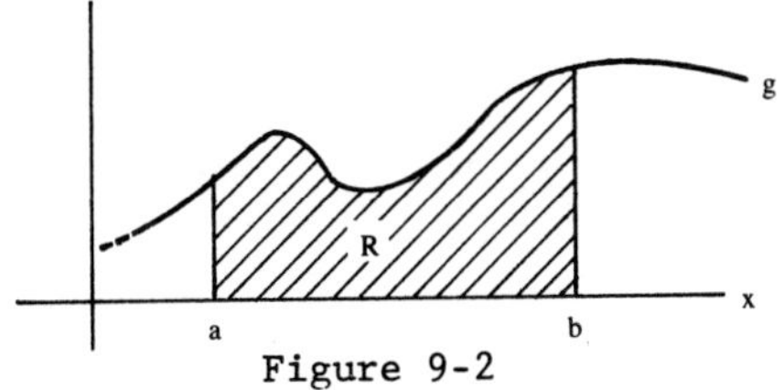

Figure 9-2

Case I: (Constant density) For a partition P_n of [a,b) , an approximation to the weight of R would be (using notation already established)

$$R(P_n) = \sum_{i=1}^{n} g(x_{i,n}^{*}) \cdot k \cdot (x_{i,n} - x_{i-1,n}) \quad ,$$

where k is a constant specific to the material constituting R . $R(P_n)$ may be recognized as the Riemann sum approximating the value of the ordinary Riemann Integral:

$$\int_a^b k\, g(x)\, dx \quad ,$$

and indeed, $R(P_n)$ has, as its limit, the value of this integral, which yields the weight of the plate R.

Case II: (Variable density) Consider now the case where the weight of a small "strip" of height h resting on the segment $[x_{i-1,n}, x_{i,n})$ depends not only upon the length $(x_{i,n} - x_{i-1,n})$ of the segment, but also upon its location. In such a case the weight of such a small "strip" of height h above $[x_{i-1,n}, x_{i,n})$ would be approximately $h \cdot [F(x_{i,n}) - F(x_{i-1,n})]$, where F is a function that reflects the varying density. (For now, we assume that F is a 1-dimensional c.d.f.). Accordingly, for the partition P_n, an approximation to the weight of R would be

$$I(P_n) = \sum_{i=1}^{n} g(x_{i,n}^*) \, [F(x_{i,n}) - F(x_{i-1,n})] \ .$$

In view of Theorem 9.1, $I(P_n)$ possesses a limit as $n \to \infty$, independent of the particular sequence of refining partitions of $[a,b)$, and this limit, which yields the weight of R, is given by the value of the Riemann-Stieltjes Integral:

$$\int_{a^-}^{b^-} g(x)\, dF(x) \quad .$$

EXERCISE 9.2 Let g be a continuous function (not necessarily bounded) over a basic, bounded interval $[a,b)$, and let F be a 1-dimensional c.d.f. If for all (any) sequences $\{P_n\}$ of refinements partitioning $[a,b)$ for which $|P_n| \to 0$ the corresponding sequences of approximating sums are all bounded above and below by fixed constants, say U and L, then the Riemann-Stieltjes Integral of g with respect to F exists over $[a,b)$. Further generalizations of this result exist. However, prove this special case.

Note that the assumption that the integrand g be continuous over a bounded interval of the form $[a,b)$ implies that g is continuous from the right at $x = a$, that is, $g(a) = \lim_{x \downarrow a} g(x) = g(a+)$, whence $g(a)$ is finite. However, this assumption does not guarantee that g is bounded over $[a,b)$, because it is possible that, as $x \uparrow b$, g becomes unbounded. (The special assumption of Theorem 9.1 precluded this possibility, but such need not be the case in general.)

Keeping in mind the above remarks, we now expand the definition of the Riemann-Stieltjes Integral so as to include integration over a singleton point. (For Riemann Integration this need not be considered.)

Accordingly, let c be any point within the interval $[a,b)$. The Riemann-Stieltjes Integral of the continuous function g with respect to the c.d.f. F over c is denoted by and defined as follows:

$$\int_{c^-}^{c^+} g\,dF = \lim_{\epsilon \downarrow 0} \int_{c^-}^{(c+\epsilon)^-} g\,dF \ .$$

Under our assumptions, this integral (limit) always exists, is finite, and has value $g(c)[F(c+) - F(c)]$, which is clearly zero if the c.d.f. F is continuous at $x = c$.

EXERCISE 9.3 Prove in detail that:

$$\int_{c^-}^{c^+} g\,dF = g(c)[F(c+) - F(c)] \quad .$$

Having extended the definition of the Integral to singleton points, we next extend its definition to include integration over bounded intervals of the form $[a,b]$, (a,b) and $(a,b]$. Accordingly, assume that any interval of the above type is included in some bounded interval of the form $[c,d)$ over which the Riemann-Stieltjes Integral of the continuous function g with respect to the c.d.f. F exists. We then define the integral of g with respect to F over the indicated intervals as follows:

$$[a,b]: \quad \int_{a^-}^{b^+} g\,dF = \int_{a^-}^{b^-} g\,dF + \int_{b^-}^{b^+} g\,dF$$

$$(a,b): \quad \int_{a^+}^{b^-} g\,dF = \int_{a^-}^{b^-} g\,dF - \int_{a^-}^{a^+} g\,dF$$

$$(a,b]: \quad \int_{a^+}^{b^+} g\,dF = \int_{a^-}^{b^-} g\,dF - \int_{a^-}^{a^+} g\,dF + \int_{b^-}^{b^+} g\,dF \quad .$$

EXERCISE 9.4 Prove that if the c.d.f. F is continuous at an endpoint (either a or b) of such a bounded interval, then the value of the integral is the same, whether or not the end is closed or open-type.

EXERCISE 9.5 Prove that if F is constant throughout any of the four types of bounded intervals, then the corresponding integral of any continuous g with respect to F is zero over that interval.

With the Riemann-Stieltjes Integral of a continuous function g with respect to a c.d.f. F now defined over any type of bounded interval (subject to existence), we may now state some of the basic properties possessed by this Integral.

(i) $\int_{a\pm}^{b\pm} gdF = \int_{a\pm}^{c\pm} gdF + \int_{c\pm}^{b\pm} gdF \qquad a < c < b$

(ii) $\int_{a\pm}^{b\pm} [\alpha g+\beta h]dF = \alpha\int_{a\pm}^{b\pm} gdF + \beta\int_{a\pm}^{b\pm} hdF \qquad \alpha, \beta$ constants

(iii) $\int_{a\pm}^{b\pm} dF = F(b\pm) - F(a\pm)$ [1]

(iv) $\int_{a\pm}^{b\pm} gdF = \int_{a}^{b} gfdx$ (Riemann) if $F' = f$ exists and is continuous

(v) $\int_{a\pm}^{b\pm} gdF \geq 0$ if $g \geq 0$

(vi) $\int_{a\pm}^{b\pm} gdf \geq \int_{a\pm}^{b\pm} hdF$ if $g \geq h$

(vii) $[\inf_I g][F(b\pm) - F(a\pm)] \leq \int_{a\pm}^{b\pm} gdF \leq [\sup_I g][F(b\pm) - F(a\pm)]$ [2]

---

[1]Note that $F(c+) = \lim_{\epsilon \downarrow 0} F(c+\epsilon)$ and $F(c-) = F(c) = \lim_{\epsilon \downarrow 0} F(c-\epsilon)$.

[2]I denotes whichever interval of integration applies.

EXERCISE 9.6 Establish properties (i) through (vii), either by using previously proved results, or by using the basic definition of the Riemann-Stieltjes Integral.

EXERCISE 9.7 Prove that if f_1 and f_2 are continuous and integrable with respect to F over the bounded interval I, and over this interval the continuous function g satisfies: $f_1 \le g \le f_2$, then g is integrable with respect to F over I, and:

$$\int_I f_1 dF \le \int_I g dF \le \int_I f_2 dF \quad .$$

EXERCISE 9.8 As a special case of the preceding Exercise, prove that if g is continuous and $|g|$ is integrable with respect to F over the bounded interval I, then g is integrable with respect to F over I, and:

$$\left| \int_I g dF \right| \le \int_I |g| dF \quad .$$

EXERCISE 9.9 Prove that if $I_1 \subseteq I_2$ are bounded intervals, and the integral of the continuous function g exists with respect to F over I_2, then it also exists over I_1. Furthermore, prove that:

$$\int_{I_1} g dF \le \int_{I_2} g dF \quad \text{if} \quad g \ge 0 \quad .$$

It has already been established (Exercise 8.1) that if F and G are c.d.f.'s, then so is $\alpha F + \beta G$, where $\alpha, \beta \ge 0$ and $\alpha + \beta = 1$. Now if the integral of the continuous function g

exists with respect to both F and G over the interval indicated, then we have the following property:

$$\text{(viii)} \quad \int_{a^\pm}^{b^\pm} g\,d(\alpha F + \beta G) = \alpha\int_{a^\pm}^{b^\pm} g\,dF + \beta\int_{a^\pm}^{b^\pm} g\,dG \quad .$$

EXERCISE 9.10 Prove property (viii).

EXERCISE 9.11 Formulate the obvious extensions of properties (i), (ii) and (viii) to the case $k > 2$.

EXERCISE 9.12 (A Mean Value Theorem for Riemann-Stieltjes Integrals). Prove that if g is bounded and continuous over the bounded interval $[a,b]$, then there can always be found a point c within this interval (possibly depending upon the particular integration interval involved) such that:

$$\int_{a^\pm}^{b^\pm} g\,dF = g(c)\,[F(b\pm) - F(a\pm)] \quad .$$

The next step in extension of the Riemann-Stieltjes Integral is for the case of unbounded intervals of integration (which is accomplished by a simple limit argument). Now provided that the indicated integrals (and limits) of the continuous function g with respect to the c.d.f. F exist, we define the integrals over the indicated intervals as follows:

$$(-\infty,b): \quad \int_{-\infty}^{b^-} g\,dF = \lim_{a\to-\infty} \int_{a^-}^{b^-} g\,dF$$

$$(-\infty,b] \quad \int_{-\infty}^{b^+} g\,dF = \lim_{a\to-\infty} \int_{a^-}^{b^+} g\,dF$$

$$(a,+\infty):\quad \int_{a^+}^{+\infty} g\,dF = \lim_{b\to+\infty} \int_{a^+}^{b^-} g\,dF$$

$$[a,+\infty):\quad \int_{a^-}^{+\infty} g\,dF = \lim_{b\to+\infty} \int_{a^-}^{b^-} g\,dF$$

$$(-\infty,\infty):\quad \int_{-\infty}^{+\infty} g\,dF = \lim_{\substack{a\to-\infty \\ b\to+\infty}} \int_{a^-}^{b^-} g\,dF \; .$$

It is not difficult to prove that if g is bounded and continuous over any of the unbounded intervals above, then the corresponding Reimann-Stieltjes Integral of g with respect to any c.d.f. F exists over that interval.

> <u>EXERCISE 9.13</u> Prove the preceding statement for one of the five cases given above.

Finally, it is important to observe that all of the properties previously established for the Riemann-Stieltjes Integral over a bounded interval carry over (with a few minor and obvious modifications) to the case of integration over unbounded intervals.

> <u>EXERCISE 9.14</u> (Expectation: Discrete Case) For the 'step' c.d.f. $F = \Sigma p_i f_{a_i}$ defined in Exercise 8.2, and any continuous function g, prove that:
>
> $$\int_{-\infty}^{+\infty} g\,dF = \Sigma \int_{a_i^-}^{a_i^+} g\,dF = \Sigma\, g(a_i)[F(a_i+) - F(a_i)] = \Sigma\, g(a_i)p_i \; ,$$
>
> provided the latter sum converges absolutely.

<u>EXERCISE 9.15</u> (Moment Generating Function and Absolute Moments) For a given c.d.f. F, its Moment Generating Function $M(t)$ is said to exist for a value of t (real) if its defining integral:

$$M(t) = \int_{-\infty}^{\infty} e^{tx} dF(x)$$

exists (finite) for that value of t. Also, the r-th absolute moment M_r of F is said to exist for a value of $r > 0$ if its defining integral:

$$M_r = \int_{-\infty}^{\infty} |x|^r dF(x)$$

exists (finite) for that value of r. Using a previous Exercise, prove that if, for some $t_0 > 0$, $M(t)$ exists for all $|t| < t_0$, then M_r exists for any $r > 0$.

<u>EXERCISE 9.16</u> (Conditions for Existence of a Moment Generating Function) Prove that a necessary and sufficient condition that the Moment Generating Function $M(t)$ of the c.d.f. F exist in some interval $|t| < t_0$ $(t_0 > 0)$ about the origin is that there exist some $\delta > 0$ such that:

$$e^{\delta x}[F(-x) + 1 - F(x)] \to 0 \quad \text{as} \quad x \to +\infty .$$

<u>EXAMPLE 9.2</u> (Existence of Absolute Moments) We shall prove that if, for some $k > 0$,

$$F(x) = O(|x|^{-k}) \quad \text{as} \quad x \to -\infty$$

and

$$1-F(x) = O(x^{-k}) \quad \text{as} \quad x \to +\infty ,$$

then the r-th absolute moment M_r of F exists for all non-negative values of $r < k$.

It is sufficient to show that the integral of $|x|^r$ with respect to F over any bounded interval of the form $[a,b)$ is less than a constant independent of a and b.[1] Now using the hypotheses, we have for $n = 1,2,\ldots$:

$$\int_{(2^{n-1})^-}^{(2^n)^-} |x|^r dF(x) \le 2^{rn}[F(2^n) - F(2^{n-1})]$$
$$\le 2^{rn}[1 - F(2^{n-1})]$$
$$\le C/2^{n(k-r)} ,$$

where C is a constant independent of n.[2]

An analogous result holds for the integrals over the intervals $[-2^n, -2^{n-1})$ for $n = 1,2,\ldots$. Summing the integrals over all of the above intervals, and adding the integral over $[-1,1)$ (which is always ≤ 1), we have for any bounded interval $[a,b)$:

$$\int_{a^-}^{b^-} |x|^r dF(x) < 1 + 2C/(2^{k-r}-1) ,$$

which establishes the desired result.

We now consider certain properties of the Riemann-Stieltjes Integral dealing with limit operations involving the integrand, integrator c.d.f., or interval of integration. These properties are of frequent value in applications.

The first of these properties involves the case where the integrand varies. The result is contained in the following Theorem.

---

[1] The result then follows from definition of the Integral over $(-\infty, +\infty)$.

[2] See Exercise 6.9.

<u>THEOREM 9.2</u> (Varying Integrand) If $\{g_n\}$ is a sequence of continuous functions that converges uniformly to the limit function g over the closed, bounded interval $[a,b]$, then for any c.d.f. F we have:

$$\int_{a^\pm}^{b^\pm} g_n dF \to \int_{a^\pm}^{b^\pm} g dF \quad \text{as} \quad n \to \infty .$$

PROOF. Since convergence of $\{g_n\}$ is uniform, each function in the sequence, and also g, is bounded and continuous over $[a,b]$. Hence, all of the above integrals exist (finite). Now, using properties of the Integral already established, together with uniform convergence, we have for all n sufficiently large:

$$\left| \int_{a^\pm}^{b^\pm} g_n dF - \int_{a^\pm}^{b^\pm} g dF \right| = \left| \int_{a^\pm}^{b^\pm} (g_n - g) dF \right|$$

$$\le \int_{a^\pm}^{b^\pm} |g_n - g| dF$$

$$< \int_{a^\pm}^{b^\pm} \epsilon \, dF \le \epsilon ,$$

which completes the proof.

<u>EXERCISE 9.17</u> Prove by example that uniform convergence cannot be weakened to pointwise convergence in Theorem 9.2.

<u>EXERCISE 9.18</u> Extend Theorem 9.2 to unbounded intervals by making suitable assumptions on g .

EXERCISE 9.19 If $g_1, g_2, \ldots, g_n$ are continuous and integrable with respect to the c.d.f. F over an interval I, then so are $g^* = \max(g_1, \ldots, g_n)$ and $g_* = \min(g_1, \ldots, g_n)$, and:

$$\int_I g_* dF \le \min_i \int_I g_i dF \le \max_i \int_I g_i dF \le \int_I g^* dF \quad .$$

Prove the above statements. (Note that the function g^* is defined for each x in I as $g^*(x) = \max(g_1(x), g_2(x), \ldots, g_n(x))$. Analogously for g_*).

The next result is well-known, and deals with the situation where the integrator (that is, the c.d.f.) varies, as opposed to the integrand. In its many versions, it is known as a Helly-Bray (type) Theorem.

THEOREM 9.3 (Helly-Bray: Varying Integrator) Let $\{F_n\}$ be a sequence of c.d.f.'s for which $F_n \overset{c}{\to} F$. If g is any continuous function over a closed, bounded interval $[a,b]$, where a and b are continuity points of F, then:

$$\int_{a^\pm}^{b^\pm} g dF_n \to \int_{a^\pm}^{b^\pm} g dF \quad \text{as } n \to \infty .$$ [1,2]

[1] Notice that, since a and b are continuity points of F, all four of the integrals:

$$\int_{a^\pm}^{b^\pm} g dF$$

coincide in value.

[2] We prove the Theorem for the case of an interval of integration of the form $[a,b)$. The remaining three cases follow in a straightforward manner upon consideration of the definition of the Integral over $[a,b]$, (a,b), and $(a,b]$ respectively.

PROOF. Observe first that all of the above integrals exist. Now let $\{P_m\}$ be any sequence of refinements partitioning the interval $[a,b)$, such that $\|P_m\| \to 0$ as $m \to \infty$; choose the endpoints of the segments within each partition as continuity points of F. (This creates no loss of generality, and can always be done).

Let $I_n(P_m)$ [respectively, $I(P_m)$] be an approximating sum corresponding to the partition P_m and the c.d.f. F_n [respectively, F], where the points chosen within the segments of the partition are chosen as continuity points of F. This can always be done for $n = 1,2,\ldots$. According to the basic definition of the Riemann-Stieltjes Integral, we may conclude the following:

$$\lim_{n\to\infty} \lim_{m\to\infty} I_n(P_m) = \lim_{n\to\infty} \int_{a^-}^{b^-} g\,dF_n$$

and

$$\lim_{m\to\infty} \lim_{n\to\infty} I_n(P_m) = \lim_{m\to\infty} I(P_m) = \int_{a^-}^{b^-} g\,dF \quad .$$

The conditions sufficient for the two iterated limits to be equal are satisfied,[1] and accordingly,

$$\lim_{n\to\infty} \int_{a^-}^{b^-} g\,dF_n = \int_{a^-}^{b^-} g\,dF \quad .$$

This completes the proof.

---

[1]See, for example, Hobson, Vol. II, pg. 48, and/or Section 2.

<u>EXERCISE 9.20</u> (Extension to Unbounded Intervals) Prove that the preceding result is valid for unbounded intervals provided the continuous function g is bounded. Note, for example, for the interval $(-\infty, \infty)$, a proof could be based upon the following inequality, which is valid for continuity points a, b of F:

$$\left| \int_{-\infty}^{\infty} g\,dF_n - \int_{-\infty}^{\infty} g\,dF \right| \le \left| \int_{-\infty}^{\infty} g\,dF_n - \int_{a^-}^{b^-} g\,dF \right| +$$

$$\left| \int_{a^-}^{b^-} g\,dF - \int_{a^-}^{b^-} g\,dF_n \right| +$$

$$\left| \int_{a^-}^{b^-} g\,dF - \int_{-\infty}^{\infty} g\,dF \right| .$$

<u>EXERCISE 9.21</u> (Further Extension) If $F_n \stackrel{c}{\to} F$, and for some non-negative continuous function g we have:

$$\int_I g\,dF_n \to \int_I g\,dF \quad \text{as } n \to \infty ,$$

then for any continuous function h for which $|h(x)| \le g(x)$ for all x in I,[1] we have:

$$\int_I h\,dF_n \to \int_I h\,dF \quad \text{as } n \to \infty .$$

[1] If the interval of integration I is bounded, the result is immediate. If I is unbounded, the condition need only hold for all $|x|$ sufficiently large.

<u>EXERCISE 9.22</u> (Application: Moment Convergence) Suppose $F_n \xrightarrow{c} F$, and for all $|t| < t_0$ $(t_0 > 0)$ the corresponding Moment Generating Functions converge, that is,

$$M_n(t) \to M(t) \quad \text{for} \quad |t| < t_0 \quad \text{as} \quad n \to \infty .$$

Then, the corresponding r-th absolute moments converge for any $r > 0$, that is,

$$\int_{-\infty}^{\infty} |x|^r dF_n \to \int_{-\infty}^{\infty} |x|^r dF \quad \text{as} \quad n \to \infty .$$

The next result in this series deals with the case where both the integrand g and the integrator F remain unchanged, but the interval of integration varies. This result is contained in the following Theorem.

<u>THEOREM 9.4</u> (Varying Interval of Integration) If $\{I_n\}$ is a sequence of intervals converging to the interval I, and g is a bounded and continuous function over $\cup I_n$, then

$$\int_{I_n} g dF \to \int_I g dF \quad \text{as} \quad n \to \infty .$$

PROOF. First note that all of the above integrals exist. To begin, we define the expression: "$x \trianglelefteq I_n$" to mean $x \le y$ for all y in I_n; the statement "$x \triangleright I_n$" is defined similarly with $x \le y$ replaced by $x > y$. In terms of this notation, we define the c.d.f.'s F_n^* $(n = 1,2,\ldots)$

and F^* as follows:

$$F_n^*(x) = \begin{cases} 0 & \text{if } x \trianglelefteq I_n \\ 1 & \text{if } x \vartriangleright I_n \\ F(x) & \text{otherwise} \end{cases}$$

and

$$F_n^*(x) = \begin{cases} 0 & \text{if } x \trianglelefteq I \\ 1 & \text{if } x \vartriangleright I \\ F(x) & \text{otherwise .} \end{cases}$$

Because $F_n^* \xrightarrow{c} F^*$ the required result follows directly from Theorem 9.3 since

$$\int_{I_n} g\,dF = \int_{\cup I_n} g\,dF_n^* \to \int_{\cup I_n} g\,dF^* = \int_I g\,dF$$

as $n \to \infty$. This completes the proof.

<u>EXERCISE 9.23</u> Referring to the preceding proof, verify that F_n^* $(n = 1,2,\ldots)$ and F^* are indeed c.d.f.'s, and that $F_n^* \xrightarrow{c} F$ as $n \to \infty$.

<u>EXERCISE 9.24</u> Verify the result contained in the last step of the proof of Theorem 9.4.

We next consider the so-called "integration-by-parts" formulas that apply to the Riemann-Stieltjes Integral. The first of two results of this type is contained in the following Theorem.

THEOREM 9.5 (Integration-by-Parts: Formula A) If F and G are continuous c.d.f.'s over the closed, bounded interval $[a,b]$ then:

$$\int_a^b FdG = FG\Big|_a^b - \int_a^b GdF \text{ ,}^1$$

where

$$FG\Big|_a^b = F(b)G(b) - F(a)G(a) \quad .$$

PROOF. First note that both of the above integrals exist. Now let $\{P_n\}$ be any sequence of refinements partitioning the interval $[a,b)$ for which $\|P_n\| \to 0$ as $n \to \infty$.

Since G is uniformly continuous[2] over $[a,b]$, given any $\epsilon > 0$ we may always find a corresponding $\delta > 0$ such that $|G(x) - G(x')| < \epsilon$ whenever the pair of points x,x' belonging to $[a,b]$ satisfies: $|x-x'| < \delta$. Accordingly, we may choose an index N sufficiently large so that $\|P_n\| < \delta$ for all $n \geq N$, whence for any pair of points within the segments $[x_{i-1,n}, x_{i,n}]$ $(i = 1,2,\ldots,n)$, $(n \geq N)$, the magnitude of the difference between the values of G at these points is always $< \epsilon$.

Next observe the following identity, which is valid for any partition P_n:

---

[1] Since F and G are continuous at $x = a$ and $x = b$, it is not necessary to indicate whether or not the interval of integration is "open-type" or "closed-type" at either end.

[2] See Footnote to Theorem 9.1.

$$\sum_1^n G(x_{i-1,n})\,[F(x_{i,n}) - F(x_{i-1,n})] + \qquad \text{(a)}$$

$$\sum_1^n F(x_{i-1,n})\,[G(x_{i,n}) - G(x_{i-1,n})] + \qquad \text{(b)}$$

$$\sum_1^n [G(x_{i,n}) - G(x_{i-1,n})][F(x_{i,n}) - F(x_{i-1,n})] \qquad \text{(c)}$$

$$\equiv F(b)G(b) - F(a)G(a) \quad .$$

By definition of the Riemann-Stieltjes Integral, and choosing $x^*_{i,n} = x_{i-1,n}$ in the approximating sums, the expressions in (a) and (b) tend, respectively, to:

$$\int_a^b GdF \quad \text{and} \quad \int_a^b FdG$$

as $n \to \infty$, whereas the expression in (c) is less than ϵ in value for any index $n \geq N$. Upon allowing n to tend to infinity, the required result then follows from the identity. This completes the proof.

EXERCISE 9.25 Verify the statement made about the expression marked (c) in the preceding proof.

EXERCISE 9.26 (Extension: Unbounded Intervals) Prove that the result of the preceding Theorem remains valid when $a = -\infty$ and/or $b = +\infty$.

EXAMPLE 9.3 If F and G are continuous c.d.f.'s, then FG is a continuous c.d.f. and, in view of Property (iii) of the Riemann-Stieltjes Integral, the conclusion of

Theorem 9.5 could alternately have been stated as follows:

$$\int_a^b d(FG) = \int_a^b FdG + \int_a^b GdF$$ [1]

<u>EXERCISE 9.27</u> (Application: Convolution Formula) First prove that, if F and G are continuous c.d.f.'s, then so are the c.d.f.'s F_z and G_z defined for all real x by:

$$F_z(x) = F(x-z)$$

and

$$G_z(x) = G(x-z) ,$$

for any fixed real number z. Next, prove the function $F*G$ of the real variable z defined by:

$$F*G(z) = \int_{-\infty}^{\infty} F(x-z)dG(x) = \int_{-\infty}^{\infty} G(x-z)dF(x) ,$$

is a bona-fide continuous c.d.f., termed the <u>convolution</u> of F with G. Do not fail to prove equality of the two integrals. Also, establish that $F*G \equiv G*F$.

The next result is the second of a pair of integration-by-parts formulas for the Riemann-Stieltjes Integral.

<u>THEOREM 9.6</u> (Integration-by-Parts: Formula B) If the function g has a continuous derivative over the closed, bounded interval $[a,b]$, and both $x = a$ and $x = b$ are

[1] See the Footnote to Theorem 9.5.

continuity points of the c.d.f. F, then:

$$\int_a^b g\,dF = gF\Big|_a^b - \int_a^b g'F\,dx \quad \text{(Riemann)}\ .$$[1]

PROOF. All of the above integrals exist. First let $\{P_n\}$ be a sequence of refinements partitioning the interval $[a,b)$ for which $\|P_n\| \to 0$ as $n \to \infty$.

By the uniform continuity argument of Theorem 9.5, applied to the continuous function g over $[a,b]$, given any $\epsilon > 0$, we may always find a corresponding $\delta > 0$, and index N, for which $\|P_n\| < \delta$ for all indices $n \geq N$ and such that, for any pair of points within the segments $[x_{i-1,n}, x_{i,n}]$ $(i = 1,2,\ldots,n)$ $(n \geq N)$, the magnitude of the difference between the values of g at these points is $< \epsilon$.

Consider now the following identity, which is valid for any partition P_n :

$$\sum_1^n g(x_{i-1,n})\,[F(x_{i,n}) - F(x_{i-1,n})] + \qquad \text{(a)}$$

$$\sum_1^n [g(x_{i,n}) - g(x_{i-1,n})]F(x_{i,n}) + \qquad \text{(b)}$$

$$\sum_1^n [g(x_{i,n}) - g(x_{i-1,n})][F(x_{i,n}) - F(x_{i-1,n})] \qquad \text{(c)}$$

$$\equiv g(b)F(b) - g(a)F(a) \quad .$$

---

[1] See the Footnote to Theorem 9.5.

By definition of the Riemann-Stieltjes Integral, and choosing $x_{i,n}^{*} = x_{i-1,n}$ in the approximating sum, the expression in (a) tends to:

$$\int_a^b g\,dF$$

as $n \to \infty$.

Next, using the Mean Value Theorem for Derivatives, the expression in (b) can be written as:

$$\sum_1^n g'(x_{i,n}^{**})F(x_{i,n})(x_{i,n}-x_{i-1,n}) \quad ,$$

where $x_{i,n}^{**}$ is some point within the interval $[x_{i-1,n}, x_{i,n}]$, for $i = 1,2,\ldots,n$. By continuity of g', it can be seen that this approximating sum tends in value to:

$$\int_a^b g'F\,dx \quad \text{(Riemann)}$$

as $n \to \infty$.

Finally, the expression in (c) is less than ϵ in value for any $n \geq N$, and the required result follows from the identity upon allowing $n \to \infty$. This completes the proof.

EXERCISE 9.28 Prove that the formula remains valid in Theorem 9.6 for $a = -\infty$ and/or $b = +\infty$ provided g remains bounded as $x \to -\infty$ and/or $x \to +\infty$, in the sense that if both integrals exist they are related in value by the stated formula.

EXERCISE 9.29 (Absolute Moments) Let F be a c.d.f. such that $F(x) = 0$ for $x \le 0$. Prove that for any $r > 0$ we have:

$$\int_0^\infty |x|^r dF(x) = r \int_0^\infty |x|^{r-1}[1-F]dx \quad \text{(Riemann)},$$

in the sense that if one integral exists (finite), then both do, and they are related as stated above. Strictly speaking, use of absolute value is unnecessary here.

EXERCISE 9.30 Prove that a sufficient condition for the r-th $(r > 0)$ absolute moment of F to exist is that, for some $\delta > 0$, $x^{r+\delta}[F(-x)+1-F(x)] \to 0$ as $x \to +\infty$.

EXERCISE 9.31 The c.d.f. F defined as follows: $F(x) = 0$ for $x \le 2$, $1-F(x) = C/x(\log x)^r$ for $x > 2$ (where $C > 0$ and $r > 1$ are constants), demonstrates that the condition of Exercise 9.30 is not, in general, necessary for the r-th absolute moment to exist. Verify this statement.

EXERCISE 9.32 (Absolute Moments: Continuation) Let F be any c.d.f. and $r > 0$. Prove that the r-th absolute moment of F exists (that is, $|x|^r$ is Riemann-Stieltjes Integrable with respect to F over the real axis) if and only if $|x|^{r-1}[1-F(x)+F(-x)]$ is Riemann Integrable over the real axis. Derive the formula relating the two integrals, thereby generalizing the results of Exercise 9.29.

<u>EXERCISE 9.33</u> Prove that for any c.d.f. F:

$$\text{(i)} \quad \lim_{x\to\infty} x\int_{x^-}^{\infty} (1/z)dF(z) = \lim_{x\to-\infty} x\int_{-\infty}^{x^-} (1/z)dF(z) = 0$$

and

$$\text{(ii)} \quad \lim_{x\downarrow 0} x\int_{x^-}^{\infty} (1/z)dF(z) = \lim_{x\uparrow 0} x\int_{-\infty}^{x^-} (1/z)dF(z) = 0.$$

<u>Extension: the Riemann-Stieltjes Integral with a B.V.F. Integrator</u>

We are now able to extend definition of the Riemann-Stieltjes Integral to the case where the integrator function is a left-continuous b.v.f. The extension is <u>immediate</u>, in view of the fact that any left-continuous b.v.f. U can be expressed as: $U = aF + bG + c$, where F,G are c.d.f.'s and a,b,c are constants. Property (viii) of the Riemann-Stieltjes Integral with respect to a c.d.f., along with the results of Exercise 9.5, provide the justification.

Accordingly, the following results are valid for left-continuous b.v.f. integrators: Properties (i), (ii), (iii), (iv), (viii); Theorems 9.1 through 9.6; Example 9.3; Exercises 9.3, 9.4, 9.7, 9.8, 9.11, 9.12, 9.20, 9.22. With an added boundedness condition, Exercises 9.21, 9.27, 9.30 are also valid.

<u>Extension: the Riemann-Stieltjes Integral with Complex-Valued Integrand</u>.

If the integrand g of the Riemann-Stieltjes Integral with respect to a left-continuous b.v.f. U is a continuous, complex-valued function of the real variable x, for example, e^{ix}, Sin(ix), or $\Gamma(ix)$, we define the integral of g with respect to U over an interval I as follows:

$$\int_I g(x)dU(x) = \int_I \text{Re}[g(x)]\, dU(x) + i \int_I \text{Im}[g(x)]\, dU(x);$$

necessary and sufficient conditions for this integral to exist are that both integrals on the right-hand side exist.

Extension: the Riemann-Stieltjes Integral with Discontinuous Integrand.

In view of Exercise 9.3, the Riemann-Stieltjes Integral of a continuous integrand with respect to a left-continuous b.v.f. integrator U over a singleton point set $\{c\}$ is zero if the integrator U is continuous at $x = c$. More generally, if g is an integrand possessing at most a finite number of finite discontinuities[1] in any bounded interval, and in addition these (possible) points of discontinuity do not coincide with (possible) points of discontinuity of the integrator U, then all results obtained previously for the Riemann-Stieltjes Integral with a continuous integrand g, remain valid in this more general case. Little more than partitioning the interval of integration, using Property (i), is required for proof.

Note that the various extensions of the Riemann-Stieltjes Integral can be combined in various manners.

[1] g need not be left-continuous; left-hand and right-hand limits, as well as the value of f at a point of discontinuity, exist and are finite, however.

We conclude this Section with some Examples of the preceding generalizations of the Riemann-Stieltjes Integral, as well as with the following "change-of variables" formula.

<u>THEOREM 9.7</u> (Change-of-Variables: Riemann-Stieltjes Version) Suppose g is a continuous integrand and U is a left-continuous b.v.f. integrator over the interval $[a,b)$. If $x = \varphi(t)$ is a continuous, monotone[1] function of t, then:

$$\int_{a^-}^{b^-} g(x)dU(x) = \int_{\alpha^-}^{\beta^-} g(\varphi(t))dU(\varphi(t)) ,$$

where $\alpha = \varphi^{-1}(a)$ and $\beta = \varphi^{-1}(b)$. If, in addition, φ and u are differentiable, then:

$$\int_{a^-}^{b^-} g(x)dU(x) = \int_{\alpha}^{\beta} g(\varphi(t))U'(\varphi(t))\varphi'(t)dt \quad \text{(Riemann)}.$$

A proof of Theorem 9.7 can be accomplished by a now-familiar appeal to the basic definition of the Riemann-Stieltjes Integral as the limit of approximating sums. If case of differentiability, an appeal to the Mean Value Theorem for Derivatives (from Calculus) yields the second part. (Similar reasoning was needed to establish Property (iv) of the Riemann-Stieltjes Integral).

[1]That is, either non-decreasing or non-increasing.

<u>EXAMPLE 9.4</u> (Change-of-Variables: Non-Parametric Theory) Let F be any continuous c.d.f., m any positive integer, and k any non-negative integer. Using the change-of-variables accomplished by: $x = F^{-1}(t)$, we have:

$$\int_{-\infty}^{\infty} F^k d(F^m) = \int_{-\infty}^{\infty} [F(x)]^k \, d([F(x)]^m)$$

$$= \int_0^1 t^k \, d(t^m)$$

$$= \int_0^1 t^k m t^{m-1} dt = m/(m+k) \quad .$$

<u>EXAMPLE 9.5</u> (Change-of-Variables) Consider evaluation of:

$$\int_0^{\pi^2} x^4 d(\text{Sin } x^2).$$

We choose the monotone function (change-of-variable) $x = t^{\frac{1}{2}}$, which is differentiable (except at $t = 0$). Thus,

$$\int_0^{\pi^2} x^4 d(\text{Sin } x^2) = \int_0^{\pi} t^2 \, (2t^{\frac{1}{2}})\text{Cos}t \cdot (\tfrac{1}{2}t^{-\frac{1}{2}})dt = \int_0^{\pi} t^2 \text{Cos}t \; dt.$$

(The last integral can be evaluated directly, using successive Riemann Integrations-by-Parts, or by "look-up" in any standard Table of Integrals).

<u>EXAMPLE 9.6</u> (Complex-valued Integrand) Consider evaluation of:

$$c(t) = \int_0^1 e^{itx} dU(x),$$

where t is a real number, and U is the b.v.f. integrator: $U(x) = 6x(1-x)$. Accordingly, since $e^{itx} = \text{Cos}(tx) + i\,\text{Sin}(tx)$,

$$c(t) = \int_0^1 \text{Cos}(tx)dU(x) + i \int_0^1 \text{Sin}(tx)dU(x) .$$

Evaluation (and simplification) yields:

$$c(t) = \begin{cases} -\frac{6}{t^2}[\text{Cos}t + i\,\text{Sin}t][1 + \frac{i}{t}], & t \neq 0 \\ 1 & t = 0 \end{cases}$$

<u>EXAMPLE 9.7</u> (Integrand and Integrator Possessing Discontinuities) Consider evaluation of the Riemann-Stieltjes Integral:

$$\int_{0^-}^{\infty} g(x)dU(x)$$

where

$$g(x) = \begin{cases} 0, & \text{for } 2n < x \leq 2n+1 \quad (n = -1, 0, 1, 2, \ldots) \\ 1, & \text{for } 2n+1 < x \leq 2n+2 \quad (n = -1, 0, 1, 2, \ldots) \end{cases}$$

and

$$U(x) = \begin{cases} 0, & \text{for } 2n-\tfrac{1}{2} < x \le 2n+\tfrac{1}{2} \quad (n = 0,1,2,\ldots) \\ (\tfrac{1}{2})^n, & \text{for } 2n+\tfrac{1}{2} < x \le 2n+\tfrac{3}{2} \quad (n = 0,1,2,\ldots) \end{cases}$$

Portions of g (solid line) and U (dashed line) are pictured below:

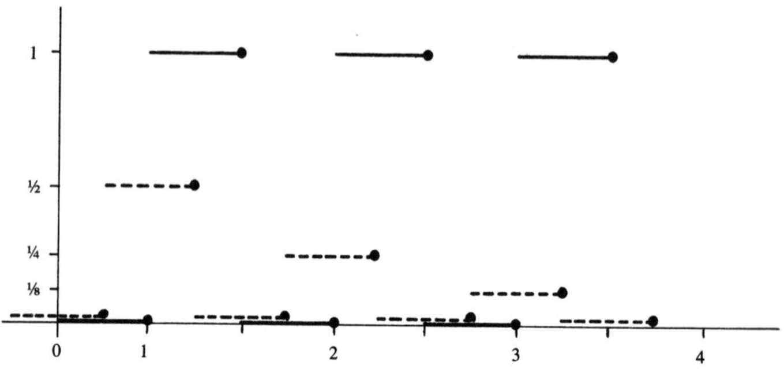

Since discontinuities of g and U do not coincide, we use the extended definition of the Riemann-Stieltjes Integral appropriate in this case to obtain:

$$\int_0^\infty g\,dU = \sum_{n=0}^{\infty} \int_{(2n+1)^-}^{(2n+1)^+} g\,dU = \sum_{n=0}^{\infty} (\tfrac{1}{2})^{2n+1} = 2/3$$

This result also gives a glimpse of the close connection between 'summation' and 'integration' that usually is reserved only for those who can study Measure Theory.

EXAMPLE 9.8 (Variable limit of Integration: Riemann-Stieltjes Case) If g is integrable with respect to the b.v.f. U over $[a,b)$, then as x varies within this interval,

$$I(x) = \int_{a^-}^{x^-} g(z)\,dU(z)$$

defines a function of x. It is not difficult to see that since $I(x+h) - I(x) = K[U(x+h) - U(x)]$ for some K between the sup and inf of g over the interval $[x,x+h)$. Thus, $I(x)$ is continuous at every point x at which either U or g is continuous. Furthermore, if g is continuous, the difference quotient:

$$\frac{I(x+h) - I(x)}{h} = g(c)\left[\frac{U(x+h) - U(x)}{h}\right], \quad x \le c \le x+h,$$

possesses the limit $g(x)U'(x)$ as $h \to 0$ for every x at which U' exists and g is continuous. At such points, gU' represents the derivative I' of I. Finally, if φ is a differentiable function of x, and provided the indicated integral exists,

$$J(x) = \int_{a^-}^{\varphi(x)^-} g(z)\,dU(z)$$

defines a function of x as x varies over $[a,b)$. By reasoning similar to that presented above, and appealing to the Mean Value Theorem for Derivatives, it is not difficult to establish that, at each point x where g is continuous and U' exists, J' exists and:

$$J'(x) = g(\varphi(x))U'(\varphi(x))\varphi'(x) .$$

Notice the similarity to analogous results for the Riemann Integral. (See, for example, Brand, pages 261 and 271.)

Hints and Answers to Exercises: Section 9

9.1 Hint: Let $\{P_n\}$ and $\{Q_n\}$ be two sequences of refinements partitioning the interval $[a,b)$. Consider the sequence $\{R_n\}$ of refinements obtained from them by 'meshing together' $\{P_n\}$ and $\{Q_n\}$ for $n = 1,2,\ldots$. This framework can be used to establish the desired result.

9.2 Hint: Assume first that g is non-decreasing and prove that if P_n and P_{n+1} are successive members of a sequence of refinements partitioning $[a,b)$, then $I(P_{n+1}) \leq I(P_n)$. Thus, according to the hypothesis and Exercise 1.37, the sequence $\{I(P_n)\}$ of approximating sums converges, whence the integral exists. Next assume that g is non-increasing and prove an analogous result. Finally, prove that any continuous function g over $[a,b)$ can always be expressed as: $g = f+h$, where f is non-decreasing and h is non-increasing over $[a,b)$. Now combine the results.

9.3 Hint: Approximate the integral over $[c,c+\epsilon)$ by a suitable approximating sume, and then consider the limit as $\epsilon \downarrow 0$, using continuity of g and left-continuity of F at the point $x=c$.

9.4 Hint: Use the result of Exercise 9.3.

9.5 Hint: Use the basic definition of the Riemann-Stieltjes Integral as the limit of approximating sums.

9.6 Hint: (i) Consider one case, say $[a,c)$, $[c,b)$, and partition the approximating sums into two parts.

(ii) Partition the approximating sums.

(iii) Use the basic definition of the Riemann-Stieltjes Integral.

(iv) Use approximating sums, and the Mean Value Theorem for Derivatives on the differences $[F(x_{i,n}) - F(x_{i-1,n})]$ in the approximating sums.

(v) Use approximating sums.

(vi) This follows from (v) by considering the function $g-h$.

(vii) Use approximating sums.

9.7 Hint: This follows from Property (vi).

9.8 Hint: Observe that $-|g| \leq g \leq |g|$.

9.10 Hint: Partition the approximating sums in a suitable manner.

9.12 Hint: Since g is continuous over the closed, bounded interval $[a,b]$, Exercise 1.43 guarantees that there exist points x_o and x^o for which $g(x_o) = \min_{a \leq x \leq b} g(x)$ and $g(x^o) = \max_{a \leq x \leq b} g(x)$. Furthermore, since g is continuous, every value lying between $g(x_o)$ and $g(x^o)$ is taken on at least once by $g(x)$ as x varies over $[a,b]$.

(This is the so-called "Intermediate Value" property.) Finally, Property (vii) yields:

$$g(x_o)\ [F(b\pm) - F(a\pm)] \le \int_{a^\pm}^{b^\pm} g\,dF \le g(x^o)\ [F(b\pm) - F(a\pm)]\ .$$

Accordingly, these results assure existence of a point c within [a,b] for which the required equality holds. Note: depending upon which of the four possible intervals is considered, a different point c may be required.

9.13 Hint: Consider, for example, the case $(-\infty, b)$. Express the integral as a suitable double limit (using an approximating sum sequence), then use boundedness of g, along with known properties of a c.d.f.

9.15 Hint: Suppose $t_0 > 0$. Then for any fixed $r > 0$ and all $|x|$ sufficiently large we have: $|x|^r \le e^{t_0 x} + e^{-t_0 x}$. Now use Property (vi).

9.16 Hint: (Necessity) If M(t) exists for $|t| < t_0$, choose a δ such that $0 < \delta < t_0$. Then prove that:

$$0 \le e^{\delta x}[1-F(x)] \le \int_x^\infty e^{\delta z}\,dF(z) \to 0 \quad \text{as} \quad x \to +\infty$$

and

$$0 \le e^{\delta x}[F(-x)] \le \int_{-\infty}^{-x} e^{\delta z}\,dF(z) \to 0 \quad \text{as} \quad x \to +\infty\ .$$

(Sufficiency) It is enough to prove that for all $|t| < \delta$:

$$\int_N^{\infty} e^{tx}dF(x) \to 0 \quad \text{and} \quad \int_{-\infty}^{-N} e^{tx}dF(x) \to 0 \quad \text{as} \quad N \to \infty .$$

Integration-by-parts can be used.

9.17 Hint: Construct an example of a pointwise convergent sequence of continuous functions whose limit function is, say, discontinuous at a countable infinity of points.

9.18 Hint: Boundedness of g would be sufficient. Would existence of the integral of the limit function g with respect to F be sufficient?

9.19 Hint: See the Hint to Exercise 8.6. In view of this, integrability need be proved only in the case I is unbounded.

9.20 Hint: Suppose $I = [b, \infty)$. Divide the interval into two parts: $I_1 = [b,c)$ and $I_2 = [c, \infty)$. Show that for a sufficiently large value of c the hypotheses guarantee that for all indices n sufficiently large,

$$\left| \int_{c^-}^{\infty} hdF_n - \int_{c^-}^{\infty} hdF \right| < \epsilon .$$

Note: It is not assumed that g or h is bounded.

9.22 Hint: Use the results of Exercise 9.20 and 9.13.

9.23 Hint: The four defining properties of a c.d.f. are easily verified, as is complete convergence.

9.24 Hint: For simplicity, assume that $\cup I_n$ is an interval. Then, actual convergence follows from Exercise 9.18 or Theorem 9.3. The two equalities follow from the definitions of F_n^* and F^*.

9.25 Hint: Use uniform continuity of F and assume N sufficiently large so that for any partition P_n with $n \geq N$ we have $|F(x) - F(x')| < \epsilon/(b-a)$ for any pair of points x, x' within the same segment of any such partition.

9.26 Hint: Note that we may take the limit as $b \to +\infty$ or $a \to -\infty$ of both sides of the formula, since all of these limits exist and are finite.

9.27 Hint: Apply the results of Exercise 9.26 first using F_z as an integrator then using G_z as an integrator.

9.28 Hint: This follows upon taking limits of both sides of the formula in Theorem 9.6.

9.29 Hint: Apply the results of Exercise 9.28.

9.30 Hint: Show that this guarantees existence of the integral defining the r-th absolute moment.

9.31 Hint: For the given c.d.f., the r-th absolute moment exists (which requires proof), yet it can be shown that the condition of Exercise 9.30 cannot be satisfied for any $\delta > 0$.

References to Additional and Related Material: Section 9

1. Bartle, R., "The Elements of Integration", John Wiley and Sons, Inc. (1966).

2. Brand, L., "Advanced Calculus", Wiley and Sons, Inc. (1958).

3. Cramér, H., "Mathematical Methods of Statistics", Princeton University Press (1958).

4. Gunther, N., "Sur les Intégrales de Stieltjes", Chelsea Publishing Company (1949).

5. Henstock, R., "Theory of Integration", Butterworths (1963).

6. Hobson, E., "Theory of Functions of a Real Variable", Vol. I, Dover Publications, Inc.

7. Kestelman, K., "Modern Theories of Integration", Dover Publications, Inc.

8. Pesin, I., "Classical and Modern Integration Theories", Academic Press (1970).

9. Zaanen, A., "An Introduction to the Theory of Integration", North-Holland Publishing Co. (1958).

10. n-Dimensional Cumulative Distribution Functions and Bounded Variation Functions

The natural generalization of 1-dimensional c.d.f.'s and b.v.f.'s (see Section 8) are n-dimensional $(n > 1)$ c.d.f.'s and b.v.f.'s. We shall develop some of their properties in this Section, and then apply these results in the following Section dealing with the n-dimensional Riemann-Stieltjes Integral. We shall follow the general pattern set in Section 8.

Before proceeding with the definition of an n-dimensional c.d.f. we have need for the following generalization of the 1-dimensional notion of "non-decreasing".

DEFINITION 10.1 (n-Monotonicity) For any bounded real-valued function $g = g(\ldots, x_k, \ldots)$ including the variable x_k, and for any pair of values $a_k < b_k$ of that variable, we define the following difference, which is a bounded function of the remaining variables only (if any):

$$D_k(a_k, b_k)\ [g] = g(\ldots, b_k, \ldots) - g(\ldots, a_k, \ldots)\ .$$

Now if $\underline{F} = F(\underline{x}')$ is a bounded real-valued function of the n real variables $\underline{x}' = (x_1, \ldots, x_n)$, and $[\underline{a}', \underline{b}') = \{\underline{x}' : a_i \le x_i < b_i\ (i = 1, \ldots, n)\}$ is a bounded non-degenerate n-dimensional "rectangle", then we define the quantity $D\{[\underline{a}', \underline{b}')\}[\underline{F}]$ as follows:

$$D\{[\underline{a}', \underline{b}')\}[\underline{F}] = D_n(a_n, b_n)[\ldots D_1(a_1, b_1)[F]\,].$$

which is obtained first by applying $D_1(a_1,b_1)$ to $\underline{F}$, then next applying $D_2(a_2,b_2)$ to the preceding quantity, and so on for n steps. The function $\underline{F}$ is termed <u>n-monotone</u> over a given domain D if and only if the constant $D\{[\underline{a},\underline{b})\}[\underline{F}] \geq 0$ for every such rectangle $[\underline{a},\underline{b})$ in D.

<u>EXERCISE 10.1</u> (Characterization of n-Monotonicity) Prove that if the bounded function $\underline{F} = F(\underline{x}')$ of the n real variables $\underline{x}' = (x_1,\ldots,x_n)$ possesses the mixed partial derivative function:

$$f(\underline{x}') = \frac{\partial^n}{\partial x_1 \ldots \partial x_n} F(\underline{x}')$$

throughout a domain D, then n-monotonicity of F over D is equivalent to the condition that $f(\underline{x}') \geq 0$ for all $\underline{x}'$ in D. (The case n=2 is sufficient.)

In terms of n-monotonicity, we now define the n-dimensional c.d.f.

<u>DEFINITION 10.2</u> (n-Dimensional Cumulative Distribution Function) A real-valued function $\underline{F} = F(\underline{x}')$, defined for all values of the n real variables $\underline{x}' = (x_1,\ldots,x_n)$, is termed an n-dimensional cumulative distribution function (n-c.d.f.) if and only if it satisfies the following four properties:

(i) $\underline{F}$ is n-monotone over E_n

(ii) $F(+\infty,\ldots,+\infty) = 1$

(iii) $F(\ldots,-\infty,\ldots) = 0$ for any given coordinate

(iv) $F(x_1-\epsilon_1,\ldots,x_n-\epsilon_n) \to F(x_1,\ldots,x_n)$ as $\epsilon_i \downarrow 0$

$(i = 1,\ldots,n)$ at each point $\underline{x}'$ of E_n .

A simple comparison of the above definition with that of a 1-dimensional c.d.f. will show that the above four conditions are natural generalizations of the four conditions required of a 1-dimensional c.d.f.

EXERCISE 10.2 (Properties of an n-C.D.F.) Prove that any n-c.d.f. is non-decreasing in each separate coordinate, whence for any $\underline{x}'$ in E_n : $0 \le F(\underline{x}') \le 1$. Show by example, however, that a function may be non-decreasing in each separate coordinate without being n-monotone (the case n=2 is sufficient).

EXERCISE 10.3 (Generating Other C.D.F.'s From an n-C.D.F.) Prove that if $\underline{F}$ is an n-c.d.f. then both $F(x,x,\ldots,x,x)$ and $F(+\infty,\ldots,x,\ldots,+\infty)$ are bona-fide 1-dimensional c.d.f.'s. Generalize this by proving that for any choice $1 \le i_1 <\ldots< i_k \le n$ $(1 \le k \le n)$ of coordinates, the function defined by: $F(+\infty,\ldots,x_{i_1},\ldots,x_{i_k},\ldots,+\infty)$ is a bona-fide k-dimensional c.d.f.

Attempt to formulate further generalizations of the above results. (For example, what can be said about $F(x,\ldots,x,y,\ldots,y)$?).

<u>EXERCISE 10.4</u> Prove that if $F_1, F_2, \ldots, F_n$ are 1-dimensional c.d.f.'s then the function $\underline{F}$ of the n real variables $\underline{x}' = (x_1, \ldots, x_n)$ defined as follows: $\underline{F}(\underline{x}') = \prod_1^n F_i(x_i)$ is a bona-fide n-dimensional c.d.f. In the process show that $D\{[\underline{a}', \underline{b}')\}[\underline{F}] = \prod_1^n [F_i(b_i) - F_i(a_i)]$. The case $n=2$ is sufficient.

It requires little additional argument beyond the case $n=1$ (see Exercise 8.7) to prove that if $\underline{F}$ is an n-c.d.f. and φ is any continuous, non-decreasing function over $[0,1]$, with $\varphi(0) = 0$ and $\varphi(1) = 1$, then $\varphi\underline{F}$ is a bona-fide n-c.d.f. It also follows easily that the following are n-c.d.f.'s:

(i) $\underline{F} = \Sigma p_i \underline{F}_i$ where $p_i \geq 0$, $\Sigma p_i = 1$, and each $\underline{F}_i$ is an n-c.d.f.

(ii) $\underline{F} = \prod_1^n \underline{F}_i$ (finite product) where each $\underline{F}_i$ is an n-c.d.f.

(iii) $\underline{F} = \min(\underline{F}_1, \underline{F}_2, \ldots, \underline{F}_n)$ where each $\underline{F}_i$ is an n-c.d.f.

(iv) $\underline{F} = \max(\underline{F}_1, \underline{F}_2, \ldots, \underline{F}_n)$ where each $\underline{F}_i$ is an n-c.d.f.

<u>EXERCISE 10.5</u> Prove that the functions defined in (ii) and either (iii) or (iv) are n-c.d.f.'s.

<u>EXERCISE 10.6</u> (n-Dimensional Discrete Distributions) For any fixed point $\underline{a}' = (a_1, a_2, \ldots, a_n)$ of E_n prove that the function $\underline{F}_{\underline{a}'}$ of $\underline{x}' = (x_1, \ldots, x_n)$ defined by:

$$\underline{F}_{\underline{a}'}(\underline{x}') = \begin{cases} 1 & \text{if } x_i > a_i \ (i = 1, \ldots, n) \\ 0 & \text{otherwise} \end{cases}$$

is a bona-fide n-c.d.f. Then prove that for any choice $\underline{a}_1', \underline{a}_2', \underline{a}_3', \ldots$ of distinct points in E_n, the function of $\underline{x}'$ defined by $\underline{F} = \Sigma p_i \underline{F}_{\underline{a}_i'}$ (where $p_i \geq 0$ and $\Sigma p_i = 1$) is an n-c.d.f. termed a 'plateau' n-c.d.f. Illustrate such an n-c.d.f. with a diagram for a simple case in $n=2$.

EXERCISE 10.7 (Riemann-Continuous Case) Prove that if

$$f(\underline{x}') = \frac{\partial^n}{\partial x_1 \ldots \partial x_n} F(\underline{x}')$$

exists and is continuous for all $\underline{x}'$ within the closed rectangle $[\underline{a}', \underline{b}'] = \{\underline{x}' : a_i \leq x_i \leq b_i \ (i = 1, \ldots, n)\}$, then $\underline{f}$ is non-negative throughout $[\underline{a}', \underline{b}']$, and for the n-c.d.f. in this special case:

$$D\{[\underline{a}', \underline{b}')\}[\underline{F}] = \int_{a_1}^{b_1} \int_{a_2}^{b_2} \cdots \int_{a_n}^{b_n} f(x_1, x_2, \ldots, x_n)\, dx_1 dx_2 \ldots dx_n \quad \text{(Riemann)}.$$

The n-c.d.f. $\underline{F}$ is termed Riemann-Continuous in this case. (Certainly $\underline{f}$ need not exist and/or be continuous in general. However, to generalize the above result to these cases would require more background than is assumed here.)

EXERCISE 10.8 (Discontinuities of an n-C.D.F.) Prove that by fixing the values of $n-1$ of the variables of an n-c.d.f., we obtain a function of one variable that has at most a countable number of points of discontinuity. Is this true for the n-c.d.f. $\underline{F}$ as a function of n variables?

We may also consider types of convergence of sequences of n-c.d.f.'s analogous to the case $n=1$. For example, the sequence $\{\underline{F}_m\}$ of n-c.d.f.'s is said to converge completely if and only if there exists an n-c.d.f. $\underline{F}$ such that $F_m(\underline{x}') \to F(\underline{x}')$ as $m \to \infty$ at every continuity point of $\underline{F}$. In this case we write: $\underline{F}_m \xrightarrow{c} \underline{F}$.

If, however, in the preceding Definition, we require only that the function $\underline{F}$ be n-monotone, bounded between zero and one, and left-continuous throughout E_n, then we say that the sequence of n-c.d.f.'s converges weakly to $\underline{F}$ (which is not necessarily an n-c.d.f.), and we write: $\underline{F}_m \xrightarrow{w} \underline{F}$ as $m \to \infty$.

> EXERCISE 10.9 (Relationship Between Weak and Complete Convergence) Clearly complete convergence implies weak convergence. What additional conditions on a weakly convergent sequence would guarantee complete convergence?

We conclude this Section by generalizing the 1-dimensional notion of Bounded Variation. This generalization is important, since many of the results that are valid for n-c.d.f.'s are also valid for the broader class of n-b.v.f.'s. A function $\underline{U} = U(\underline{x}')$ of the n real variables $\underline{x}' = (x_1, \ldots, x_n)$ is termed an n-dimensional Bounded Variation Function (an n-b.v.f.) over the closed rectangle $[\underline{a}', \underline{b}']$ if and only if it can be expressed as a difference, say $\underline{U} = \underline{H} - \underline{G}$, of two bounded, left-continuous and n-monotone functions $\underline{H}$ and $\underline{G}$ throughout $[\underline{a}', \underline{b}']$. As in the case $n=1$, there are alternate characterizations of n-b.v.f.'s.

For example, if for every partition $\{\underline{P}_m\}$ of the rectangle $[\underline{a}',\underline{b}')$ into a finite number (say k) of rectangles $[\underline{a}_i',\underline{b}_i')$ $(i = 1,2,\ldots,k)$, the "variation sum"

$$\sum_1^k |D\{[\underline{a}_i',\underline{b}_i')\}[\underline{U}]|$$

remains bounded above by some positive number B, then $\underline{U}$ is an n-b.v.f. Furthermore, a sufficient (but not necessary) condition that $\underline{U}$ be an n-b.v.f. over $[\underline{a}',\underline{b}']$ is that the mixed partial derivative function:

$$u(\underline{x}') = \frac{\partial^n}{\partial x_1 \ldots \partial x_n} U(\underline{x}')$$

exists and be bounded over $[\underline{a}',\underline{b}']$. Continuity of $\underline{U}$ alone is not sufficient for $\underline{U}$ to be an n-b.v.f.

<u>EXERCISE 10.10</u> Construct an example in the case $n=2$ to illustrate the concluding statement.

<u>EXERCISE 10.11</u> Generalize Exercise 8.12 by proving that a left-continuous n-b.v.f. $\underline{U}$ can always be expressed as: $\underline{U} = a\underline{F} + b\underline{G} + c$, where $\underline{F}$, $\underline{G}$ are n-c.d.f.'s, and a, b, c are constants.

<u>EXERCISE 10.12</u> Review the results of Section 8 and note which of these generalize to the case $n > 1$. Formulate a statement of the generalization (if one exists).

Hints and Answers to Exercises: Section 10

10.1 Hint: The case $n=2$ is sufficient. Express the mixed partial derivative as a double limit (assumed to exist) of a difference quotient.

10.2 Hint: The case $n=2$ is sufficient. For the first part use the property of n-monotonicity.

10.3 Hint: What is required is a straightforward verification of the four defining properties of a k-c.d.f. (for the appropriate value of k).

10.4 Hint: The four defining properties of a 2-c.d.f. are easily verified using the known properties of the 1-c.d.f.

10.5 Hint: Note that a proof of (i) would require an n-dimensional parallel of uniform convergence. The remaining three parts are easily established. See also the Hint to Exercise 7.6.

10.6 Hint: See the Hint for Exercise 7.2. Assume that the limit and summation operations may be interchanged.

10.7 Hint: The case $n=2$ is sufficient.

10.8 Hint: The function so determined is a non-decreasing function of one real variable, bounded between zero and one. In the case $n=2$, for instance, there may be entire lines of discontinuity.

10.9 Hint: See Exercise 7.10 and generalize.

10.10 Hint: See the Hint for Exercise 7.14.

References to Additional and Related Material: Section 10

1. Feller, W., "An Introduction to Probability Theory and its Applications", Vol. II, John Wiley and Sons, Inc. (1966).

2. Hobson, E., "Theory of Functions of a Real Variable", Vol. I, Dover Publications, Inc.

11. n-Dimensional Riemann-Stieltjes Integral

We now generalize the results of Section 9 to consider the n-Dimensional $(n > 1)$ Riemann-Stieltjes Integral with respect to an n-dimensional c.d.f., then, more generally, with respect to left-continuous n-dimensional b.v.f.'s. The development closely parallels that of the 1-dimensional case, and for this reason we will generally be briefer with proofs and descriptions than before. However, this by no means indicates that the n-dimensional Integral is any less important in applications.

Accordingly, let $\underline{F} = F(\underline{x}')$ be an n-c.d.f. and begin by assuming[1] that $\underline{g} = g(\underline{x}')$ is a continuous, real-valued function of the n-real-variables $\underline{x}' = (x_1, x_2, \ldots, x_n)$ over the bounded and non-degenerate rectangle $[\underline{a}', \underline{b}') = \{\underline{x}': a_i \leq x_i < b_i$ $(i = 1,2,\ldots,n)\}$ of E_n. Consider now a sequence $\{\underline{P}_m\}$ of partitions of the rectangle $[\underline{a}', \underline{b}')$ into disjoint rectangles $[\underline{a}_i', \underline{b}_i')$ $(i = 1,2,\ldots,m)$ of the same type, each partition being a refinement of its predecessor, and for which the n-dimensional volume of the 'largest' rectangle in a partition tends to zero as $m \to \infty$. For each partition in such a sequence, form the following sum:

$$\underline{I}(\underline{P}_m) = \sum_1^m g(\underline{x}_{i,m}'^{*})\, D\{[\underline{a}_{i,m}', \underline{b}_{i,m}')\}[\underline{F}],$$

termed an <u>approximating sum</u>, where for each index i, $\underline{x}_{i,m}'^{*}$ is a

---

[1]We are considering a special case as an introduction. The results generalize considerably.

point chosen somewhere within the rectangle $[\underline{a}'_{i,m}, \underline{b}'_{i,m})$ of the partition $\underline{P}_m$.

If, independent of any particular sequence $\{\underline{P}_m\}$ of refinements partitioning $[\underline{a}', \underline{b}')$, all such approximating sums approach the same finite limit as $m \to \infty$, then this limit is termed the n-dimensional Riemann-Stieltjes Integral of $\underline{g}$ with respect to the n-c.d.f. $\underline{F}$ over $[\underline{a}', \underline{b}')$, and is denoted by:

$$\text{(a)} \quad \int_{[\underline{a}', \underline{b}')} \underline{g} d\underline{F}$$

or

$$\text{(b)} \quad \int_{[\underline{a}', \underline{b}')} g(\underline{x}') dF(\underline{x}')$$

or

$$\text{(c)} \quad \int_{a_1^-}^{b_1^-} \int_{a_2^-}^{b_2^-} \cdots \int_{a_n^-}^{b_n^-} g(x_1, x_2, \ldots, x_n) dF(x_1, x_2, \ldots, x_n),$$

whichever form is appropriate in a given situation.

It is not difficult to generalize previous results for the case $n=1$, and prove that a sufficient condition for the above integral to exist is that the function $\underline{g}$ be continuous over the 'closed' rectangle $[\underline{a}', \underline{b}']$.

Having defined the n-dimensional Integral over the basic bounded rectangle of the form $[\underline{a}', \underline{b}')$, we now extend its definition to other types of subsets of E_n. We consider first <u>degenerate rectangles</u>. For any non-empty subset S of the indices $\{1,2,\ldots,n\}$, a bounded, degenerate rectangle R is a subset of E_n of the following form:

$$R = \{\underline{x}': x_i = c_i \ (i \in S) \quad \text{and} \quad a_i \le x_i < b_i \ (i \notin S)\}.$$

The number of degenerate 'sides' of R is the number of elements in the subset S. Thus, there may be as few as one degenerate 'side' or as many as n (in which case R reduces to a singleton point).

As a first step in extending the Integral to degenerate rectangles, we define, for any degenerate rectangle R, a corresponding bounded, non-degenerate rectangle $R_{\underline{\epsilon}}$ (of the type originally introduced) by:

$$R_{\underline{\epsilon}} = \{\underline{x}': c_i \le x_i < c_i + \epsilon_i \ (i \in S) \quad \text{and} \quad a_i \le x_i < b_i \ (i \notin S)\}.$$

<u>EXERCISE 11.1</u> For any n-c.d.f. $\underline{F}$ and non-degenerate rectangle $R_{\underline{\epsilon}}$ of the type just defined, the quantity $D\{R_{\underline{\epsilon}}\}[\underline{F}]$ is well-defined. (See Definition 10.1). Extend the definition of D to include degenerate rectangles of the form R by showing that the following limit always exists and is ≤ 1:

$$D\{R\}[\underline{F}] = \lim_{\substack{\epsilon_i \downarrow 0 \\ i \in S}} D\{R_{\underline{\epsilon}}\}[\underline{F}] .$$

(The case n=2 is sufficient).

Now let R be a degenerate rectangle. If $\underline{g}$ is continuous and integrable with respect to the n-c.d.f. $\underline{F}$ over some non-degenerate rectangle $R_{\underline{\epsilon}}$ containing R, then we define the Riemann-Stieltjes Integral of $\underline{g}$ with respect to $\underline{F}$ over R as follows:

$$\int_R \underline{g} d\underline{F} = \lim_{\substack{\epsilon_i \downarrow 0 \\ i \in S}} \int_{R_{\underline{\epsilon}}} \underline{g} d\underline{F} .$$

<u>EXERCISE 11.2</u> Prove that, under the present assumptions, the Integral of $\underline{g}$ with respect to the n-c.d.f. $\underline{F}$ always exists over the (degenerate) rectangle R. (The case n=2 is sufficient).

In the special case where <u>all</u> sides are degenerate, the degenerate rectangle R reduces to a single point $\underline{c}' = (c_1, c_2, \ldots, c_n)$. Then, it is not difficult to see that the integral of $\underline{g}$ with respect to $\underline{F}$ over $\{\underline{c}'\}$ is:

$$\int_{\{\underline{c}'\}} \underline{g} d\underline{F} = g(\underline{c}') \, D\{\underline{c}'\}[\underline{F}] .$$

<u>EXERCISE 11.3</u> Prove the preceding result.

In other cases, where the number of degenerate coordinates (sides) is $r < n$, it is possible to re-express the n-dimensional Riemann-Stieltjes Integral already defined over such a rectangle, as suitable combinations of (n-r)-Dimensional Riemann-Stieltjes

Integrals over a non-degenerate rectangle in E_{n-r}. We shall develop this result in the following paragraphs.

Accordingly, consider the n-Dimensional Riemann-Stieltjes Integral of a continuous function $\underline{g}$ with respect to an n-c.d.f. $\underline{F}$ over a bounded, degenerate rectangle R with exactly $r < n$ degenerate sides. Let $1 \leq i_1 < i_2 < \ldots < i_r < n$ be the indices associated with the degenerate coordinates, and $1 < j_1 < j_2 < \ldots j_{n-r} < n$ those associated with the non-degenerate sides.

For each of the 2^r different choices of $(d_{i_1}, d_{i_2}, \ldots, d_{i_r})$ where $d_{i_k} = \pm 1$ $(k = 1,2,\ldots,r)$, consider the 2^r functions:

$$\underline{F}_{d_{i1}, d_{i2}, \ldots, d_{ir}} \text{ of } (n-r) \text{ variables } (x_{j_1}, x_{j_2}, \ldots, x_{j_{n-r}})$$

determined as follows: for each fixed value of the (n-r) variables $(x_{j_1}, x_{j_2}, \ldots, x_{j_{n-r}})$, the value of the function $\underline{F}_{d_{i_1}, d_{i_2}, \ldots, d_{i_r}}$ for those values of the variables is defined to be the value of the limit of $\underline{F} = F(x_1, x_2, \ldots, x_n)$ taken with the values of the variables $(x_{j_1}, x_{j_2}, \ldots, x_{j_{n-r}})$ fixed as above, and the remaining variables $(x_{i_1}, x_{i_2}, \ldots, x_{i_r})$ independently approaching $(c_{i_1}, c_{i_2}, \ldots, c_{i_r})$ in the following manner:

$x_{i_k} = c_{i_k} + d_{i_k} \epsilon_{i_k}$ as $\epsilon_{i_k} \downarrow 0$ $(k = 1,2,\ldots,r)$.

The 2^r functions defined above need not all be distinct; this depends upon continuity properties of $F(\underline{x}')$ over R. Also, these functions satisfy all the properties required of an (n-r)-c.d.f. <u>except perhaps</u> that $F_{d_{i_1},d_{i_2},\ldots,d_{i_r}}(+\infty,\ldots,+\infty)$ may not equal one. Assuming that this constant <u>is</u>, however, positive[1], we may construct an (n-r)-c.d.f. from this function by 'normalizing' it by this constant. The (n-r)-c.d.f. so constructed will be denoted by: $\underline{F}^*_{d_{i_1},d_{i_2},\ldots,d_{i_r}}$

> <u>EXERCISE 11.4</u> Verify that the functions defined above are indeed (n-r)-c.d.f.'s.

Finally, let $\underline{g}^*$ denote that function of the (n-r) variables $(x_{j_1},x_{j_2},\ldots,x_{j_{n-r}})$ obtained from $\underline{g} = g(\underline{x}')$ by setting the remaining r variables $(x_{i_1},x_{i_2},\ldots,x_{i_r}) = (c_{i_1},c_{i_2},\ldots,c_{i_r})$. The function $\underline{g}^*$ so defined is continuous over the non-degenerate, bounded rectangle R* of (n-r) dimensions defined by:

$$R^* = \{(x_{j_1},x_{j_2},\ldots,x_{j_{n-r}}): a_{j_k} \le x_{j_k} < b_{j_k} \ (k = 1,2,\ldots,n-r)\}.$$

We are now in a position to formalize the assertion made at the beginning of this discussion. In the case of degeneracy being considered here, we may express:

$$\int_R \underline{g}d\underline{F}$$

[1]If it should occur that for some $(d_{i_1},d_{i_2},\ldots,d_{i_r})$ the associated constant $F_{d_{i_1},d_{i_2},\ldots,d_{i_r}}(+\infty,\ldots,+\infty) = 0$, do not attempt to construct the subsequent (n-r)-c.d.f., and drop the terms corresponding to these values in subsequent formulae.

as follows:

$$\sum_{\substack{d_{i_k} = \pm 1 \\ (k = 1,\ldots,r)}} (-1)^{\frac{1}{2}(\Sigma d_{i_k}+2r)} F_{d_{i_1},\ldots,d_{i_r}}(+\infty,\ldots,+\infty) \int_{R^*} \underline{g}^* d\underline{F}_{d_{i_1},\ldots,d_{i_r}} ,$$

where the Riemann-Stieltjes Integrals occurring in this expression are, as asserted, (n-r)-dimensional, whereas the original integral is an n-dimensional Riemann-Stieltjes Integral.

> <u>EXERCISE 11.5</u> Verify the preceding formula for the special case where $n=2$ and $r=1$.

For completeness, we shall now use the preceding results to extend the definition of the Integral to include integration of a continuous function $\underline{g}$ with respect to an n-c.d.f. $\underline{F}$ over bounded, non-degenerate rectangles other than the basic 'left-closed and right-open' rectangle of the form $[\underline{a}',\underline{b}')$. Specifically, we shall extend the definition to include any bounded 'rectangle' whose sides are any combination of the following forms:

(i)	$a_i \le x_i < b_i$	left-closed, right-open
(ii)	$a_i \le x_i \le b_i$	left-closed, right-closed
(iii)	$a_i < x_i < b_i$	left-open, right-open
(iv)	$a_i < x_i \le b_i$	left-open, right-closed.

We shall, in fact, demonstrate that the integral over any such rectangle can always be expressed as a suitable combination of integrals over rectangles[1] for which the Integral has already been defined.

---
[1]Possibly degenerate.

Accordingly, for any bounded, non-degenerate rectangle, let $w_i = 0$ or -1 according as the left end of the i-th side is of the form $a_i \leq x_i$ (left-closed) or of the form $a_i < x_i$ (left-open). Analogously, let $z_i = 0$ or 1 according as the right end of the i-th side is of the form $x_i < b_i$ (right-open) or of the form $x_i \leq b_i$ (right-closed).

For each of the 2^{2^n} different choices for $(\underline{w}';\underline{z}') = (w_1,\ldots,w_n; z_1,\ldots,z_n)$, we have a different type of non-degenerate rectangle, denoted $R(\underline{w}';\underline{z}')$. For example, the choice of $\underline{w}' = \underline{0}'$ and $\underline{z}' = \underline{0}'$ leads to the basic 'left-closed and right-open' rectangle, namely $R(\underline{0}',\underline{0}') = [\underline{a}',\underline{b}')$.

As a final notational convenience in presenting the proposed result, we shall adopt the following designation for 'basic' rectangles (both non-degenerate and degenerate) of a type previously considered: for any subset (possibly empty) S of indices (sides) $\{1,2,\ldots,n\}$, and corresponding choices for $d_i = 0$ or 1 $(i \in S)$, we denote a 'basic' rectangle $R_S(d_i; i \in S)$, which is non-degenerate or degenerate according as $S = \emptyset$ or $S \neq \emptyset$, as follows:

$$R_S(d_i; i \in S) =$$

$$\{\underline{x}' : x_i = (1-d_i)a_i + d_i b_i \ (i \in S) \ ; \ a_i \leq x_i < b_i \ (i \notin S)\}.$$

The Integral over any rectangle of the above type has already been defined.

Now, provided that the continuous function $\underline{g} = g(\underline{x}')$ is integrable with respect to the n-c.d.f. $\underline{F} = F(\underline{x}')$ over the above rectangles, we have the following relationship referred to at the outset of the present discussion:

$$\int_{R(\underline{w}';\underline{z}')} g d\underline{F} =$$

$$\sum_{r=0}^{n} \sum_{S \in \$_r} \sum_{\substack{d_i = 0,1 \\ i \in S}} \left[\prod_{i \in S} q_i\right] \int_{R_S(d_i; i \in S)} g d\underline{F}$$

where $q_i = (1-d_i)w_i + d_i z_i$ and $\$_r$ is the set of all subsets of size r of the indices $\{1,2,\ldots,n\}$.

EXERCISE 11.6 Verify the preceding relationship for a special case where $n=3$.

EXERCISE 11.7 Explain how the Integral over any 'non-basic' degenerate rectangle can be expressed as a suitable combination of Integrals over 'basic' degenerate rectangles.

There is a natural extension of the Integral to (i) unbounded rectangles, and (ii) certain sets other than rectangles. The first extension is a straightforward analogue of the 1-dimensional case, and is left as an Exercise.

EXERCISE 11.8 First, define what would be meant by an unbounded rectangle (take into account the various possibilities for both bounded and unbounded sides). Having done this, extend the definition of the Riemann-Stieltjes Integral of a continuous function $\underline{g}$ with respect to an n-c.d.f. $\underline{F}$ over such rectangles by means of an appropriate limit process.

As far as extending the definition of the Integral to sets other than rectangles (of all types) we shall proceed as follows: if R_1 and R_2 are two disjoint rectangles in E_n over which the Integral of the continuous function $\underline{g}$ exists with respect to the n-c.d.f. $\underline{F}$, then the natural definition of the Integral over $R_1 \cup R_2$ is the sum of the individual Integrals over R_1 and R_2. (Note, however, that the set $R_1 \cup R_2$ need not be a rectangle by our definition.)

EXERCISE 11.9 Verify the last statement. Prove, however, that for a general pair of rectangles, R_1 and R_2, $R_1 \cap R_2$ *is* a rectangle.

If the two rectangles R_1 and R_2 are not necessarily disjoint, then a simple diagram of the situation suggests that a natural extension of the definition of the Integral to $R_1 \cup R_2$ is:

$$\int_{R_1 \cup R_2} \underline{g} d\underline{F} = \int_{R_1} \underline{g} d\underline{F} + \int_{R_2} \underline{g} d\underline{F} - \int_{R_1 \cap R_2} \underline{g} d\underline{F} .$$

Note that each of the sets occurring on the right-hand side is a rectangle, whereas $R_1 \cup R_2$ (as noted before) need not be a rectangle.

The preceding two extensions each can be carried one step further. First, if $R_1, R_2, R_3, \ldots$ is a countable (finite or infinite) collection of disjoint rectangles, over each of which the Integral of the continuous function $\underline{g}$ with respect to the n-c.d.f. exists, then the natural extension of the definition of the Integral to $\cup R_i$ is:

$$\int_{\cup R_i} g d\underline{F} = \sum_1 \int_{R_i} g d\underline{F} \quad \text{(provided convergent)} .$$

Thus, in principle, the definition of the Integral has been extended (subject to existence) to include integration over any subset of E_n that can be represented as a countable union of disjoint rectangles.

Finally, if $R_1, R_2, \ldots, R_m$ is a finite collection of rectangles (not necessarily disjoint), then using the classical principle of "Inclusion and Exclusion", we may extend the definition of the Integral to $\cup R_i = R_1 \cup R_2 \cup \ldots \cup R_m$ as follows:

$$\int_{R_1 \cup R_2 \cup \ldots \cup R_m} g d\underline{F} = \sum_i \int_{R_i} g d\underline{F} - \sum_{1 \le i < j \le m}\sum \int_{R_i \cap R_j} g d\underline{F} +$$

$$\sum_{1 \le i < j < k \le m}\sum\sum \int_{R_i \cap R_j \cap R_k} g d\underline{F} - \ldots\ldots\ldots + (-1)^{m+1} \int_{R_1 \cap R_2 \cap \ldots \cap R_m} g d\underline{F} .$$

Note again that sets occurring on the right-hand side of the above formula are all rectangles.

<u>EXERCISE 11.10</u> Using Induction, establish the above formula.

Having generalized the Integral to the present point, we shall now consider its algebraic and limit properties. These are analogous to the 1-dimensional case, yet we shall state them in detail never-the-less.

Accordingly, over sets of the type already considered, the n-Dimensional Riemann-Stieltjes Integral of a continuous function with respect to an n-c.d.f. $\underline{F}$ (subject to existence) possesses the following algebraic properties:

(i) $\int_R [\alpha\underline{g} + \beta\underline{h}]d\underline{F} = \alpha\int_R \underline{g}d\underline{F} + \beta\int_R \underline{h}d\underline{F}$ $\quad \alpha, \beta$ constants

(ii) $\int_R d\underline{F} = D\{R\}[\underline{F}]$

(iii) $\int_R \underline{g}d\underline{F} = \int\int_R\ldots\int \underline{g}\underline{f}dx_1\ldots dx_n$ (Riemann), provided

$$\underline{f} = \frac{\partial^n}{\partial x_1\ldots\partial x_n}\underline{F} \text{ is continuous over } R$$

(iv) $\int_R \underline{g}d\underline{F} \geq 0$ if $\underline{g} \geq 0$ over R

(v) $\int_R \underline{g}d\underline{F} \geq \int_R \underline{h}d\underline{F}$ if $\underline{g} \geq \underline{h}$ over R .

EXERCISE 11.11 (Properties (vi) and (vii)) State and prove the n-dimensional generalizations of properties seven and eight of the 1-Dimensional Integral developed in Section 9.

EXERCISE 11.12 (Generalization: A Mean Value Theorem for the n-Dimensional Case) State and prove a Mean Value Theorem for the n-dimensional Riemann-Stieltjes Integral with respect to an n-c.d.f. This generalizes Exercise 9.12.

EXERCISE 11.13 State the generalizations of Exercises 9.7, 9.8 and 9.9 for the n-dimensional case being considered here.

EXERCISE 11.14 (Marginal C.D.F.'s) Prove that, for any choice $1 \le i_1 < \ldots < i_k \le n$ of indices, and fixed values $x_{i_1}, \ldots, x_{i_k}$ of the corresponding coordinates,

$$\int_{-\infty}^{+\infty} \cdots \int_{-\infty}^{\bar{x}_{i_1}} \cdots \int_{-\infty}^{\bar{x}_{i_k}} \cdots \int_{-\infty}^{+\infty} d\underline{F} = F(+\infty, \ldots, x_{i_1}, \ldots, x_{i_k}, \ldots, +\infty) .$$

The function of k real variables so determined has already been shown to be a k-c.d.f. (see Exercise 10.3). For apparent reasons, it is termed the marginal k-c.d.f. on indices $(i_1, \ldots, i_k)$ obtained from $\underline{F}$ by "integrating out" the other variables. In particular, for $n = 1,2,\ldots,n$ the one-dimensional marginal c.d.f.'s F_i are given by:

$$F_i(x_i) = \int_{-\infty}^{\infty} \cdots \int_{-\infty}^{\bar{x}_i} \cdots \int_{-\infty}^{\infty} d\underline{F} = \underline{F}(+\infty, \ldots, x_i, \ldots, +\infty) .$$

EXERCISE 11.15 (Continuity of Marginal C.D.F.'s) Let R be a degenerate rectangle with (at least one) degenerate coordinate (side) $x_i = c_i$, say. Prove that, if the 1-dimensional marginal c.d.f. F_i obtained from $\underline{F}$ is continuous at $x_i = c_i$, then $D\{R\}[\underline{F}] = 0$. What can be concluded about the n-dimensional Integral of a continuous function g with respect to $\underline{F}$ over R? The case $n=2$ is sufficient.

<u>EXERCISE 11.16</u> (Special Case of Factoring) Prove that if the n-c.d.f. $\underline{F}$ factors as follows:

$$\underline{F} = F(x_1,\ldots,x_n) = F_1(x_1)\ldots F_n(x_n)$$

into (its) 1-dimensional marginal c.d.f.'s, and $\underline{g}$ is a function of $(x_1,\ldots,x_n)$ such that:

$$\underline{g} = g(x_1,\ldots,x_n) = g_1(x_1)\ldots g_n(x_n),$$

where each g_i is continuous and bounded over the i-th side S_i of an n-dimensional rectangle R, then:

$$\int_R \underline{g}d\underline{F} = \prod_1^n \int_{S_i} g_i dF_i \quad .$$

It is sufficient to consider the case n=2, with R being a 'basic' bounded, non-degenerate rectangle in E_2. Generalize the result.

<u>EXERCISE 11.17</u> (Conditional C.D.F.'s: Iterated Integrals) Let $g(x_1,x_2)$ be bounded and continuous over the non-degenerate 2-dimensional rectangle R with 'sides' S_1 and S_2. Suppose that, for each fixed x_i in S_i, the marginal c.d.f. $F_i(x_i)$ obtained from the 2-c.d.f. $F(x_1,x_2)$, is <u>strictly positive</u>, (i = 1,2).

First prove that for each fixed x_1 in S_1 the function $F(x_2|x_1)$ of x_2 defined by:

$$F(x_2|x_1) = F(x_1,x_2)/F_1(x_1), \quad \text{for all real } x_2$$

is a bona-fide 1-dimensional c.d.f. termed the <u>conditional c.d.f. in index 2 given</u> x_1 . (Note that the preceding results are also valid if the indices 1 and 2 are interchanged.)

Next, establish the following formula, indicating how the integral of g with respect to $\underline{F}$ over R may be 'iterated':

$$\int_R g\,d\underline{F} = \int_{S_1} \left[\int_{S_2} g(x_1,x_2)\,dF(x_2|x_1)\right] dF_1(x_1)$$

$$= \int_{S_2} \left[\int_{S_1} g(x_1,x_2)\,dF(x_1|x_2)\right] dF_2(x_2) .$$

How might this result be generalized for $n > 2$?

Next we consider limit properties related to the n-Dimensional Integral. It should come as no surprise that the n-dimensional case closely parallels that of the 1-dimensional case considered in Section 9. The properties are contained in the following three Theorems.

<u>THEOREM 11.1</u> (Varying Integrand) If $\{g_m\}$ are continuous and $g_m \to g$ (uniformly) over the closed, bounded n-dimensional rectangle $[\underline{a}',\underline{b}']$, then for any rectangle[1] R within this rectangle,

$$\int_R g_m d\underline{F} \to \int_R g\,d\underline{F} \quad \text{as} \quad m \to \infty$$

for any n-c.d.f. $\underline{F}$.

---

[1] R may be any set for which the Integral has been defined.

EXERCISE 11.18 Review the proof of the result for the case n=1 (Theorem 9.2), and indicate what modifications would be necessary for the case $n > 1$.

THEOREM 11.2 (Varying Integrator: Helly-Bray) Let $\{\underline{F}_m\}$ be a sequence of n-c.d.f.'s such that $\underline{F}_m \xrightarrow{c} \underline{F}$ as $m \to \infty$. If $\underline{g} = g(\underline{x}')$ is bounded and continuous over the bounded rectangle R whose 'sides' contain no discontinuities of $\underline{F}$, then

$$\int_R \underline{g}d\underline{F}_m \to \int_R \underline{g}d\underline{F} \quad \text{as} \quad m \to \infty \ .$$

EXERCISE 11.19 Review the proof of the analogous result for the case n=1 (Theorem 9.3), and indicate what modifications and/or additions are required to make it valid for the present case of $n > 1$.

THEOREM 11.3 (Varying Set of Integration)

EXERCISE 11.20 Formulate and prove the n-dimensional analogue of Theorem 9.4.

EXERCISE 11.21 Prove the following n-dimensional counterpart of the 'integration-by-parts' formula found in Theorem 9.5. If $\underline{F}$ and $\underline{G}$ are continuous n-c.d.f.'s over the closed, bounded rectangle $[\underline{a}',\underline{b}']$, then:

$$\int_R \underline{F}d\underline{G} = D\{R\}[\underline{FG}] - \int_R \underline{G}d\underline{F} \ ,$$

where R is any rectangle contained within $[\underline{a}',\underline{b}']$.

EXERCISE 11.22 Can Theorem 9.6 be generalized to $n(> 1)$ dimensions? If so, state this version.

As an important conclusion to this Section, we indicate how these results may be generalized. That is, we have been dealing with an n-dimensional Riemann-Stieltjes Integral of a continuous integrand g with respect to an n-c.d.f. integrator. Most, if not all, of the results of this Section have immediate, obvious and useful extensions to the case where:

(i) the integrator is more generally an n-b.v.f.

(ii) the integrand is a complex-valued function of the real variables $\underline{x}' = (x_1,\ldots,x_n)$

(iii) the integrand possesses certain discontinuities, provided they do not coincide with possible discontinuities of the integrator.

Changes required in statements and/or proofs should be attempted with a little precaution, since an added assumption or two (such as boundedness of an n-b.v.f. integrator, for example) may be required.

EXAMPLE 11.1 In view of the definition of a left-continuous n-b.v.f. given in Section 10, verify that the following results remain valid for n-b.v.f. integrators: Exercises 11.1, 11.2, 11.3, 11.6, 11.7, 11.8 (with added conditions on g), 11.10, Property (i) through (iii), 11.12, 11.13 (except 9.9 part), 11.16, 11.17, Theorems 11.1, 11.18, Theorem 11.2, 11.19, Theorem 11.3, 11.21, 11.22.

Finally, it is to be noted that, in applying the n-dimensional Reimann-Stieltjes Integral, the case $n=1$ (treated in detail in Section 9) is a good guideline.

Hints and Answers to Exercises: Section 11

11.1 Hint: Since $0 \le D\{R_{\underline{\epsilon}}\}[\underline{F}] \le 1$ for any standard, non-degenerate rectangle of the form $R_{\underline{\epsilon}}$, it is sufficient to show that $D\{R_{\underline{\epsilon}}\}[\underline{F}]$ is monotone non-increasing as the $\epsilon_i \downarrow 0$ $(i \in S)$ independently.

11.2 Hint: It is sufficient to show that $\int_{R_{\underline{\epsilon}}} g\,d\underline{F}$ is a bounded, continuous function of the variables ϵ_i $(i \in S)$.

11.3 Hint: Use approximating sums and continuity of $\underline{g}$ at $\underline{x}' = \underline{c}'$.

11.4 Hint: The four defining properties - can be established in a straightforward manner using the property of $\underline{F}$ as an n-c.d.f.

11.7 Hint: Explain how to combine the preceding formulas.

11.8 Hint: Consider the non-degenerate case first. Use independent limit processes for the unbounded sides analogous to the 1-dimensional case. Next, consider the case of degenerate sides.

11.9 Hint: The first part is straightforward. The second part follows by considering the intersection of each "side" of the two rectangles.

11.10 Hint: A Venn Diagram is helpful in establishing the cases $n=2$ or $n=3$.

11.11 Hint: Use approximating sums.

(i) $\inf_R \underline{g} \cdot D\{R\}[\underline{F}] \le \int_R \underline{g}d\underline{F} \le \sup_R \underline{g} \cdot D\{R\}[\underline{F}]$

(ii) $\int_R \underline{g}d(\alpha\underline{F} + \beta\underline{G}) = \alpha \int_R \underline{g}d\underline{F} + \beta \int_R \underline{g}d\underline{G}$.

11.12 Hint: Refer to the Hint for Exercise 9.12 which corresponds to the 1-dimensional case.

11.13 Hint: Use approximating sums, analogous to the case $n=1$.

11.14 Hint: Use Property (ii) and the extension of D to unbounded rectangles. Try the case $n=2$ first.

11.15 Hint: Using established properties of D, consider the limit of $D\{R_{\epsilon_i}\}[\underline{F}]$ as $\epsilon_i \downarrow 0$, where R_{ϵ_i} is the rectangle obtained from R upon replacing its degenerate side $x_i = c_i$ by $c_i \le x_i < c_i + \epsilon_i$ $(\epsilon_i > 0)$.

11.16 Hint: For the first part, consider the approximating sums and factor the terms appropriately. An illustration of one generalization would be the case where both $\underline{g}$ and $\underline{F}$ factor into, say, two functions, one involving the variables x_i $(i \in S)$ only, and the other involving the variables x_i $(i \notin S)$ only, where S is some subset of r $(1 \le r < n)$ indices.

11.17 Hint: The four defining properties of a 1-dimensional c.d.f. can be verified easily for the conditional c.d.f.'s. Next, consider the approximating sums for the Integrals.

References to Additional and Related Material: Section 11

1. Bartle, R., "The Elements of Integration", John Wiley and Sons, Inc. (1966).

2. Gunther, N., "Sur les Integrales de Stieltjes", Chelsea Publishing Company (1949).

3. Henstock, R., "Theory of Integration", Butterworths (1963).

4. Hobson, E., "Theory of Functions of a Real Variable", Vol. I, Dover Publications, Inc.

5. Kestelman, K., "Modern Theories of Integration", Dover Publications, Inc.

6. Pesin, I., "Classical and Modern Integration Theories", Academic Press (1970).

7. Zaanen, A., "An Introduction to the Theory of Integration", North-Holland Publishing Co. (1958).

12. Finite Differences and Difference Equations

In Applied Mathematics we frequently encounter functions, relationships or equations that somehow depend upon one or more integer variables. There is a body of Mathematics, termed the Calculus of Finite Differences, that frequently proves useful in treating such situations.

First, consider a function $u(x)$, where x is thought of as a real variable. We begin by defining two unit operators "E" and "Δ" on such a function, in accordance with the following definitions:

$$E\, u(x) = u(x+1)$$

$$\Delta\, u(x) = u(x+1) - u(x)$$

Thus we have the symbolic relationship:

$$\Delta = E-1$$

connecting the two operators.[1]

By an expression of the form $E^n u(x)$ or $\Delta^n u(x)$ is meant that the operator E (resp. Δ) operates upon the function $u(x)$ iteratively n times. In view of the fact that $\Delta = E-1$ and, accordingly, $E = 1+\Delta$, we have the following two formal relationships:

$$\Delta^n = (E-1)^n = E^n - \binom{n}{1} E^{n-1} + \ldots + (-1)^{n-1}\binom{n}{n-1} E + (-1)^n$$

[1]More generally, we could consider differences of order $h (h>0)$, namely, $\Delta_h\, u(x) = u(x+h) - u(x)$ and the associated operator $E_h\, u(x) = u(x+h)$. Here we consider the case $h=1$. For a more general treatment, and its Applications, see the References.

$$E^n = (1+\Delta)^n = \Delta^n + \binom{n}{1} \Delta^{n-1} + \ldots + \binom{n}{n-1} \Delta + 1 .$$

In view of the obvious fact that $E^j u(x) = u(x+j)$ for $j = 0,1,2,\ldots$ we have the following relationship:

$$\Delta^n u(x) = u(x+n) - \binom{n}{1} u(x+n-1) + \ldots + (-1)^{n-1} \binom{n}{n-1} u(x+1) + (-1)^n u(x).$$

The operator "Δ" is called the unit difference operator; $\Delta u(x) = u(x+1) - u(x)$ is termed the first difference of $u(x)$, $\Delta^2 u(x) = \Delta[\Delta u(x)] = u(x+2) - 2u(x+1) + u(x)$ is termed the second difference of $u(x)$, and so on. The above formula for the n-th difference of $u(x)$ was derived by the so-called symbolic operator technique, that is, by formal manipulation of the operator symbols themselves, as if they were real numbers.

EXERCISE 12.1 Using the general formula for E^n, express $u(x+n)$ in terms of successive differences of various orders of $u(x)$.

EXERCISE 12.2 Using the definition, find the first and second differences of the following functions: $u(x) = x^2 - 2x - 1$ and $v(x) = x^3$. Simplify.

The previous Exercise should illustrate the fact that obtaining various differences of functions involving ordinary powers (specially large powers) of the variable x proves quite tedious. As it turns out, in the Calculus of Finite Differences (in contrast with ordinary, or Infinitesimal, Calculus) it is natural to work with a different "power" definition, known as the factorial powers. We now consider these definitions.

For any positive integer m we define the functions $x^{(m)}$ and $x^{|m|}$ (where x is considered a real variable) as follows:

$$x^{(m)} = x(x-1)\dots(x-m+1) \qquad \underline{\text{m-th descending factorial}}$$

$$x^{|m|} = x(x+1)\dots(x+m-1) \qquad \underline{\text{m-th ascending factorial}},$$

where, by convention, each function is defined to be 1 for $m = 0$.

To illustrate why ascending and descending factorial powers are more natural to work with in the Calculus of Finite Differences than are ordinary powers, consider the following illustrations:

$$\Delta x^{(m)} = (x+1)^{(m)} - x^{(m)} = mx^{(m-1)}, \quad m = 1,2,\dots$$

$$\Delta \frac{1}{x^{|m|}} = \frac{1}{(x+1)^{|m|}} - \frac{1}{x^{|m|}} = \frac{-m}{x^{|m+1|}}, \quad m = 1,2,\dots$$

Here we have an analog with ordinary Calculus; the "Δ" operator replacing the derivative, or "D" operator, and $x^{(m)}$ $(m > 0)$ behaves like x^m $(m > 0)$, with $1/x^{|m|}$ or $x^{-|m|}$ behaving like $1/x^m$ or x^{-m} $(m > 0)$. Simple formulas do not exist for the first difference of ordinary powers of x.

There are, however, simple connections between ordinary powers of x and the descending factorial powers of x. It enables us to convert an expression involving ordinary powers of x into an equivalent expression involving only descending factorial powers, and vice versa, as the situation might require. These relationships involve the so-called Sterling Numbers of First and Second Kinds. We now consider these relationships.

Clearly $x^{(m)}$ is a polynomial of degree m in ordinary powers of x. If it were expanded, we would have, say,

$$x^{(m)} = \sum_{j=1}^{m} S_j^m x^j ,$$

where the constant coefficients S_j^m in the expansion are known as <u>Sterling Numbers of the First Kind</u>.

<u>EXERCISE 12.3</u> Derive the recursion formula: $S_j^{m+1} = -mS_j^m + S_{j-1}^m$ for Sterling Numbers of the First Kind.

By direct expansion and inspection, we obtain the following (partial) table of Sterling Numbers of the First Kind:

S_j^m

m \ j	1	2	3	4	5	6	7	8	9
1	1								
2	-1	1							
3	2	-3	1						
4	-6	11	-6	1					
5	24	-50	35	-10	1				
6						1			
7							1		
8								1	
9									1

EXERCISE 12.4 Fill in the remaining rows of the above table using the recursion formula.

EXERCISE 12.5 Using the above table, express $3x^{(4)} - x^{(2)}$ and $x^{(7)} + 5x^{(3)} + 1$ as polynomials in x.

Analogously, we could consider the reverse procedure of expressing an ordinary power of x in terms of a polynomial in descending factorial powers. Thus, we would have, say,

$$x^m = \sum_{j=1}^{m} \mathcal{S}_j^m \, x^{(j)}$$

where the constant coefficients, namely $\mathcal{S}_j^m$, in this expansion are known as Sterling Numbers of the Second Kind.

EXERCISE 12.6 Derive the recursion formula: $\mathcal{S}_j^{m+1} = j\mathcal{S}_j^m + \mathcal{S}_{j-1}^m$ for Sterling Numbers of the Second Kind.

Upon direct calculation of the first few numbers, and then use of the recursion formula, we obtain the following (partial) table of Sterling Numbers of the Second Kind:

$\mathcal{S}_j^m$

m \ j	1	2	3	4	5	6	7	8	9
1	1								
2	1	1							
3	1	3	1						
4	1	7	6	1					
5	1	15	25	10	1				
6						1			
7							1		
8								1	
9									1

EXERCISE 12.7 Using the recursion formula, fill in the remaining lines of the above table.

EXERCISE 12.8 Express $x^6 + 2x^4 + x^2$ and $x^7 - 2x^3 + x + 1$ as polynomials in descending factorials.

EXERCISE 12.9 (Moments of the Poisson Distribution) The m-th moment μ_m $(m = 0,1,2,\ldots)$ of the Poisson Distribution is defined as:

$$\mu_m = \sum_{x=0}^{\infty} x^m e^{-p} p^x / x!$$

for $p > 0$. Prove that:

$$\mu_m = \sum_{j=1}^{m} p^j \mathcal{S}_j^m$$

by first proving that:

$$p^j = \sum_{x=0}^{\infty} x^{(j)} e^{-p} p^x / x! \quad .$$

The "Δ" operator of finite calculus obeys properties analogous to those of the derivative operator "D" of ordinary calculus. If we set $U_x = u(x)$ and $V_x = v(x)$, we have the following three properties as illustrations:

(i) $\Delta[k \cdot U_x] = k \, \Delta U_x$ (k a constant)

(ii) $\Delta[U_x \pm V_x] = \Delta U_x \pm \Delta V_x$

(iii) $\Delta[U_x \cdot V_x] = U_x \, \Delta V_x + V_{x+1} \, \Delta U_x = V_x \, \Delta U_x + U_{x+1} \, \Delta V_x$

These rules obviously can be combined in various manners.

EXERCISE 12.10 Verify the above three properties.

EXERCISE 12.11 Find $\Delta(5x3^x - x^2/2^x)$ by applying rules (i) through (iii).

EXERCISE 12.12 In Finite Calculus the function $u(x) = 2^x$ plays much the same role as the function $f(x) = e^x$ in ordinary Calculus. Demonstrate this statement by proving that for any constant a we have: $\Delta a^x = (a-1)a^x$, whence for $a = 2$: $\Delta 2^x = 2^x$.

Finally, we consider a natural generalization of the ascending and descending factorials, previously defined. The generalized ascending and descending factorials are defined as follows:

$$(a + bx)^{(m)} = (a + bx)(a + b\overline{x-1})\ldots(a + b\overline{x-m+1}) \quad \underline{\text{descending}}$$
$$(a + bx)^{|m|} = (a + bx)(a + b\overline{x+1})\ldots(a + b\overline{x+m-1}) \quad \underline{\text{ascending}},[1]$$

where a and b are constants and m is any positive integer. By convention, both expressions are defined as one if $m = 0$. It is not difficult to establish that:

$$\Delta(a + bx)^{(m)} = mb(a + bx)^{(m-1)}$$

$$\Delta \frac{1}{(a+bx)^{|m|}} = \frac{-mb}{(a+bx)^{|m+1|}} \quad .$$

EXERCISE 12.13 Prove the above results. In view of previous remarks, what are the analogous results in regular Calculus?

EXERCISE 12.14 Using Sterling Numbers of the First Kind, then of the Second Kind, express Δx^4 and $\Delta^2 x^4$ as polynomials in powers of x.

EXERCISE 12.15 Prove that $\Delta\binom{x}{m} = \binom{x}{m-1}$.

EXERCISE 12.16 Let $P_m(x) = a_m x^m + a_{m-1}x^{m-1} + \ldots + a_0$ be a polynomial of degree m in x. Prove that: $\Delta^m P_m(x) = m!a_m$.

There is a natural reverse process to that of finding finite differences, namely, the process of finding anti-differences or finite integrals. The parallel process in ordinary Calculus is that of finding anti-derivatives or indefinite integrals.

---

[1]The notation $b\,\overline{x-1}$ means $b(x-1)$, and $b\,\overline{x-m+1}$ means $b(x-m+1)$

Specifically, if U_x is any function such that $\Delta U_x = V_x$, then we term U_x an anti-difference or finite integral of V_x and write: $U_x = \Delta^{-1}V_x$. This relationship defines the anti-difference operator Δ^{-1}. Note also that if U_x is an anti-difference of V_x then so is $U_x + c$ for any constant c.

EXERCISE 12.17 Prove the following statement is equivalent to the preceding statement:

$$\Delta^{-1}U_x - \Delta^{-1}V_x = c \quad \text{iff} \quad U_x = V_x \ .$$

Thus, for example, $\Delta^{-1}\ 2^x = 2^x + c$, $\Delta^{-1}\ x = x^{(2)}/2! + c$, $\Delta^{-1}\ 4x^{(3)} = x^{(4)} + c$ and $\Delta^{-1}\ 3^x = 3^x/2 + c$, where c is an arbitrary constant of finite integration.

The Δ^{-1} operator of Finite Calculus possesses properties analogous to the indefinite integral operator $\int$ of regular Calculus. For example, we have the following basic properties:

(i) $\Delta^{-1}[k \cdot V_x] = k\Delta^{-1}\ V_x$

(ii) $\Delta^{-1}[V_x \pm W_x] = \Delta^{-1}\ V_x \pm \Delta^{-1}\ W_x$

(iii) $\Delta^{-1}[V_x\ \Delta\ W_x] = V_xW_x - \Delta^{-1}\ W_{x+1}\ \Delta\ V_x$.

EXERCISE 12.18 Prove the preceding three properties.

EXERCISE 12.19 Find $\Delta^{-1}[5x3^x - 2x]$ by applying properties (i) through (iii).

For the three functions already defined we have:

(a) $\Delta^{-1} a^x = \frac{a^x}{(a-1)} + c \quad (a \neq 1)$

(b) $\Delta^{-1} (a + bx)^{(m)} = \frac{(a+bx)^{(m+1)}}{b(m+1)} + c$

(c) $\Delta^{-1} \frac{1}{(a+bx)^{|m|}} = \frac{-1}{b(m-1)(a+bx)^{|m-1|}} + c \quad (m > 1)$.

<u>EXERCISE 12.20</u> Establish the preceding three formulas. State the special cases where the constants in the last two formulas are $a = 0$ and $b = 1$. What are the analogous results from regular Calculus?

<u>EXERCISE 12.21</u> Using Sterling Numbers of both kinds, express $\Delta^{-1}[2x^4 - x^3 - x]$ as a polynomial in x. Also, find $\Delta^{-1} \binom{x}{m}$.

<u>EXERCISE 12.22</u> Prove that the operators Δ and Δ^{-1} are inverses of one another in the sense that any function U_x simultaneously satisfies both of the following equations: $\Delta(\Delta^{-1} U_x) = U_x$ and $\Delta^{-1} (\Delta U_x) = U_x$.

Because obtaining the anti-difference of a given function may be the key step in solving many problems in Applied Finite Differences, it is essential to become familiar with the anti-differences of basic functions, and to develop efficient techniques for finding anti-differences of more complicated combinations of these basic functions. We now pursue this point.

The first technique deals with a systematic way of finding the anti-difference of a product of two basic functions such as $x3^x$ or $(\frac{1}{2})^x/(x+1)$, for example. The method is based upon the "integration-by-parts" formula (iii) given above, namely,

$$\Delta^{-1}[V_x \; \Delta W_x] = V_x W_x - \Delta^{-1} [W_{x+1} \; \Delta V_x] .$$

The idea involved is the following: treat the product function whose anti-difference is to be found as $V_x \; \Delta W_x$. Now we may choose either component of the product as V_x or ΔW_x. The point is, by judicious choice of what we call the components, and use of the above formula, we may obtain an "easy" anti-difference on the right-hand side, whereas the anti-difference on the left-hand side may be "difficult". The following Example illustrates this point.

EXAMPLE 12.1 To find $\Delta^{-1} x3^x$ we choose $V_x = x$ and $\Delta W_x = 3^x$. Then, of course, $\Delta V_x = 1$ and so $W_x = 3^x/2$ (we choose $c = 0$). Using formula (iii) we then have $\Delta^{-1} x3^x = x3^x/2 - \Delta^{-1} 3^{x+1}/2 = x3^x/2 - 3/2 \; \Delta^{-1} 3^x = x3^x/2 - 3^{x+1}/4 + c$.

EXERCISE 12.23 Referring to the preceding Example, why would the reverse choice, namely $V_x = 3^x$ and $\Delta W_x = x$, have been inappropriate? Also, why was it possible to set $c = 0$ when finding W_x in the Example? Experience and trial-and-error are involved in the suitable choice of what to call V_x and what to call ΔW_x .

<u>EXERCISE 12.24</u> Find $\Delta^{-1}\ (x+1)^{(2)}\ 2^x$. Note: two successive integrations-by-parts will be required.

There is another technique for finding the anti-difference of more complicated combinations of the basic functions, and of certain non-standard functions. It is termed the method of <u>Undetermined Coefficients and Functions</u>. It proves useful when an educated 'guess' can be made concerning the <u>form</u> of the required anti-difference. Naturally, a certain amount of experience in evaluating standard anti-differences is required before we can use this method effectively. It is best described by Examples.

<u>EXAMPLE 12.2</u> Find $\Delta^{-1}\ \frac{x2^x}{(x+1)(x+2)}$. Here we wish to find a function V_x such that ΔV_x equals the given function. Based upon experience and common sense, we begin with an initial 'guess' of:

$$V_x = \frac{f(x)2^x}{x+1} \ ,$$

where $f(x)$ is some function yet to be determined. Because $\Delta V_x = V_{x+1} - V_x$ we must have:

$$\frac{f(x+1)2^{x+1}}{x+2} - \frac{f(x)2^x}{x+1} = \frac{x2^x}{(x+1)(x+2)} \ .$$

Upon finding a common denominator and equating numerators, we must have:

$$2(x+1)f(x+1) - (x+2)f(x) \equiv x \ .$$

'Guessing' that $f(x) = k$ (constant) we have:

$$2(x+1)k - (x+2)k \equiv x$$

and $k = 1$ is a solution. Thus, the anti-difference

$$V_x = \frac{2^x}{x+1} + c$$

is the solution required. Verify this is indeed the anti-difference of the given function.

<u>EXAMPLE 12.3</u> Find $\Delta^{-1} \frac{2x-1}{2^{x-1}}$. As before, we begin with an initial (possibly incorrect) guess:

$$V_x = \frac{f(x)}{2^{x-2}} \quad .$$

We require that ΔV_x equal the given function and, because $\Delta V_x = V_{x+1} - V_x$, we have:

$$\frac{f(x+1)}{2^{x-1}} - \frac{f(x)}{2^{x-2}} = \frac{2x-1}{2^{x-1}} \quad .$$

After finding a common denominator and then equating numerators we find that we must have:

$$f(x+1) - 2f(x) \equiv 2x-1 \; .$$

This suggests a linear form for $f(x)$, whence we try $f(x) = ax + b$, which then yields:

$$a(x+1) + b - 2ax - 2b \equiv 2x-1 \quad .$$

Solving the two equations in a and b obtained by equating coefficients of like powers of x, we obtain: $a = -2$ and $b = 1$. Thus the anti-difference required is given by:

$$V_x = \frac{-(2x-1)}{2^{x-2}} + c \ .$$

EXERCISE 12.25 Find $\Delta^{-1} \frac{(x+2)}{x^{|2|}} (\frac{1}{2})^x$ and $\Delta^{-1} \frac{(x+8)}{x(x+1)(x+2)} (\frac{3}{4})^x$

The Calculus of Finite Differences is quite useful in problems related to summation of certain convergent series (both finite and infinite). First, consider the series $U_1 + U_2 + U_3 + \ldots + U_n = \sum_1^n U_x$. The objective is to express the sum of this series in so-called 'closed form'. With this in mind, suppose that we can obtain an anti-difference V_x of the function U_x, whence it would follow that $\Delta V_x = V_{x+1} - V_x = U_x$. Then we could write:

$$U_1 = V_2 - V_1$$

$$U_2 = V_3 - V_2$$

$$U_3 = V_4 - V_3$$

$$\vdots$$

$$U_n = V_{n+1} - V_n$$

Upon adding both sides of the above equalities, and observing the concellation that takes place, we obtain that:

$$\sum_1^n U_x = U_1 + U_2 + U_3 + \ldots + U_n = V_{n+1} - V_1 \ .$$

Thus, to obtain the sum of the series $\sum_{1}^{n} U_x$ is closed form, it is necessary only to find (if possible) an anti-difference V_x of U_x and make use of the preceding result.

EXAMPLE 12.4 To find the sum of the series $1 + 2 + 3 + \ldots + n$ in closed form we note that $U_x = x$ and $V_x = x^{(2)}/2$. Thus, the sum of the first n positive integers is: $x^{(2)}/2\Big]_1^{n+1} = n(n+1)/2.$[1] To sum the series $1^2 + 2^2 + \ldots + n^2$ note that $U_x = x^2 = x^{(1)} + x^{(2)}$ whence $V_x = x^{(2)}/2 + x^{(3)}/3$. Upon simplification it then follows that $\sum_{1}^{n} x^2 = V_{n+1} - V_n =$ $n(n+1)(2n+1)/6$.

EXERCISE 12.26 Obtain the sum $1^3 + 2^3 + \ldots + n^3$ in 'closed form'.

Recalling that an infinite 'sum' $\sum_{1}^{\infty} U_x$ is defined to be $\lim_{n\to\infty} \sum_{1}^{n} U_x$, provided the limit exists, we see that the above method may be used to find the sum of certain convergent infinite series. Assuming that the appropriate anti-difference V_x can be found, and that the indicated limits exist, we have:

$$\sum_{1}^{\infty} U_x = \lim_{n\to\infty} \sum_{1}^{n} U_x = \lim_{n\to\infty} V_{n+1} - V_1 .$$

This technique is illustrated in the following Example.

EXAMPLE 12.5 To find $\sum_{1}^{\infty} \frac{2x-1}{2^{x-1}}$ note that from Example 12.3 a suitable choice for V_x is $V_x = \frac{-(2x-1)}{2^{x-2}}$, whence,

[1] The notation $x^{(2)}/2\Big]_1^{n+1}$ means $(n+1)^{(2)}/2 - (1)^{(2)}/2$. An analogous notation is used in Integral Calculus.

$$\sum_{1}^{\infty} \frac{2x-1}{2^{x-1}} = \lim_{n\to\infty} - \frac{2n+1}{2^{n-1}} + 2 = 2.$$

<u>EXERCISE 12.27</u> Find $\sum_{1}^{\infty} \frac{(x+2)}{x^{|2|}} (\frac{1}{2})^{x}$ and also the sum

$$\frac{9}{1\cdot 2\cdot 3} (\tfrac{3}{4}) + \frac{10}{2\cdot 3\cdot 4} (\tfrac{3}{4})^{2} + \ldots \quad .$$

<u>EXERCISE 12.28</u> (Moments of the Geometric Distribibition) The m-th moment $(m = 1,2,\ldots)$ μ_m of the Geometric Distribution is defined as

$$\mu_m = \sum_{1}^{\infty} x^{m} p(1-p)^{x-1}$$

where $0 < p < 1$. Using the above method, and Sterling Numbers, find μ_1, μ_2, and μ_3 .

<u>EXERCISE 12.29</u> (Continuation) Verify that:

$$\frac{ps}{1-s(1-p)} = \sum_{1}^{\infty} s^{x} p(1-p)^{x-1}$$

is valid for $0 < p < 1$ and $|s(1-p)| < 1$. Accordingly, show that an alternate method for finding the moments of the preceding Exercise is to formally differentiate the above relation with respect to s and evaluate both sides for $s = 1$.

Another aspect of Finite Calculus that occurs often in Applied Mathematics is the solution of so-called <u>Difference Equations</u> (and certain generalizations to be introduced later). Formally, a difference equation is an equation expressing a relationship amongst successive values and/or differences of a function U_x of the real variable x. For example:

(i) $\Delta^3 U_x - \Delta^2 U_x + U_x = 0$

(ii) $U_{x+3} - U_{x+2} + xU_{x+1} = U_x$

(iii) $U_{x+2} - \Delta^2 U_x = 2^x$.

Since we may always make use of the relations: $\Delta U_x = U_{x+1} - U_x$, $\Delta^2 U_x = U_{x+2} - 2U_{x+1} + U_x$ and so on, equations involving differences (such as the first and third equations) may always be replaced by equivalent equations (such as the second) in which the differences have been eliminated. In view of this, we shall assume that all difference equations with which we work are of the latter type. General methods will be developed for solving such difference equations.

The <u>general n-th order linear difference equation with constant coefficients</u> is a difference equation of the form:

$$U_{x+n} + A_1 U_{x+n-1} + \ldots + A_n U_x = V_x$$

where the A_i's are constants and V_x is an arbitrary function of x. The above equation can also be expressed in operator notation as:

$$F(E)\, U_x = V_x$$

where

$$F(E) = E^n + A_1 E^{n-1} + \ldots + A_n \quad .$$

The homogeneous form of the general equation is given by:

$$F(E)\ U_x = 0 \quad .$$

Clearly, the general solution of the complete (non-homogeneous) equation is the sum of the general solution of the homogeneous equation plus any particular solution of the non-homogeneous equation.

> EXERCISE 12.30 Prove the preceding statement. (Note the parallel here with linear Differential Equations of ordinary Calculus.)

The following Theorem provides the general solution of the homogeneous equation (in this linear, constant coefficient case). Its proof is based upon the fact that: $a^n + A_1 a^{n-1} + \ldots + A_n = 0$ if and only if $U_x = a^x$ is a solution of the homogeneous equation. The preceding polynomial equation in a is termed the auxiliary equation.

THEOREM 12.1 (General Solution of Homogeneous Equation) If the auxiliary equation $a^n + A_1 a^{n-1} + \ldots + A_n = 0$ has n distinct roots $a_1, a_2, \ldots, a_n$, then the general solution of $U_{x+n} + A_1\ U_{x+n-1} + \ldots + U_n = 0$ is given by:

$$U_x = c_1 a_1^x + c_2 a_2^x + \ldots + c_n a_n^x \ ,$$

where the c_i's are arbitrary constants. If, however, certain roots of the auxiliary equation are multiple roots (say, for instance, that $a_1 = a_2 = \ldots = a_k = a$), then the k terms of the general solution given above and corresponding to this root should be replaced by:

$$(c_1 + c_2x + \ldots + c_kx^{k-1})\ a^x\ .$$[1]

EXERCISE 12.31 Prove the preceding Theorem in the case where all roots of the auxiliary equations are assumed to be distinct.

EXAMPLE 12.6 To solve $U_{x+2} - 5U_{x+1} + 6U_x = 0$ note that the auxiliary equation is $a^2 - 5a + 6 = 0$ which has distinct roots $a_1 = 2$ and $a_2 = 3$. Thus the general solution is $U_x = c_1 2^x + c_2 3^x$. If initial conditions $U_0 = U_1 = 1$ were to be imposed, this would uniquely determine the solution (constants) as $U_x = 2^{x+1} - 3^x$.

EXAMPLE 12.7 Solve $U_{x+3} - 3U_{x+1} - 2U_x = 0$. Here the auxiliary equation is given by: $a^3 - 3a-2 = 0 = (a+1)^2(a-2)$, whence the roots are $a_1 = a_2 = -1$, $a_3 = 2$. Thus, the general solution is $U_x = (c_1 + c_2x)(-1)^x + c_3 2^x$.

In Applied Mathematics it is sometimes possible to obtain a difference equation that must be satisfied by some unknown function we wish to determine. If this difference equation can then be solved, the unknown function can then (at least partially) be obtained. The following Exercises illustrate this point.

EXERCISE 12.32 (Probability of Gambler's Ruin) Solve the difference equation $Q_z = pQ_{z+1} + qQ_{z-1}$ (where $0 < p < 1$ and $p + q = 1$) subject to the initial conditions that

[1]The Fundamental Theorem of Algebra (Theorem 13.6) guarantees existence of the n roots.

$Q_0 = 1$ and $Q_m = 0$ (and where m is some positive integer). Note: first change variables by setting $U_z = Q_{z-1}$ to get the equation in standard form. Treat the case $p = q = \frac{1}{2}$ separately.

EXERCISE 12.33 (Geometric Distribution) Prove that if $p_n > 0$ satisfies the difference equation: $p_{n+1} = ap_n$ $(n = 1,2,\ldots,;\ 0 < a < 1)$ then $p_n = ka^{n-1}$ for some constant k. Then show that if we require $\sum_1^\infty p_n = 1$, we must have $k = 1-a$.

EXERCISE 12.34 (Expected Waiting Time in Bernoulli Trials) Solve the difference equation: $m_r = p(1+m_{r-1}) + q(1+m_r)$ $(r = 1,2,\ldots,;\ 0 < p < 1,\ p + q = 1)$ subject to the condition $m_1 = 1/p$.

EXERCISE 12.35 (Size of Waiting Line: a non-linear equation) Consider the non-linear difference equation: $(a + bx)\, U_x = a\, U_{x-1} + (x+1)\, U_{x+1}$ $(a,b > 0;\ x = 1,2,\ldots)$ subject to the conditions $a\, U_0 = b\, U_1$ and $\sum_0^\infty U_x = 1$. Solve this equation recursively (inductively) whence prove that

$$U_x = e^{-(a/b)}\, (a/b)^x/x!\ \ (x = 0,1,2,\ldots)$$

is the solution. (This inductive approach is a common-sense alternative when standard linear techniques fail to apply).

By the use of experience and trial-and-error, together with common sense, the above techniques often lend themselves to solution of (i) certain sets of simultaneous difference equations and (ii) certain non-linear difference equations (by means of inductive solution or suitable substitution(s)). General rules are difficult to state; the techniques usually must be tailored to the specifics of each individual problem. The following Examples and Exercise illustrate these comments.

EXAMPLE 12.8 Solve the simultaneous system:

$$U_{x+1} - V_x = 0$$

$$V_{x+1} - U_x = 0 \quad .$$

From the first equation we conclude that $U_{x+2} - V_{x+1} = 0$ and, upon substitution in the second equation we obtain: $U_{x+2} - U_x = 0$ which has general solution $U_x = c_1 + c_2(-1)^x$. Substituting this in the last equation we then obtain that $V_{x+1} = c_1 + c_2(-1)^x$ whence $V_x = c_1 - c_2(-1)^x$.

EXAMPLE 12.9 Solve the non-linear difference equation: $x_{n+1} = 4 - 4/x_n$. We try the substitution $x_n = y_{n+1}/y_n$, whence the original equation becomes: $y_{n+2} = 4y_{n+1} - 4y_n$, which is linear and homogeneous with general solution: $y_n = (c_1 + c_2n)\, 2^n$. Thus, the original equation has as its solution: $x_n = 2 + 2c_2/(c_1 + c_2n)$.

EXERCISE 12.36 Solve the non-linear difference equation: $U_{x+1}\, U_x + (x+2)\, U_{x+1} + x\, U_x + x^2 + 2x + 2 = 0$ by using the substitution: $U_x = V_{x+1}/V_x - (x+2)$.

Recall that, in order to provide the general solution to a general n-th order linear difference equation with constant coefficients, it is necessary to obtain a particular solution to the non-homogeneous equation. We now consider the problem of obtaining particular solutions of this type. One technique is based upon the formal use of operators, and is developed in what follows.

Consider the non-homogeneous equation $F(E)\ U_x = V_x$, where $F(E) = E^n + A_1E^{n-1} +\ldots+ A_n$. By formal manipulation of this equation, a particular solution is given by:

$$U_x = \frac{1}{F(E)} V_x \quad .$$

The difficulty here is in evaluating (even interpreting) the expression on the right-hand side. There are, however, certain results concerning algebraic manipulators with operators such as "E" that will enable us to evaluate the above expression in a variety of situations. The following result considers the case where V_x is of the form a^x.

<u>THEOREM 12.2</u> If $F(E)$ is a polynomial in E, then $F(E)\ a^x = a^x\ F(a)$, whence, if $F(a) \neq 0$ we have:

$$\frac{1}{F(E)} a^x = \frac{a^x}{F(a)} \quad .$$

PROOF. $F(E)\ a^x = (E^n + A_1E^{n-1} +\ldots+ A_n)\ a^x = a^{x+n} + A_1a^{x+n-1} +\ldots+ A_na^x = a^x(a^n + A_1a^{n-1} +\ldots+ A_n) = a^x\ F(a)$. If $F(a) \neq 0$ the result follows upon simple division.

<u>EXAMPLE 12.10</u> Solve: $U_{x+2} - 5U_{x+1} + 6U_x = 5^x$. From previous results, the homogeneous equation has general solution $c_12^x + c_23^x$. To obtain a particular solution

of the non-homogeneous equation, note that $F(E) = E^2 - 5E + 6$ and $F(5) = 6 \neq 0$. Thus, a particular solution is $5^x/6$ and so the general solution is $U_x = c_1 2^x + c_2 3^x + 5^x/6$. (Alternately, a particular solution could have been obtained by first 'guessing' that it was of the form $k5^x$, then using the equation to evaluate k.)

EXERCISE 12.37 Solve $U_{x+3} - 3U_{x+1} - 2U_x = 3^x$ subject to the conditions $U_1 = U_2 = U_3 = 0$.

The above method fails to apply when $V_x = a^x$ if $F(a) = 0$, that is, if a is a root (possibly multiple) of the polynomial equation $F(E) = 0$. If this be the case, then $F(E)$ contains the factor $(E-a)^k$ for some positive integer k. The next result enables us to overcome the difficulties encountered in this case.

THEOREM 12.3

$$\frac{1}{(E-a)^k} a^x = \frac{x^{(k)} a^{x-k}}{k!} \qquad (k = 1,2,\ldots) .$$

PROOF. (By induction on k) The result is true for $k = 1$ since $(E-a)xa^{x-1} = (x+1)a^x - xa^x = a^x$. Now assume that the result is true for all positive integers $\leq k$. We then show that this implies the truth of the result for $k + 1$, which then completes the proof. Now, by the inductive hypothesis,

$$\frac{1}{(E-a)^{k+1}} a^x = \frac{1}{E-a} \frac{1}{(E-a)^k} a^x$$

$$= \frac{1}{E-a} \frac{x^{(k)} a^{x-k}}{k!} .$$

Denoting the latter quantity on the right by $a^x V_x$, it suffices to show that $V_x = x^{(k+1)} a^{-(k+1)}/(k+1)!$. Operating on both sides of the following equation:

$$\frac{1}{E-a} \frac{x^{(k)} a^{x-k}}{k!} = a^x V_x$$

by E-a, we obtain the following relation:

$$\frac{x^{(k)} a^{x-k}}{k!} = a^{x+1} \Delta V_x . \quad \text{Whence,} \quad \Delta V_x = \frac{x^{(k)}}{k!} a^{-(k+1)} ,$$

so

$$V_x = \frac{x^{(k+1)}}{(k+1)!} a^{-(k+1)} + c .$$

The case $k = 1$ can be used to show that $c = 0$ and so the result is proved.

<u>EXAMPLE 12.11</u> Solve $U_{x+2} - 5U_{x+1} + 6U_x = 3^x$. Here, $F(E) = (E-3)(E-2)$. Combining, successively, the results of the two preceding Theorems, we obtain a particular solution:

$$\frac{1}{F(E)} 3^x = \frac{1}{E-3} [\frac{1}{E-2} 3^x] \frac{1}{E-3} 3^x = x3^x .$$

Thus, the general solution is given by:

$$U_x = c_1 3^x + c_2 2^x + x3^x .$$

<u>EXERCISE 12.38</u> Solve: $U_{x+3} - 7U_{x+2} + 16U_{x+1} - 12U_x = 2^x$.

Next, we consider an operator technique for finding a particular solution to the equation $F(E)\, U_x = V_x$ in the case where V_x is a polynomial in x. It is, perhaps, best described by example.

<u>EXAMPLE 12.12</u> Solve $U_{x+2} + U_{x+1} + U_x = x^2 + x + 1$. First, the homogeneous form has auxiliary equation: $a^2 + a + 1 = 0$ with roots $1/2 \pm i\sqrt{3}/2$, whence its general solution is: $c_1(1/2 + i\sqrt{3}/2)^x + c_2(1/2 - i\sqrt{3}/2)^x = b_1\mathrm{Cos}\, 2\pi x/3 + b_2\mathrm{Sin}\, 2\pi x/3$. Next we find a particular solution to the non-homogeneous equation. Using the relation $E = 1 + \Delta$ and Sterling Numbers of the Second Kind, we rewrite the equation as: $(3 + 3\Delta + \Delta^2)\, U_x = x^{(2)} + 2x^{(1)} + 1$. Then, at least formally, a particular solution is given by:

$$\frac{1}{3+3\Delta+\Delta^2}\,[x^{(2)} + 2x^{(1)} + 1]\ .$$

By formal long division we obtain:

$$\frac{1}{3+3\Delta+\Delta^2} = \frac{1}{3}[1 - \Delta + \frac{2}{3}\,\Delta^2 + \text{terms higher than } \Delta^2].$$

Thus, $\frac{1}{3}[1 - \Delta + \frac{2}{3}\,\Delta^2 + \ldots][x^{(2)} + 2x^{(1)} + 1] =$ $\frac{1}{3}\,[x^{(2)} + \frac{1}{3}] = x^2/3 - x/3 + 1/9$ is a particular solution. Therefore the general solution is $U_x = b_1\mathrm{Cos}\, 2\pi x/3 + b_2\mathrm{Sin}\, 2\pi x/3 + x^3/ - x/3 + 1/9$.

Note that the extent to which the formal long division procedure is carried out depends upon the degree of the polynomial in x that forms the function V_x .

EXERCISE 12.39 By combining several of the preceding results, solve: $U_{x+2} - 2U_{x+1} + U_x = 2^x + x^{(2)}$.

If the above general techniques fail to apply in finding a particular solution to the non-homogeneous equation (that is, if V_x is not a linear combination of exponentials and/or powers of x), then a common sense alternative might be to 'guess' the general form of a particular solution and then use the equation itself to identify the solution exactly. The following Example illustrates this point.

EXAMPLE 12.13 Solve: $U_{x+2} - 7U_{x+1} - 8U_x = x^{(2)}2^x$. Of course, the homogeneous equation has general solution $c_1(-1)^x + c_2 8^x$. Now we 'guess' that a particular solution to the non-homogeneous equation is of the form: $[ax^{(2)} + bx^{(1)} + c]\ 2^x$. Upon substituting this in the equation, and equating the appropriate coefficients on both sides, the values of $a, b,$ and c may be determined (provided, of course, that the correct form was chosen originally).

EXERCISE 12.40 Complete the preceding Example.

EXERCISE 12.41 Find the general solution to the following system of difference equations (k is an arbitrary constant):

$$U_{x+1} - V_x = 2k(x+1)$$

$$V_{x+1} - U_x = -2k(x+1).$$

In more complicated difference equations, the function U involved in the equation depends not only upon an integer variable but also upon another variable that may be integer or continuous. The following sample equations illustrate this point.

EXAMPLE 12.14 (Coin Tossing) Solve the difference equation:

$U_{x+1}(t) = \frac{t-1}{N} U_x(t) + \frac{N-t+1}{N} U_x(t-1)$ where N is a fixed positive integer and $x,t = 1,2,\ldots$.

EXAMPLE 12.15 (Poisson Distribution) Solve the differential-difference equation: $\frac{d}{dt} U_x(t) = -\lambda\, U_x(t) + \lambda\, U_{x-1}(t)$, where $x = 0,1,2,\ldots,$ $t > 0$, and λ is a positive constant.

Special techniques must be developed for solving such complicated systems of difference equations. Although a first step might be to use any initial conditions and attempt to solve the system recursively, general techniques are available in some cases. They are somewhat sophisticated. (See References).

EXERCISE 12.42 (Poisson Distribution) Solve the differential-difference equation of Example 12.15 subject to the initial conditions: $U_{-1}(t) = 0$, $U_x(0) = 0$ $(x = 0,1,2,\ldots)$, and $\sum_0^\infty U_x(t) = 1$ for any $t > 0$.

EXERCISE 12.43 (Moments of Standard Normal Distribution)

Show that the even moments:

$$\mu_{2n} = \int_{-\infty}^{\infty} x^{2n} \frac{1}{\sqrt{2\pi}} e^{-\frac{1}{2}x^2} dx$$

$(n = 0,1,2,\ldots)$ of the Standard Normal Distribution satisfy the difference (recursion) equation: $\mu_{2n} = (2n-1)\mu_{2n-2}$. Using the initial condition $\mu_0 = 1$, prove that $\mu_{2n} = (2n-1)(2n-3)\ldots 3 \cdot 1$ by solving the equation recursively.

Hints and Answers for Exercises: Section 12.

12.1 Answer: $u(x+n) = \Delta^n u(x) + \binom{n}{1} \Delta^{n-1}u(x) + \ldots + \binom{n}{n-1}\Delta u(x) + u($

12.2 Answer: $2x^{(1)} - 3$ and $3x^2 - 9x + 2$.

12.3 Hint: $x^{(m+1)} = (x-m)x^{(m)}$.

12.4 Answer: (Partial)

6:	-120	274	-225	85	-15	1		
7:	720	-1764	1624	-735	175	-21	1	
8:	-5040	13068	-13132	6769	-1960	322	-28	1

12.6 Hint: Note that $x^{(j+1)} + jx^{(j)} = xx^{(j)}$, and by definition:
$x^{n+1} = \sum_{1}^{n+1} \mathcal{S}_j^{n+1} x^{(j)}$.

12.7 Answer: (Partial)

6:	1	31	90	65	15	1		
7:	1	63	301	350	140	21	1	
8:	1	127	966	1701	1050	266	28	1

12.9 Hint: Note that:

$$\sum_0^\infty x^{(j)} e^{-p} p^x/x! = p^j \sum_0^\infty e^{-p} p^y/y! = p^j$$

since

$\sum_0^\infty p^y/y! = e^p$. Now express x^m in terms of descending factorials to evaluate μ_m .

12.10 Hint: Use the definition of the Δ operator.

12.11 Hint: Use properties (i) through (iii), noting that $x^2/2^x = (\frac{1}{2})^x[x^{(2)} + x^{(1)}]$.

12.12 Hint: $\Delta a^x = a^{x+1} - a^x$.

12.13 Hint: For the first part note that:

$$\begin{aligned}\Delta(a+bx)^{(m)} &= (a+b\overline{x+1})^{(m)} - (a+bx)^{(m)} \\ &= (a+bx)^{(m-1)}[a+b(x+1) - (a+b\overline{x+m-1})] \\ &= -mb(a+bx)^{(m-1)} \quad .\end{aligned}$$

In regular Calculus, $\frac{d}{dx}(a+bx)^m = mb(a+bx)^{m-1}$.

12.15 Hint: $\Delta \binom{x}{m} = \binom{x+1}{m} - \binom{x}{m}$.

12.16 Hint: $\Delta^m x^{(m)} = m!$.

12.17 Hint: For the first part, apply the Δ operator to both sides of the left-hand equality, and for the second part apply the Δ^{-1} operator to the relationship: $U_x - V_x = 0$.

12.18 Hint: Use the results of Exercise 12.17.

12.19 Hint: Use properties (i) through (iii).

12.20 Hint: Apply the Δ operator to both sides of each equality, then use the results of Exercise 12.18.

12.22 Hint: Apply the Δ^{-1} operator to both sides of the first equality, and the Δ operator to both sides of the second equality.

12.24 Hint: Apply Property (iii) choosing, in the first application, $V_x = (x+1)^{(2)}$.

12.25 Hint: Apply Property (iii) in an appropriate manner by grouping terms at first. Check the result by applying the Δ operator to it.

12.26 Hint: Use the technique of Example 12.4. Upon simplification the sum is: $[n(n+1)/2]^2$.

12.27 Hint: Use the results of Exercise 12.25.

12.28 Hint: Express x^m in terms of descending factorials, then find the anti-difference of the general term by applying Property (iii) repeatedly.

12.29 Hint: Express $s^x p(1-p)^{x-1}$ as $\frac{p}{1-p}[s(1-p)]^x$ and use the result of Example 3.11.

12.30 Hint: Prove that if W_x and Z_x are solutions of the general equation, then $W_x - Z_x$ is a solution of the homogeneous equation.

12.31 Hint: Clearly U_x satisfies the homogeneous equation and, if the roots are distinct, the n functions a_i^x $(i = 1,2,\ldots,n)$ are linearly independent functions.[1]

12.32 Hint: Use Theorem 12.1. The case $p = q = \frac{1}{2}$ corresponds to multiple roots.

12.33 Hint: Use Theorem 12.1.

12.34 Hint: Use the change of variables $n = r-1$, then Theorem 12.1.

12.37 Hint: Apply Theorem 12.2. Note that the coefficient of U_{x+2} is zero, whence the auxiliary equation becomes: $(a+1)^2(a-2) = 0$.

12.39 Answer: $U_x = (c_1 + c_2 x) + 2^x + x^{(4)}/12$.

12.40 Answer: A particular solution is: $-\frac{1}{54}[3x^{(2)} - 2x^{(1)} + 2]$.

12.41 Answer: $U_x = A + B(-1)^x + kx$

$V_x = C + D(-1)^x + kx$.

12.42 Hint: Solve the equation recursively, beginning with $x = 0$.

12.43 Hint: Use Integration-by-Parts along with properties of the respective integrands.

---

[1]That is, $b_1 a_1^x + b_2 a_2^x + \ldots + b_n a_n^x = 0$ if and only if $b_1 = b_2 = \ldots = b_n = 0$.

References to Additional and Related Material: Section 12

1. Batchelder, P., "An Introduction to Linear Difference Equations", Harvard University Press (1927).

2. Boole, G., "A Treatise on the Calculus of Finite Differences". Dover Publications, Inc.

3. Brand, L., "Differential and Difference Equations", John Wiley and Sons, Inc. (1966).

4. Chorlton, F., "Ordinary Differential and Difference Equations: Theory and Applications", Van Nostrand, Inc. (1965).

5. Fort, T., "Finite Differences and Difference Equations in the Real Domain", Clarendon Press (1948).

6. Goldberg, S., "Introduction to Difference Equations, with Illustrative Examples from Economics", John Wiley and Sons, Inc. (1958).

7. Jordan, C., "Calculus of Finite Differences", Chelsea Publishing Co. (1947).

8. Richardson, C., "An Introduction to the Calculus of Finite Differences", Van Nostrand, Inc. (1954).

13. Complex Variables

Applications of the complex number system, sometimes referred to as complex variables, form an essential tool in many areas of Applied Mathematics. The complex number system can be viewed as a useful generalization of the familiar real number system. For, if the real number system can be thought of as the familiar properties of points - called real numbers - on the real line, then the complex number system can be thought of as the yet-to-be-examined properties of points - called complex numbers - of the complex plane, of which the real line is its abscissa. Properties of the complex number system are determined by the special manner in which complex numbers are combined, that is, added, multiplied, etc.

Here we shall examine only the fundamentals of complex variables; however, the material we develop will be sufficient for a wide variety of Applications.

A complex number z is a quantity of the form $z = x + iy$, where x and y are real numbers and i is the imaginary unit which satisfies: $i^2 = -1$; x is the real component (part) of z written $x = \text{Re}(z)$, and y is termed the imaginary part of z written $y = \text{Im}(z)$. Thus, $z = \text{Re}(z) + i\text{Im}(z)$.

The complex numbers can be plotted on a complex plane much the same way points are plotted in E_2. This is shown below.

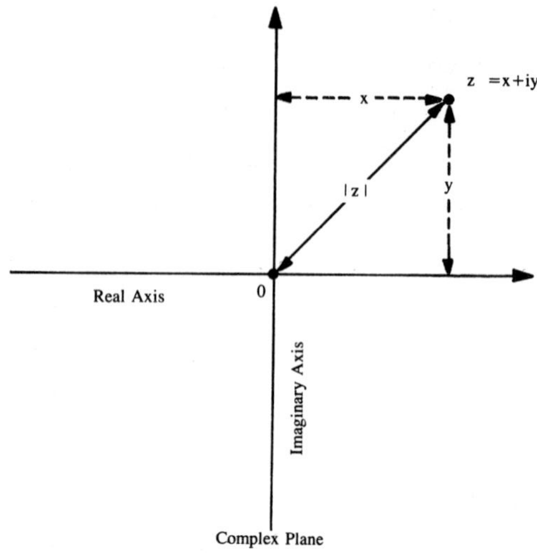

Complex Plane

Figure 13-1

The quantity $|z| = (x^2 + y^2)^{\frac{1}{2}}$ is termed the modulus of z. Geometrically it represents the straight-line distance from z to the origin of the complex plane. If z is real, that is $\text{Im}(z) = 0$, then modulus reduces to ordinary absolute value. Furthermore, since two complex numbers $z_1 = x_1 + iy_1$ and $z_2 = x_2 + iy_2$ are equal iff $x_1 = x_2$ and $y_1 = y_2$, we see that $|z| = 0$ iff $z = 0 = 0 + i0$.

Complex numbers may be combined according to the following rules:

(1) Addition: $z_1 \pm z_2 = (x_1 \pm x_2) + i(y_1 \pm y_2)$

(2) Multiplication: $z_1 \cdot z_2 = (x_1x_2 - y_1y_2) + i(x_1y_2 + y_1x_2)$

(3) Division: $z_1/z_2 = [(x_1x_2 + y_1y_2) + i(x_2y_1 - y_2x_1)]/(x_2^2 + y_2^2)$, $z_2 \neq 0$.

<u>EXERCISE 13.1</u> (Real Scalar Multiplication) Prove that if c is real then $cz = cx + icy$.

With these definitions, combinations of complex numbers satisfy the usual algebraic properties, e.g. commutativity, associativity, distributivity, that are so familiar from corresponding properties of ordinary real numbers.

The operations of addition (subtraction) and real scalar multiplication are easily demonstrated geometrically.

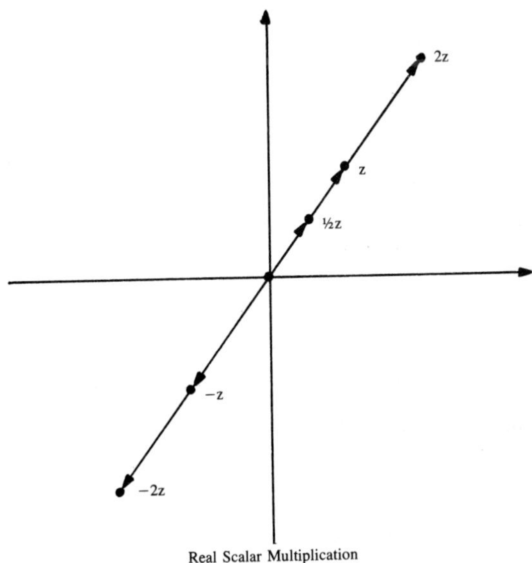

Real Scalar Multiplication

Figure 13-2

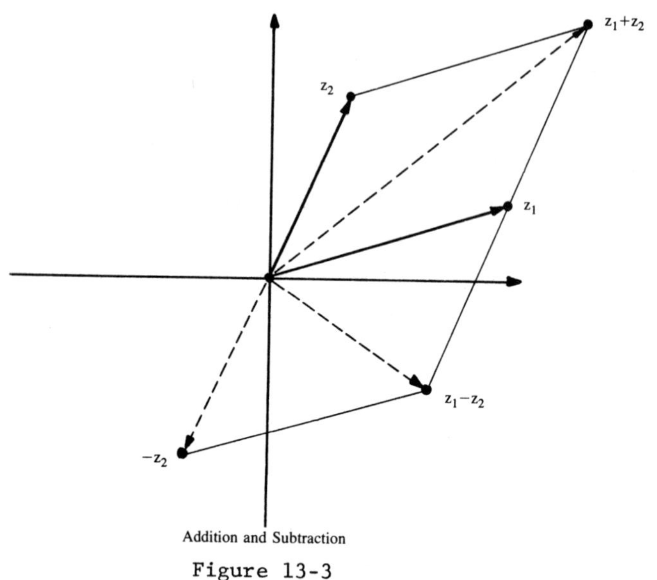

Figure 13-3

<u>EXERCISE 13.2</u> Exploit the geometry of complex addition and subtraction (that is, the so-called Parallelogram Rule) to prove: $|z_1| - |z_2| \le ||z_1| - |z_2|| \le |z_1 + z_2| \le |z_1| + |z_2|$.

The conjugate $\bar{z}$ of a complex number z is defined to be the complex number $\bar{z} = x - iy$. The operation of conjugation is pictured below; it amounts to reflection in the Real Axis.

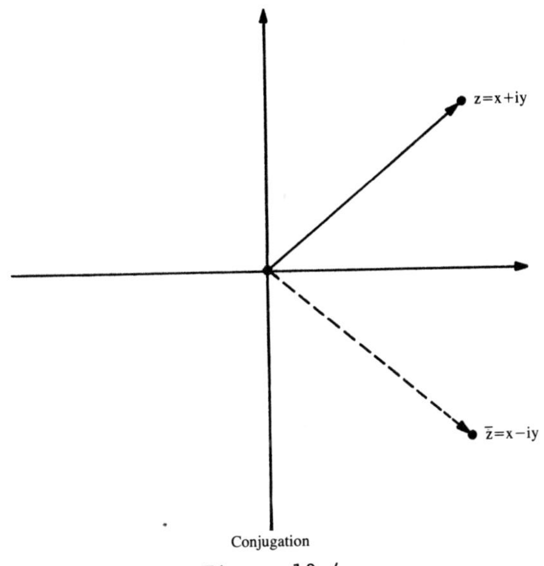

Conjugation

Figure 13-4

EXERCISE 13.3 Prove that conjugation obeys the following algebraic properties: $\overline{z_1 \pm z_2} = \bar{z}_1 \pm \bar{z}_2$, $\overline{z_1 \cdot z_2} = \bar{z}_1 \cdot \bar{z}_2$, $(\overline{z_1/z_2}) = \bar{z}_1/\bar{z}_2$.

EXERCISE 13.4 Prove that $|\bar{z}| = |z|$ and $z \cdot \bar{z} = |z|^2$.

An alternate and equivalent means of expressing complex numbers is the so-called polar form. It is most useful when products, powers and quotients of complex numbers occur. For a complex number $z = x + iy$ with modulus $r = |z|$, define the argument θ of z as follows: $\theta = \mathrm{Tan}^{-1} y/x$. Then observe that we may rewrite z in terms of r and θ as follows: $z = r(\mathrm{Cos}\ \theta + i\mathrm{Sin}\ \theta)$. Now z is said to be expressed in polar form. Geometrically we may picture locating a point in the complex plane by means of r, θ as follows:

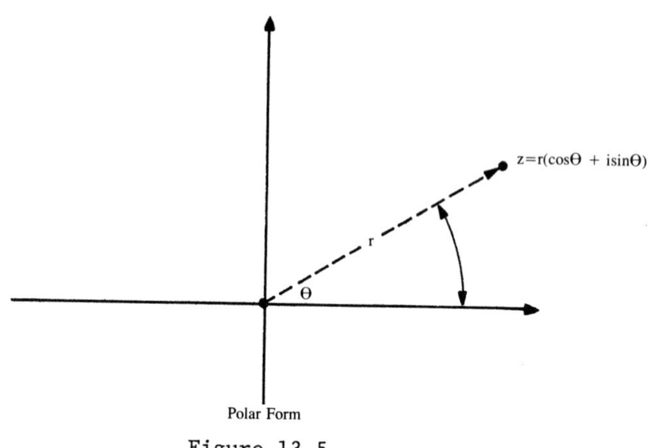

Polar Form

Figure 13-5

<u>EXERCISE 13.5</u> Express the following complex numbers in polar form: $1 + i$, $-i$, -4, $3 - i$.

<u>EXERCISE 13.6</u> Express the following complex numbers in standard form: $4(\mathrm{Cos}\ \pi/4 + i\mathrm{Sin}\ \pi/4)$, $6(\mathrm{Cos}\ \pi/12 + i\mathrm{Sin}\ \pi/12)$, $2(\mathrm{Cos}\ -\pi/6 + i\mathrm{Sin}\ -\pi/6)$.

Once complex numbers are expressed in polar form, products, powers and quotients become simple both algebraically and geometrically.

<u>EXERCISE 13.7</u> (Algebra of Polar Form Computations)
Let $z_1 = r_1(\mathrm{Cos}\ \theta_1 + i\mathrm{Sin}\ \theta_1)$ and $z_2 = r_2(\mathrm{Cos}\ \theta_2 + i\mathrm{Sin}\ \theta_2)$. Prove that $z_1 \cdot z_2 = r_1 r_2[\mathrm{Cos}(\theta_1 + \theta_2) + i\mathrm{Sin}\ (\theta_1 + \theta_2)]$ whence $z^n = r^n(\mathrm{Cos}\ n\theta + i\mathrm{Sin}\ n\theta)$ $(n = 0,1,2,\ldots)$ and $z_1/z_2 = r_1/r_2\ [\mathrm{Cos}(\theta_1 - \theta_2) + i\mathrm{Sin}(\theta_1 - \theta_2)]$, $z_2 \neq 0$.

<u>EXERCISE 13.8</u> For any complex number a define $a^{1/n}$ as any solution to the equation $z^n = a$. Prove that if $a \neq 0$, then there are n distinct solutions (called n-th roots of a). In fact, show that if $a = q(\mathrm{Cos}\ \phi + i\mathrm{Sin}\ \phi)$ then the solutions are given by $q^{1/n}[\mathrm{Cos}(\phi + 2k\pi/n) + i\mathrm{Sin}(\phi + 2k\pi/n)]$ for $k = 0,1,\ldots,n-1$.

A geometrical representation of product and quotient in polar form is given below.

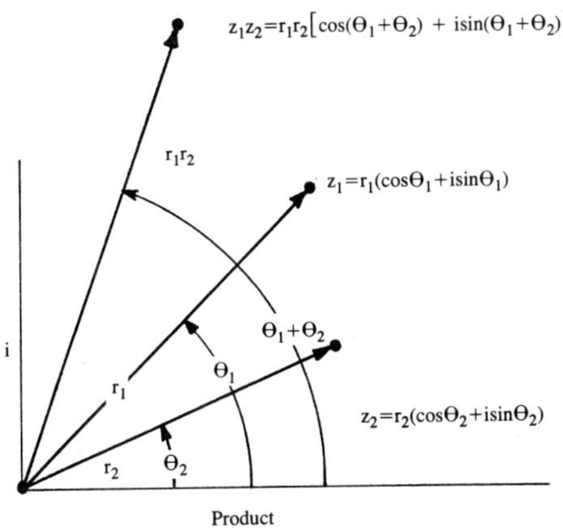

Figure 13-6

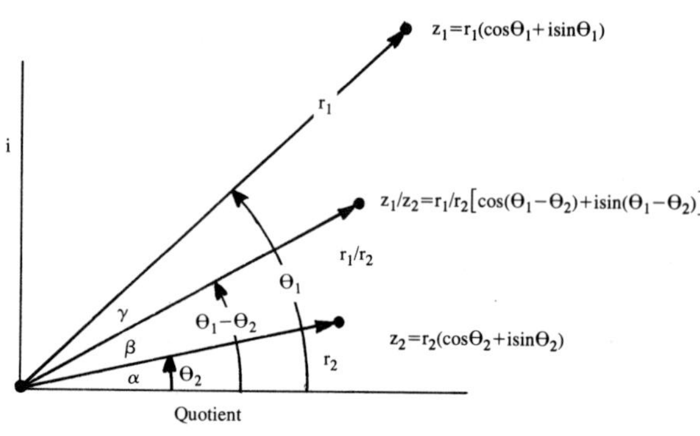

Figure 13-7

The notion of a complex function is now easy to explain. If, corresponding to every complex number z in a certain subset (perhaps all) of the complex plane there is a single complex number w, then this correspondence defines a function of the complex variable z, generally denoted $w = f(z)$. A simple example is the power function $f(z) = z^n$ $(n = 1,2,...)$. This is obviously a generalization of the notion of a real-valued function of a real variable x.

Any function of the complex variable $z = x + iy$ can be expressed in the form $f = u + iv$, where $u = u(x,y)$ is the real part of f and $v = v(x,y)$ is the imaginary part of f. Conversely, every expression of the form $u(x,y) + iv(x,y)$, in which u and v are real-valued functions of the two real variables x and y, can be viewed as a complex function of z because $x = \frac{1}{2}(z + \bar{z})$ and $y = \frac{1}{2i}(z - \bar{z})$.

<u>EXERCISE 13.9</u> Find u and v for $f(z) = z^2$. Then find $f(z)$ if $u(x,y) = 3x + y$ and $v(x,y) = 3y - x$.

Thus, in view of the above, the study of functions of a complex variable z can be viewed as the study of pairs of real functions of two real variables (linked, of course, by the special way in which complex numbers are combined).

Some elementary functions of z are as follows:

Exponential: $e^z = \exp z = e^x(\mathrm{Cos}\,y + i\,\mathrm{Sin}\,y)$

Trigonometric: $\mathrm{Cos}\,z = (e^{iz} + e^{-iz})/2$

$\mathrm{Sin}\,z = (e^{iz} - e^{iz})/2i$

These functions are defined for all values of z. Next, note that we can write $z = re^{i\theta}$, but then the expression $\log z = \ln r + i\theta$ does not define a (single-valued) function of z unless we restrict θ by $-\pi < \theta \leq \pi$, say. Similarly, the expression $f(z) = z^{1/n}$ $(n = 1,2,\ldots)$ does not, as-is, determine a (single-valued) function of z because for $z \neq 0$ there are n distinct values of w satisfying the equation $w^n = z$, namely

$r^{1/n}[\mathrm{Cos}(\theta + 2k\pi/n) + i\mathrm{Sin}(\theta + 2k\pi/n)]$ $(k = 0,1,\ldots,n-1)$. If, however, we agree to choose one of the solutions (or 'branches' as they are called), then we may define an n-th root function (in fact, n of them).

Henceforth we shall assume that the complex functions with which we deal are defined within some domain D of the complex plane; that is, within an open, connected[1] subset (possibly all) of the complex plane such as $\{z: |z| < R\}$ or $\{z: 0 < |z| < 1\}$.

The notions of existence of a limit, continuity and differentiability of a complex function have definitions parallel to those for functions of a real variable; in this sense there are few surprises when these notions are formally defined for complex functions. We begin first with the notion of a limit.

DEFINITION 13.1 (Existence of a Limit at a Point) A complex function $f(z)$ defined over D is said to possess the limit L at the point $z = a$ in D iff given any $\epsilon > 0$ there can be found a $\delta > 0$ (possibly depending upon ϵ and a) such that if $0 < |z-a| < \delta$ then $|f(z) - L| < \epsilon$. In such a case we write $\lim_{z\to a} f(z) = L$.

EXERCISE 13.10 (Uniqueness of Limit) Prove that the phrase 'the limit L' in Definition 13.1 is proper. That is, prove that if $\lim_{z\to a} f(z) = L_1$ and $\lim_{z\to a} f(z) = L_2$ then $L_1 = L_2$.

[1] A subset D of the complex plane is open iff the subset $\{(x,y): z = x + iy \in D\}$ is an open subset of E_2. D is connected iff every pair of points in D can be joined by a curve composed of straight-line segments lying completely within D.

<u>EXERCISE 13.11</u> Using the formal definition, prove that for any complex number $z = a$ we have $\lim_{z \to a} z^3 = a^3$. Display δ explicitly.

The following diagram helps to illustrate the idea of existence of a limit at a point.

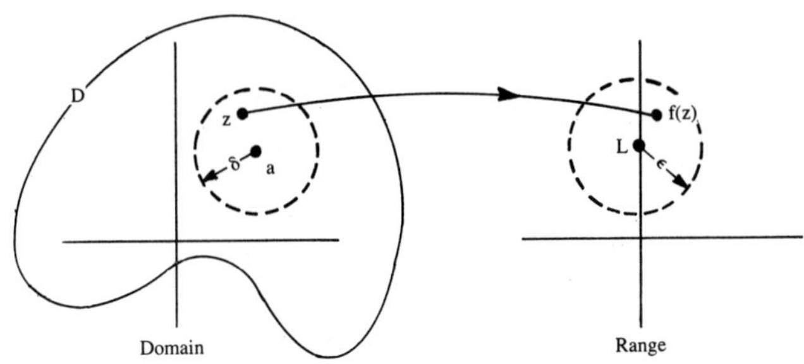

Figure 13-8

It is not difficult to prove that the definition of the limit in the complex case satisfies the same algebraic properties as that of the real case. For example, if $\lim_{z \to a} f(z) = L$ and $\lim_{z \to a} g(z) = M$ then $\lim_{z \to a} [f(z) \pm g(z)] = L \pm M$. We shall not pursue this point further.

In terms of this basic definition of a limit, we can now define what is meant by continuity and differentiability for complex functions. Again, the similarity with the real case is to be noted, although it will be seen later that, for example, for a complex function to possess a derivative is a more far-reaching property than for a real function to possess a derivative.

<u>DEFINITION 13.2</u> (Continuity at a Point) A complex function $f(z)$ defined over D is said to be continuous at a point $z = a$ within D iff given any $\epsilon > 0$ there can be found a $\delta > 0$ (possibly depending upon on ϵ and a) such that if $0 < |z-a| < \delta$ then $|f(z) - f(a)| < \epsilon$. Alternately, iff $\lim_{z \to a} f(z) = f(a)$.

<u>EXERCISE 13.12</u> (Combinations of Complex Functions) Let $f(z)$ and $g(z)$ be continuous at $z = a$. Prove that $\alpha f(z) + \beta g(z)$ (α, β complex constants), $f(z) \cdot g(z)$ and $f(z)/g(z)$ ($g(a) \neq 0$) are all continuous at $z = a$.

<u>EXERCISE 13.13</u> (Compositions of Continuous Functions) Let f and g be two functions which are continuous at each point of their respective domains. If the domain of g is contained within the range of f then the so-called composite function $g \circ f$ defined by: $g \circ f(z) = g(f(z))$ is continuous at each point of its domain. Prove the last statement.

<u>EXERCISE 13.14</u> Using the basic definition, prove that the function $e^z = e^x(\text{Cos}y + i\text{Sin}y)$ is continuous at $z = 0$, then apply the results of the previous Exercise to prove that $\text{Cos}z$ is continuous at $z = 1$ and $\text{Sin}z$ is continuous at $z = 0$.

<u>DEFINITION 13.3</u> (Differentiability at a Point) A complex function $f(z)$ defined over D is said to be differentiable at a point $z = a$ within D with derivative A iff given any $\epsilon > 0$ there can be found a $\delta > 0$

(possibly depending on ϵ and a) such that if

$0 < |z-a| < \delta$ then $\left| \frac{f(z)-f(a)}{z-a} - A \right| < \epsilon$. In such a case, the dervative A of f at $z = a$ is usually denoted $f'(a)$.

If a complex derivative exists at a certain point, its value is in general a complex number. Furthermore note that $f(z)$ possesses a derivative $f'(a)$ at $z = a$ iff the following is true:

$$\lim_{z \to a} \frac{f(z)-f(a)}{z-a} = f'(a) \quad .$$

<u>EXERCISE 13.15</u> (Differentiability Implies Continuity) Prove that if $f(z)$ is differentiable at a point $z = a$ within D then $f(z)$ is continuous there. Provide a counterexample to prove that the converse is false.

<u>EXERCISE 13.16</u> Are the functions $f(z) = |z|$ and $f(z) = \text{Re}(z)$ differentiable at certain points?

<u>EXERCISE 13.17</u> Consider the complex function defined by:

$$f(z) = \begin{cases} \dfrac{x^3 y(y-ix)}{x^6+y^2} & \text{for } z \neq 0 \\ 0 & \text{for } z = 0 \quad . \end{cases}$$

Prove that the difference quotient $\frac{f(z)-f(0)}{z-0}$ tends to zero as z approaches the complex origin along any straight line (or ray), but not as z approaches the complex origin in <u>any</u> manner. What conclusion may be drawn?

Determining differentiability properties of complex functions always by applying the basic definition of the derivative as a limit of a difference quotient would prove quite laborious. The following Theorem is a first step in providing alternate and simpler conditions for existence of the derivative. This result, known as the Cauchy-Riemann Equations, provides a simple necessary condition for a complex derivative to exist; it is based upon the behavior of the ordinary partial derivatives of the real and imaginary parts of the complex function involved. Later, this necessary condition will be augmented by a sufficient condition for existence of the complex derivative.

<u>EXERCISE 13.18</u> Let P be any path approaching the point $z = a$ of the complex plane. Suppose $f(z)$ possesses a limit, written $\lim_P f$, as z approaches a along P. Prove that if u and v are the real and imaginary parts of f, then $\lim_P u$ and $\lim_P v$ both exist and $\lim_P f = \lim_P u + i \cdot \lim_P v$.

<u>THEOREM 13.1</u> (Cauchy-Riemann Equations: Necessary Condition) Let $f(z)$ be a complex function with real and imaginary parts $u = u(x,y)$ and $v = v(x,y)$ respectively. A necessary condition that $f(z)$ be differentiable at a point $z = a$ within D is that the partial derivatives of u and v exist and satisfy the following equations at $z = a$: $u_x = v_y$ and $u_y = -v_x$.

PROOF. Let $a = x + iy$ and $z = x + t + iy$, where t is real. Then,

$$\frac{f(z)-f(a)}{z-a} = \frac{u(x+t,y) + iv(x+t,y) - u(x,y) - iv(x,y)}{t}.$$

Since $f'(a)$ is assumed to exist, the limit of the left side approaches this value as $t \to 0$. In view of the previous Exercise, both:

$$\lim_{t\to 0} \frac{u(x+t,y)-u(x,y)}{t} = u_x$$

and

$$\lim_{t\to 0} \frac{v(x+t,y)-v(x,y)}{t} = v_x$$

exist and $f'(a) = u_x + iv_x$. Similarly, if we take $z = x+i(y+t)$, where t is real, and proceed as above, we obtain the relation: $f'(a) = (u_y + iv_y)/i$. Upon equating the real and imaginary parts of the two relations we find that $u_x = v_y$ and $u_y = -v_x$. This completes the proof.

EXERCISE 13.19 (Cauchy-Riemann Condition Alone is Not Sufficient) Consider the complex function $f(z) = |xy|^{\frac{1}{2}}$; thus, $u = |xy|^{\frac{1}{2}}$ and $v \equiv 0$. Show that the Cauchy-Riemann Equations are satisfied at $z = 0$ yet $f(z)$ is not differentiable at $z = 0$.

EXERCISE 13.20 Show that the following functions satisfy the Cauchy-Riemann Equations at every point:
$f(z) = z^n$ $(n = 0,1,2,\ldots)$, $f(z) = e^z$, $f(z) = \text{Sin}z$.
Does it then follow that these functions are differentiable at every point?

<u>EXERCISE 13.21</u> Use the Cauchy-Riemann Theorem to prove that the following complex functions are not differentiable at any point: $f(z) = \bar{z}$, $f(z) = 2x + xy^2 i$, $f(z) = e^x$ (Cosy - iSiny), $f(z) = z - \bar{z}$.

<u>EXERCISE 13.22</u> (Cauchy-Riemann Equations in Polar Form) As usual, let $r = |z|$ and $\theta = \mathrm{Tan}^{-1} y/x$, whence $(z \neq 0)$, $x = r\mathrm{Cos}\ \theta$ and $y = r\mathrm{Sin}\ \theta$. Prove that if f is expressed in the Polar form, then the Cauchy-Riemann Equations become:

$$u_r = v_\theta / r \quad \text{and} \quad u_\theta / r = -v_r \quad .$$

Then, apply this result by demonstrating that the function $f(z) = z^n$ $(n = 0,1,2,\ldots)$ satisfies these equations.

As was proved in Exercise 13.19, for a function to satisfy the Cauchy-Riemann Equations at a point is, by itself, not sufficient to guarantee that the function is differentiable at that point. However, with some added conditions, we can obtain a partial converse of the Cauchy-Riemann Theorem that does provide sufficient conditions that a complex function possess a derivative at a point. The result is contained in the following Theorem.

THEOREM 13.2 (Sufficient Conditions for Differentiability) Let f be a complex function with real and imaginary parts u and v respectively. If:

(1) the partial derivatives of u and v exist in a neighborhood[1] of $z = a$, and

(2) the partial derivatives of u and v are continuous functions of x and y at $z = a$, and

(3) the Cauchy-Riemann Equations are satisfied at $z = a$, then f is differentiable at $z = a$, and at this point $f' = u_x + iv_x = v_y - iu_y$.

PROOF. Let $f(z) - f(a) = \Delta f \equiv \Delta u + i\Delta v$ and $z - a = \Delta z \equiv \Delta x + i\Delta y$. We must show that

$\lim_{\Delta z \to 0} \frac{\Delta f}{\Delta z} = u_x + iv_x = v_y - iu_y$ at $z = a$. Now, within a neighborhood of $z = a$ we have:

$\Delta u = u_x \Delta x + u_y \Delta y + \epsilon_1 \Delta x + \epsilon_2 \Delta y$, and

$\Delta v = v_x \Delta x + v_y \Delta y + \epsilon_1' \Delta x + \epsilon_2' \Delta y$, where

$\epsilon_1, \epsilon_2, \epsilon_1', \epsilon_2'$, tend to zero as $|\Delta x| + |\Delta y| \to 0$.

Thus we may write $\Delta f = u_x(\Delta x + i\Delta y) + iv_x(\Delta x + i\Delta y) + \rho$, where $\rho = \epsilon_1 \Delta x + \epsilon_2 \Delta y + i\epsilon_1' \Delta x + i\epsilon_2' \Delta y$, whence

$\frac{\Delta f}{\Delta z} = u_x + iv_x + \rho/\Delta z$ because $\Delta z = \Delta x + i\Delta y$. Now $\rho/\Delta z \to 0$ as $\Delta z \to 0$, thus we have that $\lim_{\Delta z \to 0} \frac{\Delta f}{\Delta z} = u_x + iv_x$. The remaining part of the equality follows easily, thus completing the proof.

[1]A neighborhood of $z = a$ is an open set containing $z = a$; for example, $\{z: |z-a| < \delta\}$.

EXERCISE 13.23 Review the proof of the preceding Theorem and establish where each of the three assumptions is needed.

EXERCISE 13.24 Prove that $f(z) = e^z$ is differentiable everywhere and $f'(z) = e^z$; prove that $f(z) = z^n$ $(n = 0, \pm 1, \ldots)$ is differentiable everywhere, except perhaps at $z = 0$, and $f'(z) = nz^{n-1}$.

A function $f(z)$ that possesses a derivative at each point of a domain D is said to be Regular over D. A function that is Regular over the entire complex plane, such as $f(z) = e^z$, is usually called an Entire function. The next Exercise stresses the algebraic similarity between the real and complex cases when dealing with derivatives of combinations of functions.

EXERCISE 13.25 (Algebra of Derivatives of Regular Functions) Let f and g be regular over D with derivatives f' and g' respectively. Then $af + bg$ (a, b complex constants), $f \cdot g$ and f/g $(g \neq 0)$ are regular over D with derivatives $af' + bg'$, $fg' + gf'$ and $(gf' - fg')/g^2$ respectively. Thus, for example, the ratio of two polynomials in z is regular at every complex point for which the demominator is non-zero. Prove the product rule only.

Important Note: All definitions and properties of Sections 1 through 5 that did not specifically depend upon order (i.e. largest, smallest) properties of real numbers, but rather only on magnitude (that is, absolute value), are automatically valid for complex numbers and functions when we replace absolute value by modulus. Thus, for example, sequences and series of complex numbers and functions and their convergence properties have, effectively, already been developed.

Thus, without any additional comments, we will refer to notions such as convergent (absolutely convergent) sequences and series of complex numbers and functions, and uniform convergence of sequences and series (including Power Series) of complex functions. Most of the Theorems proved in Sections 3, 4 and 5 remain valid - almost as stated - for the complex case; most proofs require minor modifications, if any.

As an illustration of the preceding remarks, consider Complex Power Series. A series of the form $S(z) = \sum_0^\infty a_n (z-a)^n$ with complex coefficients is termed a complex Power Series about $z = a$. The series converges (absolutely) inside its <u>circle of convergence</u> $|z-a| < R$, diverges outside its circle of convergence $|z-a| > R$ (with no conclusion if $|z-a| = R$), where $1/R = \limsup \sqrt[n]{|a_n|}$. This is to be compared with Theorem 5.1, which considered the radius of convergence of Real Power Series. As in the real case, for the complex case R may equal zero, a positive number, or $+\infty$. The complex series converges at least for $z = a$ and convergence is uniform for $|z-a| \le R' < R$.

A result of great importance is that the sum function of a Complex Power Series is a regular function within its circle of convergence. This is contained in the following Theorem.

<u>THEOREM 13.3</u> (Regularity of Sum Function) The sum function $S(z) = \sum_0^\infty a_n (z-a)^n$ of a Complex Power Series is a regular function within its circle of convergence and, therein, $S'(z) = \sum_0^\infty na_n (z-a)^{n-1}$.

PROOF. Suppose that $R > 0$. From preceding results we know that $g(z) = \sum_0^\infty na_n (z-a)^{n-1}$ is also (absolutely) convergent for $|z-a| < R$. Now we must show that within $|z-a| < R$, $S'(z)$ exists and equals $g(z)$. Without loss of generality we assume $a = 0$. Then, it suffices to prove that:

$$\left| \frac{S(z+w)-S(z)}{w} - g(z) \right| \to 0$$

as the complex number w approaches zero, where z is any fixed complex number within $|z| < R$. Choose p such that $|z| + |w| < p < R$. Then,

$$\left| \frac{S(z+w)-S(z)}{w} - g(z) \right| = \left| \sum_0^\infty a_n \left\{ \frac{(z+w)^n - z^n}{w} - nz^{n-1} \right\} \right|$$

$$\leq \sum_0^\infty \frac{c}{p^n} \left| \frac{(z+w)^n - z^n}{w} - nz^{n-1} \right| \quad \text{(for some constant } c\text{)}$$

$$= \sum_0^\infty \frac{c}{p^n} \left| \sum_2^n \binom{n}{r} z^{n-r} w^{r-1} \right|$$

$$\leq \sum_0^\infty \frac{c}{p^n} \sum_2^n \binom{n}{r} |z|^{n-r} |w|^{r-1}$$

$$= \sum_0^\infty \frac{c}{p^n} \left\{ \frac{(|z|+|w|)^n - |z|^n}{|w|} - n|z|^{n-1} \right\}$$

$$= c \left\{ \frac{p}{(p-|z|-|w|)(p-|z|)} - \frac{p}{(p-|z|)^2} \right\}$$

$$= c \left\{ \frac{p|w|}{(p-|z|-|w|)(p-|z|)^2} \right\} \to 0 \quad \text{as} \quad w \to 0,$$

and the proof is completed.

EXERCISE 13.26 Verify the algebraic steps in the preceding Theorem. The Complex Geometric Series is used.

Before pursuing further properties of Regular functions, we shall first define the complex integral of a continuous complex function. The reason for this particular approach will become clearer later on.

DEFINITION 13.4 A contour in the complex plane is a continuous curve with a continuously turning tangent (except perhaps at a finite number of points). It can be expressed parametrically as follows: $\{z: z = z(t) = x(t) + iy(t),\ t_1 \le t \le t_2\}$, where $x'(t)$ and $y'(t)$ are continuous functions of the real variable t (except perhaps at a finite number of points).

A simple closed contour (s.c.c.) is a contour which does not cross itself and is closed, that is, $x(t_1) = x(t_2)$ and $y(t_1) = y(t_2)$. In addition, a closed convex contour is a s.c.c. whose interior forms a convex set (see Section 18).

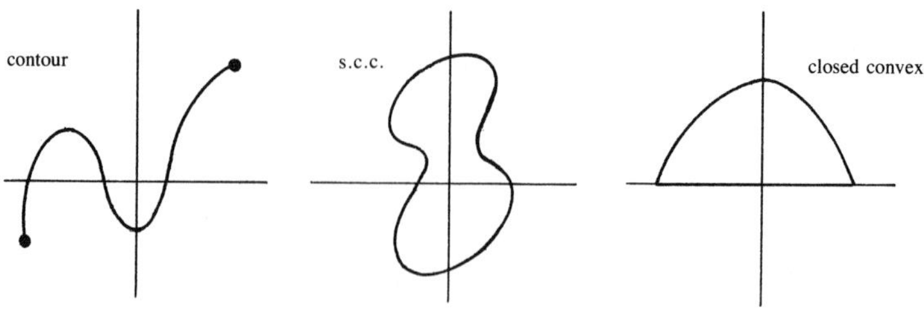

Figure 13-9

Let $f(z)$ be a continuous (but not necessarily Regular) complex function of z along the contour $\mathcal{C}$ determined by the parametric equations $x = x(t), y = y(t)$: $T_0 \le t \le T_1$. Without loss of generality we assume that $x'(t)$ and $y'(t)$ are everywhere continuous within $T_0 \le t \le T_1$.

Let $P_n = \{T_0 = t_0 < t_1 < t_2 <\ldots< t_n = T_1\}$ be a partition of the interval $T_0 \le t \le T_1$ with $|P_n| = \max\limits_{j=1,\ldots,n} |t_j - t_{j-1}|$; set $z_i = z(t_i) = x(t_i) + iy(t_i) = x_i + iy_i$ for $i = 1,2,\ldots,n$. The following diagram illustrates this notation and the notation to be used in the following discussion.

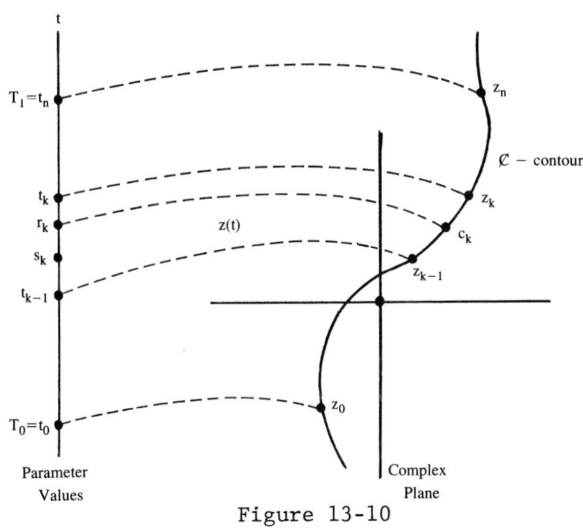

Figure 13-10

We begin by forming the following approximating sum corresponding to the partition P_n: $J_n = \sum_1^n f(c_k)(z_k - z_{k-1})$, where $c_k = z(r_k)$, and r_k is any point satisfying $t_{k-1} \le r_k \le t_k$ $(k = 1,2,\ldots,n)$. Writing $f(z) = u(x,y) + iv(x,y)$, we have: $f(c_k) = u[x(r_k), y(r_k)] + iv[x(r_k), y(r_k)]$, or briefly, $f(c_k) = u(r_k) + iv(r_k)$. In these terms,

$$J_n = \sum_1^n [u(r_k) + iv(r_k)]\,[(x_k - x_{k-1}) + i(y_k - y_{k-1})] .$$

By the Mean Value Theorem for derivatives, we may write $x_k - x_{k-1} = x(t_k) - x(t_{k-1}) = x'(s_k)(t_k - t_{k-1})$ where s_k is some point satisfying $t_{k-1} \le s_k \le t_k$ $(k = 1,2,\ldots,n)$. Recall now that a function, such as $x'(t)$, that is continuous over a closed, bounded interval, such as $T_0 \le t \le T_1$, is uniformly continuous over that interval; thus, given $\epsilon > 0$ there exists a $\delta > 0$ such that $x_k - x_{k-1} = [x'(r_k) + \epsilon_k][(t_k - t_{k-1})]$, where $|\epsilon_k| < \epsilon$, $k = 1,2,\ldots,n$ provided $|P_n| < \delta$.

Thus, for example, $\sum_1^n u(r_k)(x_k - x_{k-1}) = \sum_1^n u(r_k)x'(r_k)(t_k - t_{k-1}) + \sum_1^n \epsilon_k u(r_k)(t_k - t_{k-1}) = A_n + B_n$, say. Now as $n \to \infty$ and $|P_n| \to 0$ we have, by definition of the Riemann Integral, $A_n \to \int_{T_0}^{T_1} u(t)x'(t)dt$, whereas $|B_n| \le \epsilon\, |T_1 - T_0|\, M$, where $M = \max_{T_0 \le t \le T_1} u(t)$.

Since ϵ is arbitrary, we see that $B_n \to 0$ as $n \to \infty$.

Treating the other (three) parts of J_n in a similar manner we can see that as $n \to \infty$ and $|P_n| \to 0$,

$$J_n \to \int_{T_0}^{T_1} [u(t) + iv(t)][x'(t) + iy'(t)]\, dt,$$

which is defined to be:

$$\int_{\mathcal{C}} f(z)dz \quad .$$

In general, the value of a complex integral will be a complex number, and may depend upon the specific contour joining two points, and the direction in which the contour is traversed.

EXAMPLE 13.1 Evaluate $\int_{\not{C}} z dz$ where $\not{C}$ is the contour defined by the parametric equations: $x(t) = t^2$, $y(t) = t$, $0 \le t \le 1$. Here, $f(z) = x + iy$ so that $\int_{\not{C}} z dz = \int_0^1 (t^2 + it)(2t + i)dt = i$. The contour is a segment of a parabola.

EXERCISE 13.27 Evalute $\int_{\not{C}} z^2 dz$ for the contour given above.

EXERCISE 13.28 (Dependence of Integral Upon Contour) Evaluate $\int_{\not{C}} f(z)dz$ where $f(z) = y - x - 3ix^2$ and: (i) in the first case $\not{C}$ is the straight line segment from $z = 0$ to $z = 1+i$, (ii) in the second case $\not{C}$ is formed of two straight line segments: one from $z = 0$ to $z = i$ and the next from $z = i$ to $z = 1+i$. The endpoints of both contours are the same; are the values of the two integrals the same?

The complex integral as defined enjoys many of the same algebraic properties as the ordinary Riemann Integral of a real function of a real variable. Some of these properties are established in the following Note and Exercise.

NOTE (Algebraic Properties of the Complex Integral) Since the very definition of $\int_{\not{C}} f(z)dz$ is based upon the ordinary Riemann Integral of a real function of a real variable, it is not surprising to find that both types of integrals enjoy the same basic algebraic properties.

For example, subject to existence,

$$\int_{\mathcal{C}} af(z)dz = a\int_{\mathcal{C}} f(z)dz \qquad \text{(a complex)}$$

and

$$\int_{\mathcal{C}} \{f(z) \pm g(z)\}dz = \int_{\mathcal{C}} f(z)dz \pm \int_{\mathcal{C}} g(z)dz \quad .$$

Furthermore, since the conditions required for interchange of the operations of $\lim_{n\to\infty}$, $\frac{d}{dz}$ and $\sum_0^\infty$ with the integral sign $\int_{\mathcal{C}}$ involve only magnitude and not order (e.g. the condition of uniform convergence), it is not surprising to find that most of the results of previous Sections dealing with this topic find almost word-for-word parallels for the complex integral. We shall not specifically state or prove, but rather assume and use such results later on.

Finally, if a contour $\mathcal{C}$ is defined by the parametric equations: $x = x(t)$, $y = y(t)$, $T_0 \le t \le T_1$, then by "$-\mathcal{C}$" we mean that contour traversed in reverse direction. Thus, $-\mathcal{C}$ is defined by the parametric equations: $x = x(T_1) - x(t) + x(T_0)$, $y = y(T_1) - y(t) + y(T_0)$, $T_0 \le t \le T_1$. Then,

$$\int_{-\mathcal{C}} f(z)dz = -\int_{\mathcal{C}} f(z)dz \quad .$$

EXERCISE 13.29 Prove the last statement.

The first general property of the complex integral is contained in the following Theorem; the result provides a bound on the modulus of the value of the integral.

<u>THEOREM 13.4</u> (Bound) If $f(z)$ is continuous and $|f(z)| \le M$ on the contour $\mathcal{C}$: $x = x(t)$, $y = y(t)$, $T_0 \le t \le T_1$, then $|\int_{\mathcal{C}} f(z)dz| \le M \cdot L(\mathcal{C})$, where $L(\mathcal{C})$ is the arc-length of the contour defined by:

$$L(\mathcal{C}) = \int_{T_0}^{T_1} \sqrt{x'(t)^2 + y'(t)^2} \, dt \quad .$$

PROOF. For a partition P_n of $T_0 \le t \le T_1$ consider the approximating sum J_n, and note that, given $\epsilon > 0$ we can write $|J_n| = |\sum_1^n f(c_k)(z_k - z_{k-1})| \le M \cdot \sum_1^n |z_k - z_{k-1}| = M \cdot \sum_1^n |x_k - x_{k-1} + i(y_k - y_{k-1})| = M \cdot \Sigma\, |t_k - t_{k-1}| \cdot |x'(t_{k-1}) + iy'(t_{k-1}) + \epsilon_k|$, where $|\epsilon_k| < \epsilon$ $(k = 1,2,\ldots,n)$ if $|P_n| < \delta$. The latter result follows from uniform continuity of both $x'(t)$ and $y'(t)$ over $T_0 \le t \le T_1$. Thus,

$$|J_n| \le M \sum_1^n |t_k - t_{k-1}|\,|x'(t_{k-1}) + iy'(t_{k-1})| +$$

$$M\epsilon \sum_1^n |t_k - t_{k-1}| = M \cdot \sum_1^n (t_k - t_{k-1}) \cdot \sqrt{x'(t_{k-1})^2\; y'(t_{k-1})^2}$$

$+ M\epsilon(T_1 - T_0) = A_n + B_n$, say. Now by definition of the Riemann integral, $A_n \to M \cdot L(\mathcal{C})$ as $n \to \infty$ and $|P_n| \to 0$, whereas since ϵ is arbitrary, $B_n \to 0$ as $\epsilon \to 0$. This completes the proof.

<u>EXERCISE 13.30</u> Upon selecting a suitable M, use the preceding Theorem to find an upper bound on the modulus of the integrals in Exercise 13.28.

EXAMPLE 13.2 (Evaluating Certain Integrals Over S.C.C.'s)
Consider a countour $\mathcal{C}$ originating at $z = a$ and terminating at $z = b$ in the complex plane. We wish to consider both $\int_{\mathcal{C}} 1\,dz$ and $\int_{\mathcal{C}} z\,dz$. For any partition P_n of the parameter interval, the approximating sum J_n for $\int_{\mathcal{C}} 1\,dz$ is:

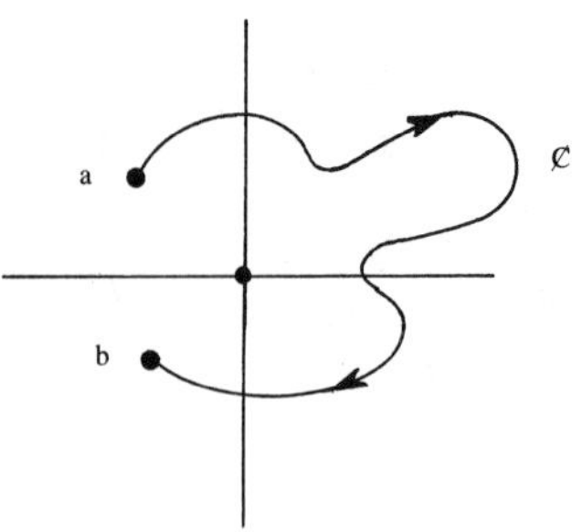

Figure 13-11

$J_n = \sum_1^n (z_k - z_{k-1}) \equiv b - a$, whence $\int_{\mathcal{C}} 1\,dz = b - a$. Consider now the second integral. Choosing $c_k = z_{k-1}$, we have as approximating sum: $J_n' = \sum_1^n z_{k-1}(z_k - z_{k-1})$, which tends to $\int_{\mathcal{C}} z\,dz$, whereas choosing $c_k = z_k$, we have as approximating sum: $J_n'' = \sum_1^n z_k(z_k - z_{k-1})$, which also tends to $\int_{\mathcal{C}} z\,dz$ as $n \to \infty$ and $|P_n| \to 0$. Therefore $\frac{1}{2}(J_n' + J_n'') \to \int_{\mathcal{C}} z\,dz$. But the expression on the left has value $\sum_1^n (z_k^2 - z_{k-1}^2) = \frac{1}{2}(b^2 - a^2)$, which must then be the value of the integral. Note that if $\mathcal{C}$ is a simple closed contour (s.c.c.), in which case $a = b$, the value of both integrals is zero.

The preceding Example demonstrates a special case of a far-reaching property of the integral of a Regular function over a s.c.c. The property (two versions of it) is contained in the following Theorem.

<u>THEOREM 13.5</u> (Cauchy's Theorem: Strong Form) If $f(z)$ is Regular at every point interior to a s.c.c. $\mathcal{C}$ and continuous at every point on $\mathcal{C}$ then $\int_{\mathcal{C}} f(z)dz = 0$.[1]

<u>THEOREM 13.6</u> (Cauchy's Theorem: Weak Form) If $f(z)$ is Regular at every point within and on a s.c.c. $\mathcal{C}$ then $\int_{\mathcal{C}} f(z)dz = 0$.

We shall prove only the weak version of Cauchy's Theorem. It is sufficient for many Applications.

Before proving Cauchy's Theorem, we need the following Lemma of existence due to Goursat.

<u>LEMMA 13.1</u> (Goursat's Lemma) Let $f(z)$ be Regular at all points z in the closed region R consisting of the points interior to and on a s.c.c. $\mathcal{C}$. Given any $\epsilon > 0$ it is always possible to subdivide R into a finite number N of squares and partial squares - whose boundaries are denoted by $\mathcal{C}_j$, $j = 1,2,\ldots,N$ - such that a point z_j exists within or on each $\mathcal{C}_j$ for which the inequality:

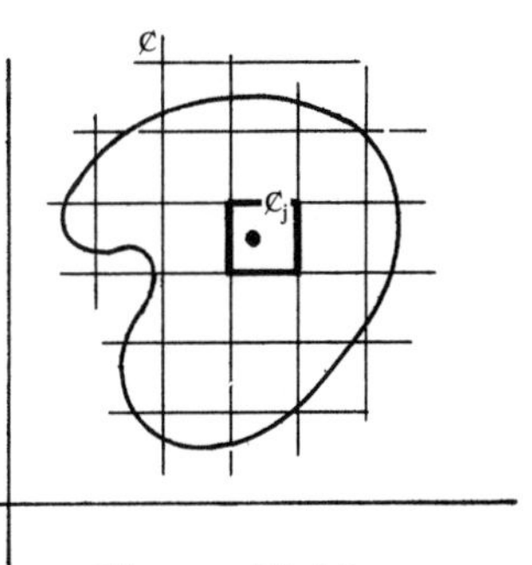

Figure 13-12

[1]By convention we shall henceforth assume that all s.c.c.'s are so parametrized that they are traversed in a counter-clockwise fashion.

$|f(z) - f(z_j) - (z-z_j)f'(z_j)| < \epsilon |z-z_j|$ is satisfied at every point $z \neq z_j$ within or on $\mathcal{C}_j$, $j = 1,2,\ldots,N$.

PROOF. Let the region R be covered with a set of equal squares formed by drawing equally-spaced lines parallel to Real and Imaginary axes. The portion of any square lying outside of R is to be removed, leaving R subdivided into squares and 'partial' squares. See above.

Now, suppose to the contrary that for some $\epsilon > 0$ there is at least one of these regions in which no point z_j exists so that the stated inequality is satisfied. If this region is a square let it be divided into four equal squares; if it is a 'partial' square let the whole original square be so subdivided and let the portions lying outside of R be discarded. If any one of the smaller regions so obtained has no point z_j so that the stated inequality is satisfied, let that region be subdivided as above, etc.

After a finite number of steps of subdividing every region that requires it, we may arrive at a subdivision such that the stated inequality is satisfied for some z_j in each subregion. In this case the Lemma is true. However, suppose to the contrary that points z_j do not exist such that the stated inequality is satisfied after having subdivided some one of the original subregions any finite number of times.

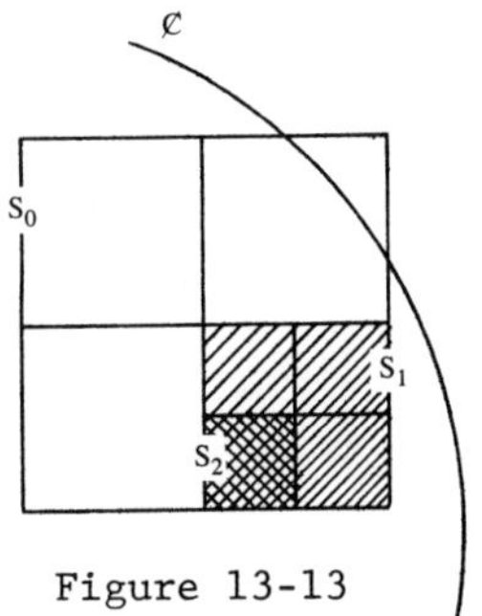

Figure 13-13

Let S_0 denote the original subregion if it is an entire square, and if it is a 'partial' square let S_0 denote the whole square before deletion. Accordingly, after the square S_0 is subdivided into four equal squares, at least one of the four smaller squares, say S_1, contains point of R but no appropriate points z_j after any finite number of subdivisions. See above. After S_1 is subdivided as above, at least one of the four smaller squares, say S_2, fails to qualify, etc. If at any step more than one of the smaller squares could be chosen in this process, let it be chosen as that square lowest and farthest to the left, thus making the choice unique.

Each square S_k in the infinite sequence $S_0, S_1, \ldots, S_{k-1}, S_k, \ldots$ is contained in its predecessor S_{k-1}, and its side is half as long. Also, each square S_k contains points of R. Thus, by Exercise 3.34, there exists a point z_0 contained in (common to) each square in the infinite sequence. Also, each circular neighborhood $|z-z_0| < \delta$ of z_0 contains a square of this sequence. This is clearly so whenever the length of the diagonal of a square is less than δ. Consequently, each neighborhood of z_0 contains an infinite number of points of R and, because R is a closed set, the limit point z_0 of R must belong to R.

Now because z_0 belongs to R, $f(z)$ is Regular at $z = z_0$ and so, from the definition of the derivative, given any $\epsilon > 0$ there exists a $\delta > 0$ such that $|f(z) - f(z_0) - (z-z_0)f'(z_0)| < \epsilon\ |z-z_0|$ whenever $|z-z_0| < \delta$. But the circular neighborhood $|z-z_0| < \delta$ of z_0 involved in this condition contains the entire square S_K whenever the index K is large enough so that the diagonal of the square does not exceed δ. Consequently the point z_0 does serve as the point z_j so that the stated inequality is satisfied in the subregion consisting of S_K or the part of R in S_K. Thus, contrary to our hypothesis, it was not necessary to subdivide S_K in the first place; with this contradiction the proof is completed.

We are now in a position to prove the weak form of Cauchy's Theorem.

PROOF. (Cauchy's Theorem: weak form)

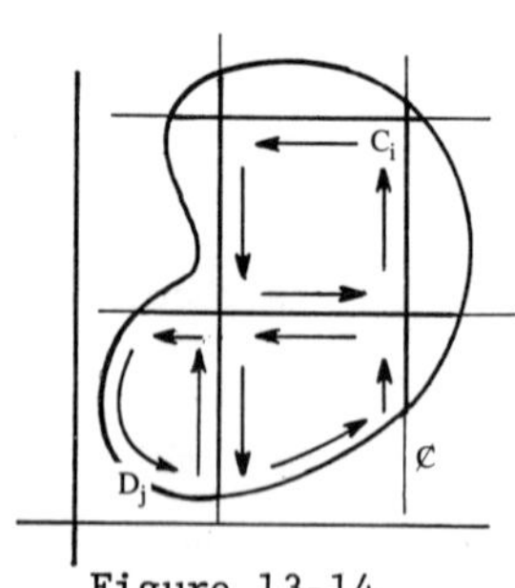

Figure 13-14

First subdivide R into squares and 'partial' squares as in Goursat's Lemma. Denote the complete squares by $C_1,\ldots,C_n$ and the 'partial' squares by $D_1,\ldots,D_m$ $(m+n = N)$. Next note that:

$$\int_{₵} f(z)dz = \sum_1^n \int_{C_i} f(z)dz + \sum_1^m \int_{D_j} f(z)dz\ .$$

Now by Goursat's Lemma we may write:

$f(z) = f(z_i) + (z-z_i)f'(z_i) + g_i(z)$, where $g_i(z)$ is some function of z such that $|g_i(z)| < \epsilon\,|z-z_i|$ for all z within and on C_i -- where in addition, z_i is some fixed point within or on C_i. This is true for $i = 1,2,\ldots,n$ and so:

$$\int_{C_i} f(z)dz = \int_{C_i} f(z_i)dz + \int_{C_i} (z-z_i)f'(z_i)dz + \int_{C_i} g_i(z)dz,$$

which equals $\int_{C_i} g_i(z)dz$, because in view of Example 13.2, the first two integrals vanish. Now $|\int_{C_i} g_i(z)dz| \le \epsilon\sqrt{2}\, L_i(4L_i)$, where L_i is the side-length of C_i. Then we have:

$$\left|\sum_1^n \int_{C_i} f(z)dz\right| \le \sum_1^n\left|\int_{C_i} g_i(z)dz\right| \le \sum_1^n \epsilon\sqrt{2}\; 4L_i^2 \le 4\;\epsilon\sqrt{2}\; d^2,$$

where d is the side-length of some square that completely encloses $\mathcal{C}$.

The other portion of the integral follows similarly because $\int_{D_j} f(z)dz = \int_{D_j} g_j(z)dz$ and $|\int_{D_j} g_j(z)dz| \le \epsilon\sqrt{2}\, L_j'(4L_j' + s_j)$, where L_j' is the side-length of the complete square associated with the 'partial' square D_j, and s_j is the arc-length of $\mathcal{C}$ lying therein. Now we have that:

$$\left|\sum_1^m \int_{D_j} f(z)dz\right| \le \sum_1^m \epsilon\sqrt{2}\; 4L_j'^2 + \sum_1^m \epsilon\sqrt{2}\; L_j'\, s_j \le 4\;\epsilon\sqrt{2}\; d^2 + \epsilon\sqrt{2}\; d\; L(\mathcal{C}).$$

Combining the two results:

$$\left|\int_{\mathcal{C}} f(z)dz\right| \leq \epsilon\sqrt{2}\ (8d^2 + dL(\mathcal{C})).$$

Because ϵ is arbitrarily small, and d and $L(\mathcal{C})$ are fixed, it follows that:

$$\int_{\mathcal{C}} f(z)dz = 0,$$

and the proof is completed.

Several useful results concerning the integral of Regular functions now follow immediately from Cauchy's Theorem. They are included in the results below.

THEOREM 13.7 (Independence of Path)
Let $\mathcal{C}_1$ and $\mathcal{C}_2$ be two contours from $z = a$ to $z = b$ in the complex plane. If $f(z)$ is Regular within and on the s.c.c. formed by these two contours, then:

$$\int_{\mathcal{C}_1} f(z)dz = \int_{\mathcal{C}_2} f(z)dz .$$

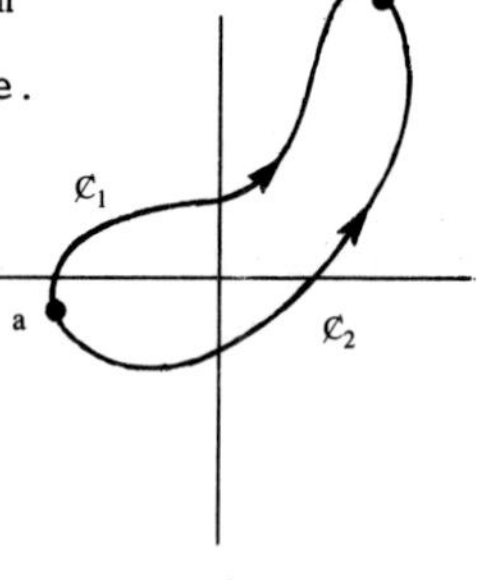

Figure 13-15

PROOF. EXERCISE 13.31

Note that if the function $f(z)$ above does not satisfy the stated regularity condition, then the values of the two integrals may very well be different. This point was illustrated in Exercise 13.28.

THEOREM 13.8 (Future Foundation of Residue Theory)

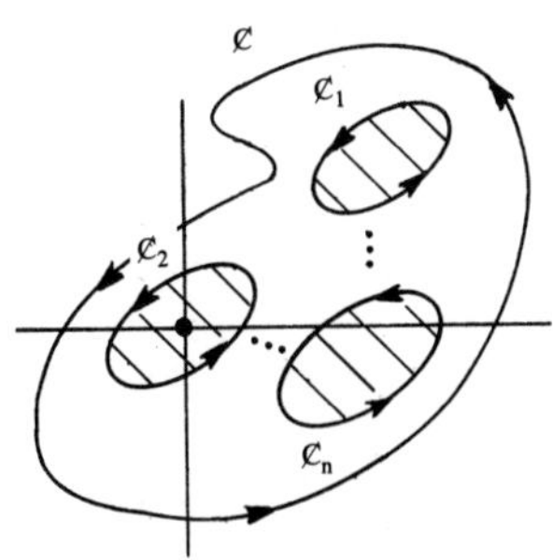

Figure 13-16

Let $₵_1, ₵_2, \ldots, ₵_n$ and $₵$ be s.c.c.'s which together with their boundaries form closed regions $D_1, D_2, \ldots, D_n$ and D such that $D_1, D_2, \ldots, D_n$ are all pair-wise disjoint and D contains $\cup D_i$. If $f(z)$ is Regular within $D \setminus \cup D_i$, as well as on its boundary, then:

$$\int_{₵} f(z)dz = \sum_{1}^{n} \int_{₵_i} f(z)dz .$$

PROOF. EXERCISE 13.32

EXAMPLE 13.3 (A Special Contour Integral) Evaluate $\int_{₵} (z-a)^n dz$, where $₵$ is any s.c.c. enclosing the point $z = a$ and n is an integer. Now by the previous Theorem we may replace $₵$ by $₵'$, where the latter is the s.c.c. $|z-a| = 1$. Writing a as $a = b + ic$, we can express $₵'$ parametrically as: $x(t) = b + \text{Cos}t$, $y(t) = c + \text{Sin}t$, $0 \le t \le 2\pi$. Thus, in accordance with the definition of the complex integral, we have

$$\int_{₵} (z-a)^n dz = \int_0^{2\pi} e^{int}(ie^{it})dt = i \int_0^{2\pi} e^{i(n+1)t}dt =$$

$$i \int_0^{2\pi} \{\text{Cos}(n+1)t + i\text{Sin}(n+1)t\}\, dt = \begin{cases} 0, & n \ne -1 \\ 2\pi i, & n = -1 \end{cases} .$$

This particular integral will be important later in studying Residue Theory.

<u>EXAMPLE 13.4</u> Evaluate $\int_{\not{C}} dz/(2z-1)$ where $\not{C} = \{z: |z| = 1\}$. If $\not{C}'$ is the contour $\not{C}' = \{z: |z-\frac{1}{2}| = r < \frac{1}{2}\}$, then from the previous Theorem, the integrals over $\not{C}$ and $\not{C}'$ are identical in value. From the preceding Example we have: $\int_{\not{C}'} dz/(2z-1) = \frac{1}{2} \int_{\not{C}'} dz/(z-\frac{1}{2}) = \frac{1}{2}(2\pi i) = \pi i$.

<u>EXERCISE 13.33</u> Apply the previous Theorem and Examples to evaluate $\int_{\not{C}} [(z-2)/z]dz$ where (i) in the first case $\not{C}$ is the unit circle centered at the complex origin and (ii) in the second case $\not{C}$ is the unit circle centered at $z = 2i$.

<u>EXAMPLE 13.5</u> Evaluate $\int_{\not{C}} [(z-2)/(z^2-3z)]dz$ where $\not{C}$ is the elliptical contour $\not{C} = \{z: |z+1|+|z-1| = 4\}$. Now, using the standard partial fraction expansion, $(z-2)/(z^2-3z) = 2/3z+1/3(z-3)$, whence the integral equals $2/3 \int_{\not{C}} dz/z + 1/3 \int_{\not{C}} dz/(z-3)$, which equals $(2/3)(2\pi i) + (1/3)(0) = 4\pi i/3$. If, however, the elliptical contour were $\{z: |z+1|+|z-1| = 8\}$ instead, then the value of the integral would have been $(2/3)(2\pi i) + (1/3)(2\pi i) = 2\pi i$.

A point at which a complex function is not Regular (differentiable) is termed a point of singularity (or simply a singularity) of the function. In view of the preceding results, it is seen that the value of the integral of a complex function over a s.c.c. depends upon the (possible) singularities of the function within the contour. If no singularities exist, then the integral is clearly zero.

EXERCISE 13.34 Evaluate $\int_{\mathcal{C}} f(z)dz$ where $\mathcal{C} = \{z: |z| = 2\}$, and (i) $f(z) = 1/(4z^2 + 1)$, (ii) $f(z) = z^2/(z-3)$, (iii) $f(z) = 1/(z^2 + 2z + 2)$, (iv) $f(z) = ze^{-z}$.

We now begin to explore the rather powerful properties of Regular functions. Our first result asserts that if $f(z)$ is Regular within and on a s.c.c. $\mathcal{C}$ then the value of $f(z)$ at any point interior to $\mathcal{C}$ can be determined from the values of $f(z)$ on $\mathcal{C}$ alone. This result is known as Cauchy's Integral Formula.

THEOREM 13.9 (Cauchy's Integral Formula)

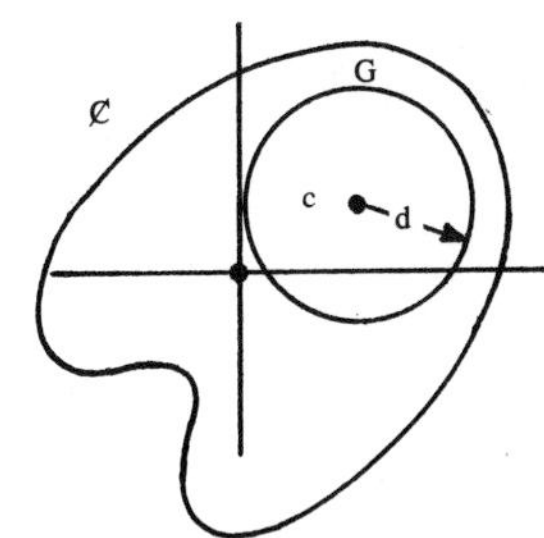

If $f(z)$ is Regular within and on the s.c.c. $\mathcal{C}$ and c is any point inside $\mathcal{C}$ then:

$$f(c) = \frac{1}{2\pi i} \int_{\mathcal{C}} \frac{f(z)}{z-c} dz .$$

PROOF. Surround c by a circular s.c.c. G centered at c and having radius d, so that for prescribed $\epsilon > 0$ we have $|f(z) - f(c)| < \epsilon$ whenever $|z-c| < d$. From previous results the stated integral over $\mathcal{C}$ may be replaced by the integral over G. Now since $f(z) = f(c) + f(z) - f(c)$ we may write:

the integral as:

$$\frac{1}{2\pi i}\int_{G}\frac{f(z)}{z-c}\,dz = \frac{1}{2\pi i}\int_{G}\frac{f(c)}{z-c}\,dz + \frac{1}{2\pi i}\int_{G}\frac{f(z)-f(c)}{z-c}\,dz = A + B,$$

say. But clearly $A = f(c)$, whereas $|B| \le |\frac{1}{2\pi i}|\ \frac{\epsilon 2\pi d}{d} = \epsilon$, so that, since ϵ is arbitrarily small, and $|\frac{1}{2\pi i}\int_{\mathcal{C}}\frac{f(z)}{z-c}\,dz - f(c)| \le \epsilon$, the Theorem is established.

Thus, according to the preceding Theorem, we can theoretically find $f(c)$ at any interior point c knowing only the values of $f(z)$ on the contour $\mathcal{C}$.

EXAMPLE 13.6 (Application) Evaluate $\int_{\mathcal{C}}[z/(9-z^2)(z+i)]dz$ where $\mathcal{C}$ is the s.c.c. given by $\{z: |z| = 2\}$. This could be done using the technique of Example 13.5. However, consider this alternate approach employing Cauchy's Formula. Let $g(z) = z/(9-z^2)$; since g is Regular within and on $\mathcal{C}$, it follows that for any point c within $\mathcal{C}$ we have:

$$g(c) = \frac{1}{2\pi i}\int_{\mathcal{C}}\frac{g(z)}{z-c}\,dz\ .$$

In particular this is true for $c = -i$, whence:

$$g(-i) = \frac{1}{2\pi i}\int_{\mathcal{C}}\frac{g(z)}{z+i}\,dz = \frac{1}{2\pi i}\int_{\mathcal{C}}\frac{z}{(9-z^2)(z+i)}\,dz\ .$$

Clearly from direct evaluation we have that $g(-i) = -i/10$, thus the value of the original integral is $+\pi/5$.

The following result is an almost immediate Corollary of the previous Theorem.

<u>THEOREM 13.10</u> If $f(z)$ and $g(z)$ are both Regular within and on the s.c.c. $\mathcal{C}$ and $f(z) = g(z)$ at all points on $\mathcal{C}$, then $f(z) \equiv g(z)$ within $\mathcal{C}$.

PROOF. <u>EXERCISE 13.35</u>

The next general result concerning Regular functions is that a Regular function possesses derivatives of <u>all</u> orders, that is, existence of the first derivative implies existence of derivatives of all orders. This is certainly a strong result.

<u>THEOREM 13.11</u> (Representation Theorem: Derivatives) If $f(z)$ is Regular within and on the s.c.c. $\mathcal{C}$, then derivatives of all orders of $f(z)$ exist at every point c inside $\mathcal{C}$ and:

$$f^{(n)}(c) = \frac{n!}{2\pi i} \int_{\mathcal{C}} \frac{f(z)}{(z-c)^{n+1}}\, dz$$

for $n = 1,2,\dots$.

In order to prove the preceding Representation Theorem, we shall need the following Lemma.

<u>LEMMA 13.2</u> Let N be any integer ≥ 2 and suppose $|z| \leq B < \infty$. Then, given any $\epsilon > 0$ there can be found a $\delta > 0$ such that:

$$\left| \frac{z^N-(z-h)^N}{h} - N(z-h)^{N-1} \right| < \epsilon$$

for all complex numbers h such that $|h| < \delta$.

PROOF. The quantity within the modulus sign above can be expressed as a polynomial of degree N-2 in z having h as a factor, because:

$$\left| \frac{z^N - (z-h)^N}{h} - N(z-h)^{N-1} \right| =$$

$$\left| \sum_2^N \left[N \binom{N-1}{j-1} - \binom{N}{j} \right] z^{N-j} (-h)^{j-1} \right| =$$

$$\left| \sum_2^N a_j z^{N-j} (-h)^{j-1} \right| \le \sum_2^N |a_j| B^{N-j} |h|^{j-1} \le \frac{aB^N |h|}{1-|h|} ,$$

where $a = \max|a_j|$ and B (≥ 1) was given. Clearly the upper bound can be made $< \epsilon$ by choosing $|h| < \delta$ for δ sufficiently small. This proves the Lemma.

We are now in a position to prove the so-called Representation Theorem for derivatives of Regular functions.

PROOF. (Representation Theorem: Derivatives)
First let

$$g_n(c) = \frac{n!}{2\pi i} \int_{\mathbb{C}} \frac{f(z)}{(z-c)^{n+1}} dz .$$

If we can establish that:

$$\left| \frac{g_n(c+h) - g_n(c)}{h} - g_{n+1}(c) \right| \to 0 \quad \text{as} \quad h \to 0$$

then we have proved that:

$$g_n'(c) = g_{n+1}(c) \quad \text{for} \quad n = 1,2,\ldots$$

and we are done. Accordingly, note that:

$$\left| \frac{g_n(c+h)-g_n(c)}{h} - g_{n+1}(c) \right| =$$

$$\frac{n!}{2\pi} \left| \int_{\mathcal{C}} f(z) \left[\frac{1}{h(z-c-h)^{n+1}} - \frac{1}{h(z-c)^{n+1}} - \frac{(n+1)}{(z-c)^{n+2}} \right] dz \right| =$$

$$\frac{n!}{2\pi} \left| \int_{\mathcal{C}} \frac{f(z)}{(z-c-h)^{n+1}(z-c)^{n+2}} \left[\frac{(z-c)^{n+2}}{h} - \right.\right.$$

$$\left.\left. \frac{(z-c)(z-c-h)^{n+1}}{h} - (n+1)(z-c-h)^{n+1} \right] dz \right| =$$

$$\frac{n!}{2\pi} \left| \int_{\mathcal{C}} \frac{f(z)}{(z-c-h)^{n+1}(z-c)^{n+2}} \left[\frac{(z-c)^{n+2}-(z-c-h)^{n+2}}{h} - \right.\right.$$

$$\left.\left. (n+2)(z-c-h)^{n+1} \right] dz \right| .$$

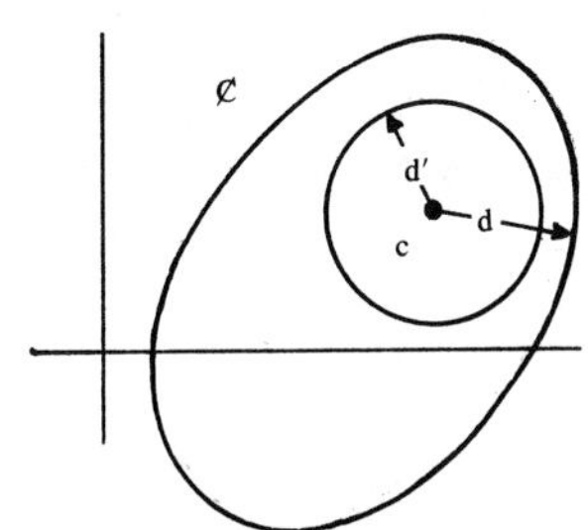

Figure 13-18

Now let d be the minimum distance from c to $\mathcal{C}$, and restrict h so that $0 < |h| < d' < d$. From the preceding Lemma, given any $\epsilon > 0$ there can be found a $\delta > 0$ such that for all $|h| < \delta$ the following relation is satisfied for all points z lying on $\mathcal{C}$:

$$\left| \frac{(z-c)^{n+2}-(z-c-h)^{n+2}}{h} - (n+2)(z-c-h)^{n+1} \right| < \epsilon .$$

Next choose δ' as $\delta' = \min(\delta, d')$ and observe hen that for all points z on $\not{C}$ we have: $|z-c| \geq d$ and $|z-c-h| \geq |z-c| - |h| > d-d'$ whenever $|h| < \delta'$.

Assembling the above results we now have for all h such that $|h| < \delta'$:

$$\left| \frac{g_n(c+h)-g_n(c)}{h} - g_{n+1}(c) \right| \leq \frac{n!}{2\pi} \frac{M\epsilon}{(d-d')^{n+1}d^{n+2}} L(\not{C}),$$

where M is a constant such that $|f(z)| \leq M$ for all z on $\not{C}$. Since all quantities in the bound are fixed, and ϵ is arbitrarily small, the Theorem is proved.

<u>EXERCISE 13.36</u>. Prove the following result which was used in the proof of the preceding Theorem: if $f(z)$ is Regular on a s.c.c. $\not{C}$ then there exists a (finite, positive) constant M such that $|f(z)| \leq M$ for all z on $\not{C}$.

We are now in a position to prove a converse of Theorem 13.5 (Cauchy's Theorem). The result is due to Morera and is contained in the following Theorem.

<u>THEOREM 13.12</u> (Morera: a Converse of Cauchy's Theorem) If $f(z)$ is continuous throughout a domain D and $\int_{\not{C}} f(z)dz = 0$ for every s.c.c. in D then $f(z)$ is Regular throughout D.

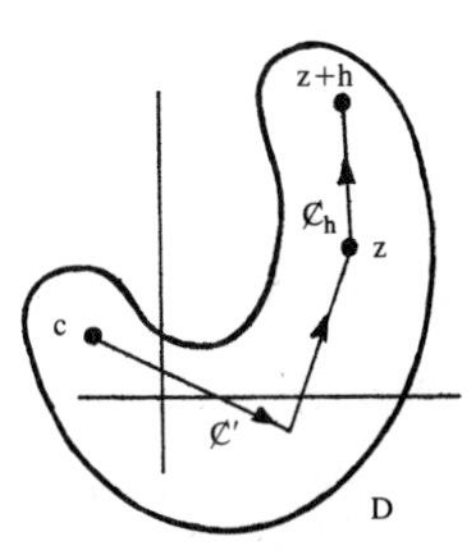

Figure 13-19

PROOF. Let c be any fixed point in D and define a function $F(z)$ of z over D as follows:
$F(z) = \int_{\mathcal{C}'} f(w)dw$, where $\mathcal{C}'$ is a contour from c to z composed of straight line segments in D. This defines $F(z)$ uniquely. Now consider the following:

$$\frac{F(z+h)-F(z)}{h} - f(z) = \int_{\mathcal{C}_h} \frac{f(w)-f(z)}{h}\,dw;$$

$\mathcal{C}_h$ is the straight line contour from z to $z+h$. By continuity of $f(z)$ at z, given $\epsilon > 0$ there exists a $\delta > 0$ such that $|f(w) - f(z)| < \epsilon$ for all w on $\mathcal{C}_h$ whenever $|h| < \delta$. Thus,

$$\left| \frac{F(z+h)-F(z)}{h} - f(z) \right| \le |h| \cdot \epsilon / |h| = \epsilon \quad \text{for all} \quad |h| < \delta .$$

Therefore we may conclude that $F'(z)$ exists and equals $f(z)$ at every point z in D, that is, $F(z)$ is Regular in D. But by Theorem 13.11 (Representation Theorem: Derivatives) the derivative of a Regular function is Regular, whence $f(z)$ is Regular over D, thus concluding the proof.

Now in what follows, we shall make use of the complex parallels of Theorems (previously proved for the real case) dealing with such things as interchange of the operations of

$\lim_{n\to\infty}$, $\sum_0^\infty$ and $\frac{d}{dz}$ with $\int_{\mathbb{C}}$. As mentioned before, statements and proofs of such Theorems usually require only minor modifications (if any) of the corresponding proofs in the real case.

Our next result resolves the (previously postponed) question concerning what functions possess a convergent Taylor Series Representation.

THEOREM 13.13 (Taylor's Theorem: General Form) If $f(z)$ is Regular in a neighborhood of $z = a$, then in that neighborhood $f(z)$ has the representation:

$f(z) = \sum_0^\infty \frac{f^{(n)}(a)}{n!} (z-a)^n$. In fact, this infinite series converges to $f(z)$ inside a circle with center $z = a$ and radius equal to the distance from $z = a$ to the nearest point at which $f(z)$ is singular.

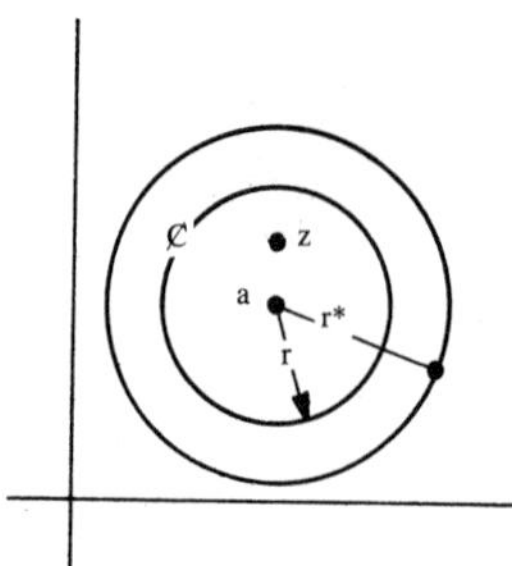

Figure 13-20

PROOF. Let r^* be the distance from $z = a$ to the nearest singularity of $f(z)$. Note that $0 < r^* \leq \infty$. Now let z be any point interior to a circle of radius r^* centered at $z = a$, and finally let $\mathbb{C}$ be a circle of radius r $(0 < r < r^*)$ centered at $z = a$. Now by the Cauchy Integral Formula we may write:

$$f(z) = \frac{1}{2\pi i} \int_{\mathbb{C}} \frac{f(w)}{w-z}\, dw \quad .$$

And, upon expressing $w - z$ as $w - z = (w-a) - (z-a)$:

$$\frac{1}{w-z} = \frac{1}{(w-a)-(z-a)} = \frac{1}{w-a}\left[\frac{1}{1-\frac{z-a}{w-a}}\right] = \frac{1}{w-a}\sum_0^\infty \left(\frac{z-a}{w-a}\right)^n ,$$

and the series on the right converges uniformly on $\mathcal{C}$, at least. Therefore we may write:

$$f(z) = \frac{1}{2\pi i}\int_{\mathcal{C}} f(w)\left[\sum_0^\infty \frac{(z-a)^n}{(w-a)^{n+1}}\right] dw = \frac{1}{2\pi i}\sum_0^\infty \int_{\mathcal{C}} \frac{f(w)}{(w-a)^{n+1}} (z-a)^n\, dw$$

$$= \sum_0^\infty \frac{f^{(n)}(a)}{n!}(z-a)^n , \quad \text{as was to be shown.}$$

Note that convergence is uniform in the region $|z-a| \le r$. This completes the proof.

In view of the preceding Theorem, a function $f(z)$ is Regular in a domain D iff it has a (convergent) Taylor Series expansion in a neighborhood of every point in D.

EXERCISE 13.37 (Complex vs. Real Case) Let $f(x)$ be a real valued function of the real variable x and let $f(z)$ be the complex function obtained from it by replacing the real variable by the complex variable z.

(1) Let p be a point on the real axis. Consider both $\lim_{x\to p} f(x)$ and $\lim_{z\to p} f(z)$.

(a) When do both exist, and in this case, are the two equal?

(b) Need both exist if at least one exists? Demonstrate by example(s).

(2) Repeat part (1) with 'limit at p' replaced first by 'continuity at p' and then by 'existence of a derivative at p'.

(3) Formulate a connection similar to the above between the complex integral over a straight line contour $\mathcal{C}$ on the real axis and the ordinary Riemann Integral over the interval.

The next topic we shall develop is the so-called Residue Theory that finds so many applications in various fields of Applied Mathematics. However, prior to this we shall pause to establish a few classical results concerning Regular functions, that follow almost immediately from the preceding Theorems. These classical results sometimes find important Applications in their own right.

The first result deals with a bound on the modulus of derivatives of Regular functions at the origin.

<u>THEOREM 13.14</u> (Inequality due to Cauchy) Suppose $f(z)$ is Regular for $|z| < R$, and let $\mathcal{C}_r$ be the circular contour centered at the origin with radius $r < R$. Finally, let $M(r)$ denote the maximum value attained by $|f(z)|$ as z varies over $\mathcal{C}_r$. Then,

$$|f^{(n)}(0)| \le \frac{n!M(r)}{r^n} .$$

PROOF. By the Representation Theorem for Derivatives (Theorem 13.11) we have:

$$f^{(n)}(0) = \frac{n!}{2\pi i} \int_{\mathcal{C}_r} \frac{f(z)}{z^{n+1}}\,dz, \quad \text{whence}$$

$$|f^{(n)}(0)| \le \frac{n!}{2\pi} \frac{M(r)}{r^{n+1}} 2\pi r, \quad \text{which completes the proof.}$$

EXERCISE 13.38 Prove that a maximum $M(r)$ is actually attained by $|f(z)|$ as z varies over $\mathcal{C}_r$ -- the fact that was used in the preceding Theorem.

A complex function that is Regular for all complex z is termed an entire function. The next classical result asserts that if an entire function has bounded modulus, then the function must reduce to a constant.

THEOREM 13.15 (Liouville's Theorem) If $f(z)$ is an entire function and for some finite, positive constant M we have: $|f(z)| \le M$ for all z, then $f(z)$ must be a constant function.

PROOF. First, $f(z)$ possesses a convergent Taylor Series Representation, $\sum\limits_0 a_n z^n$, about the origin that is valid for all complex z. Furthermore, $a_n = f^{(n)}(0)/n!$ for $n = 0,1,2,...$ But by the preceding Inequality of Cauchy, for any $r > 0$, $|a_n| \le M/r^n \to 0$ as $r \to \infty$, for $n = 1,2,...$. Thus we must have $a_n = 0$ for $n = 1,2,...$ whence $f(z) \equiv a_0$, that is, a constant. This completes the proof.

<u>THEOREM 13.16</u> (Generalization of Liouville's Theorem) If $f(z)$ is an entire function and $|f(z)| \le M|z^k|$ for all z, where M is some finite positive constant and k is a non-negative integer, then $f(z)$ must be a polynomial in z of degree at most k.

PROOF. <u>EXERCISE 13.39</u>.

<u>THEOREM 13.17</u> Suppose $f(z)$ is entire. Let $\mathcal{C}_r$ be the circle of radius r situated at the origin, and let $A(r)$ represent the maximum value of $\mathrm{Re}(f(z))$ as z varies over $\mathcal{C}_r$. If $A(r)$ is bounded above by a (finite) constant A, then $f(z)$ is a constant.

PROOF. Let $f = u + iv$. Then the function $g(z)$ defined by: $g(z) = e^{f(z)} = e^{u+iv}$ is also entire and furthermore $|g(z)| = e^u \le e^A$ for all z. Thus, by Liouville's Theorem, $g(z)$ must be a constant function, whence the same is true of $f(z)$. This completes the proof.

<u>EXERCISE 13.40</u> In the preceding Theorem, prove, as asserted, that g is entire and that g = const. implies that f = const.

Finally, we shall prove the so-called Fundamental Theorem of Algebra. This classical result is applied in all areas of Applied Mathematics. It establishes that a polynomial of degree $n \ge 1$ in z has exactly n "roots", not all necessarily real or distinct.

This result is contained in the following Theorem.

THEOREM 13.18 (Fundamental Theorem of Algebra) If $P(z) = a_0 + a_1 z + \ldots + a_n z^n$ ($a_n \neq 0$, $n \geq 1$) is a polynomial of degree n in z, then the equation $P(z) = 0$ has at least one solution.

PROOF. (By Contradiction) Suppose to the contrary that for all z, $P(z) \neq 0$. It then follows that $1/P(z)$ is an entire function and furthermore $|f(z)|$ is bounded since $|f(z)| \to 0$ as $|z| \to \infty$. Thus, by Liouville's Theorem, $f(z)$ must be a constant function. But this cannot be because $f(z) = 1/P(z)$, and $P(z)$ is not a constant function. Therefore we must conclude that $P(z) = 0$ for at least one value of z. This completes the proof.

We shall now enter into certain preliminaries that are aimed at leading to a discussion of the so-called Residue Theorem of complex variables. This particular theory finds many uses in the various fields of Applied Mathematics.

First, consider a function $f(z)$ that is Regular in the circular domain $D = \{z: |z-a| < r\}$ around the point $z = a$. From previous results we know that $f(z)$ possesses a convergent Taylor Series representation, say $f(z) = \sum_0^\infty a_n(z-a)^n$, within D. Now if, in this representation, we have say $a_0 = 0$, then we say that $f(z)$ has a zero at $z = a$. If, furthermore, $a_0 = a_1 = \ldots = a_{k-1} = 0$ and $a_k \neq 0$, then we have, say, that $f(z)$ possesses a zero of order k at $z = a$.

Note that ($f(z)$ has a zero of order k at $z = a$) implies that we can write the Taylor Series representation of $f(z)$ about $z = a$ as follows: $f(z)=(z-a)^k \sum_0^\infty a_{k+m}(z-a)^m$, where $a_k \neq 0$, and so accordingly,
$f(a) = f^{(1)}(a) = f^{(2)}(a) = \ldots = f^{(k-1)}(a) = 0$, but $f^{(k)}(a) \neq 0$.

Actually, the essential feature of a zero of order k of $f(z)$ at $z = a$ is that:

$$\lim_{z\to a} \frac{f(z)}{(z-a)^k} = L \neq 0 \text{ whereas } \lim_{z\to a} \left| \frac{f(z)}{(z-a)^p} \right| = +\infty \text{ for any } p > k.$$

<u>EXERCISE 13.41</u> Prove that the statements in the two preceding paragraphs are in fact true.

<u>EXAMPLE 13.7</u> (Zeros are Isolated) Let $f(z) \not\equiv 0$ be regular within the domain D. Then the zeros of $f(z)$ are isolated in the sense that if $z = a$ is a zero of $f(z)$, then there exists a deleted circular neighborhood, say $0 < |z-a| < \delta$, around $z = a$ within which $f(z) \neq 0$. To prove this suppose that $f(z)$ possesses a zero of order k at $z = a$. Then, throughout D we may express $f(z)$ as follows: $f(z) = (z-a)^k g(z)$ where $g(z)$ is Regular and $\lim_{z\to a} g(z) \neq 0$. Therefore, for $\delta > 0$ sufficiently small there exists an $\eta > 0$ such that if $|z-a| < \delta$ then $|g(z)| > \eta$. Therefore within $0 < |z-a| < \delta$ we have $f(z) \neq 0$ as was to be shown.

<u>EXAMPLE 13.8</u> (Agreement of Regular Functions) If both $f(z)$ and $g(z)$ are Regular within the domain D and for some infinite, bounded subset $N \subseteq D$ we have

$f(z) \equiv g(z)$ for $z \in N$, then $f(z) \equiv g(z)$ throughout D. To prove this choose a sequence $\{z_n\}$ of distinct points in N converging to z_0, say (z_0 belongs to D but not necessarily to N). This can always be done. Now consider the function $F(z)$ defined by $F(z) = f(z) - g(z)$, which is clearly Regular within D. Now each z_n and z_0 are zeros of F, but z_0 is not isolated. Since F is Regular, we must have $F(z) \equiv 0$ over D or, $f(z) \equiv g(z)$ over D. This completes the proof.

EXERCISE 13.42 In reference to the preceding Example, prove (i) that a sequence $\{z_n\}$ with the given convergence properties can always be chosen, and that its limit, z_0, is actually a zero of $F(z)$ and (ii) z_0 is not isolated.

If $f(z)$ is not Regular at the point $z = a$ within a domain D then there is no longer the guarantee of a convergent Taylor Series representation of f therein. However, the following Examples illustrate that perhaps under weaker assumptions such a function may possess a series representation, but including, perhaps, negative powers of $z - a$ (a Taylor Series consists only of non-negative powers of $z - a$).

EXAMPLE 13.9 Consider the function $f(z) = e^z/z^2$. Although not Regular at $z = 0$, $f(z)$ is Regular for all z such that $|z| > 0$, and for such values of z we have $f(z) = 1/z^2(1 + z + z^2/2! + \dots) = 1/z^2 + 1/z + 1/2! + z/3! + \dots$, that is, a series representation in both non-negative and negative powers of z (convergence considerations are postponed).

EXAMPLE 13.10 First note that $1/(1-w) = \sum_{0}^{\infty} w^n$ for $|w| < 1$. Upon formally substituting $z = 1/w$ we obtain: $z/(z-1) = f(z) = \sum_{0}^{\infty} (1/z)^n$ for $|z| > 1$, which is a series representation in non-positive powers of z. Note that $f(z)$ is Regular for $|z| > 1$.

The above Examples illustrate special cases of a general representation Theorem known as Laurent's Theorem. It sets forth the conditions under which a function may be represented as a convergent series in non-negative and negative powers of the variable over a suitable domain. This Theorem is the foundation of Residue Theory.

THEOREM 13.19 (Laurent's Theorem) Suppose $f(z)$ is Regular within the annular region $D = \{z: 0 < r \le |z-a| \le R \le \infty\}$. Then for all z satisfying: $r < |z-a| < R$, we can express $f(z)$ as a convergent series $f(z) = \sum_{-\infty}^{+\infty} a_n (z-a)^n$ of non-negative and negative powers of $(z-a)$, where the coefficients a_n are:

$$a_n = \frac{1}{2\pi i} \int_{\mathcal{C}} \frac{f(z)}{(z-a)^{n+1}}\, dz, \quad \text{and}$$

where $\mathcal{C}$ is any circular contour of the form $\mathcal{C} = \{z: |z-a| = \rho, \ r < \rho < R\}$.

PROOF. Without loss of generality, and for simplicity of algebra, we assume that $a = 0$.

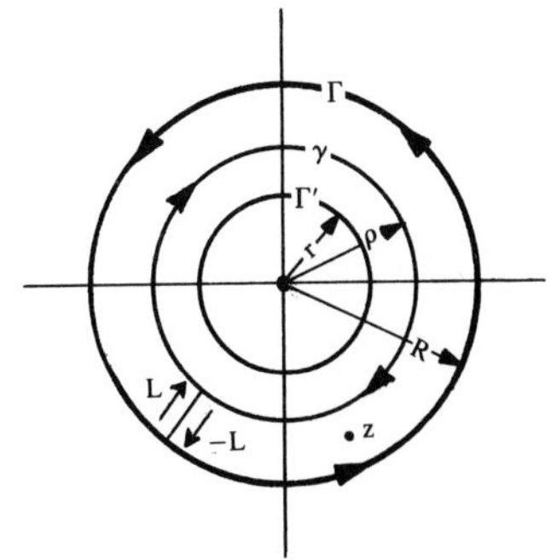

Figure 13-21

Let $\not{C}'$ be the s.c.c. composed of Γ, L, γ and -L, and note that $\gamma = -\not{C}$, where $\not{C}$ is the s.c.c. $\not{C} = \{z: |z| = \rho\}$. Now by assumption, f(z) is Regular within and on $\not{C}'$, whence for any point z interior to $\not{C}'$ we may apply Cauchy's Integral Formula and obtain:

$$f(z) = \frac{1}{2\pi i}\int_{\not{C}'}\frac{f(w)}{w-z}\,dw = \frac{1}{2\pi i}\int_{\Gamma}\frac{f(w)}{w-z}\,dw + \frac{1}{2\pi i}\int_{\gamma}\frac{f(w)}{w-z}\,dw = A + B$$

where we can then write:

$$A = \frac{1}{2\pi i}\int_{\Gamma}\frac{f(w)}{w(1-z/w)}\,dw = \frac{1}{2\pi i}\int_{\Gamma}\frac{f(w)}{w}\sum_{0}^{\infty}\left(\frac{z}{w}\right)^n dw$$

$$= \sum_{0}^{\infty} z^n \frac{1}{2\pi i}\int_{\Gamma}\frac{f(w)}{w^{n+1}}\,dw$$

$$= \sum_{0}^{\infty} z^n \frac{1}{2\pi i}\int_{\not{C}}\frac{f(w)}{w^{n+1}}\,dw$$

and $$= \sum_{0}^{\infty} a_n z^n$$

$$B = -\frac{1}{2\pi i}\int_{\gamma}\frac{f(w)}{z}\sum_{0}^{\infty}\left(\frac{w}{z}\right)^n dw = -\sum_{-\infty}^{-1} z^n \frac{1}{2\pi i}\int_{\gamma}\frac{f(w)}{w^{n+1}}\,dw$$

$$= \sum_{-\infty}^{-1} z^n \frac{1}{2\pi i}\int_{\mathcal{C}}\frac{f(w)}{w^{n+1}}\,dw$$

$$= \sum_{-\infty}^{-1} a_n z^n .$$

Thus, $f(z) = \sum_{-\infty}^{\infty} a_n z^n$ where $a_n = \frac{1}{2\pi i}\int_{\mathcal{C}}\frac{f(w)}{w^{n+1}}\,dw$

as was asserted. Since any point z satisfying $r < |z| < R$ is interior to $\mathcal{C}'$ for a suitable choice of ρ, the proof is complete.

<u>EXERCISE 13.43</u> Prove that if a Laurent expansion of $f(z)$ exists in an annulus about $z = a$, then it is unique therein. Uniqueness of a Taylor series can be viewed as a special case.

<u>EXAMPLE 13.11</u> Consider $f(z) = \dfrac{1}{z(1+z^2)}$ in the annular region $D = \{z: 0 < |z| < 1\}$. Formally we then have $f(z) = 1/z(1-z^2+z^4-\ldots) = 1/z-z+z^3-\ldots$ within D. By uniqueness, this must be the Laurent expansion of $f(z)$ within D. Next, consider the same function in the annular region $D' = \{z: 1 < |z| < \infty\}$. Formally we have $f(z) = 1/z^3\left[\dfrac{1}{1+1/z^2}\right] = 1/z^3(1-1/z^2+1/z^4-\ldots) =$ $1/z^3-1/z^5+1/z^7-\ldots$ which must be the Laurent expansion of $f(z)$ in D'.

EXERCISE 13.44 Using the above technique of formal long division and then an appeal to uniqueness, obtain the first few terms of the Laurent expansions of $f(z) = \dfrac{e^z}{z(z^2+1)}$ and $g(z) = \dfrac{1}{e^z-1}$ within the annulus $D = \{z: 0 < |z| < 1\}$.

It will be recalled that a point at which a function $f(z)$ is not Regular is termed a singular point (or singularity) of the function. It often proves useful to have a classification of the various types of singularities.

If $f(z)$ is Regular in an annular region $R = \{z: 0 < |z-a| < r\}$ about a singularity $z = a$ then $f(z)$ is said to possess an isolated singularity at $z = a$. If, furthermore, in the Laurent expansion of $f(z)$ about such an isolated singularity, viz. $f(z) = \sum_{-\infty}^{\infty} a_n (z-a)^n$, we have for some positive integer k: $a_{-n} = 0$ $(n > k)$ and $a_{-k} \neq 0$, then the isolated singularity is termed a pole of order k at $z = a$. If no such positive integer k exists, then the singularity is termed an essential isolated singularity at $z = a$.

EXAMPLE 13.12 The function $f(z) = \dfrac{1}{(z-1)^2} + \dfrac{3}{(z-1)} + (z-1)^4$ has a pole of order 2 at $z = 1$ whereas the function $e^{1/z} = \sum_0^{\infty} \dfrac{1}{n!z^n}$ has an essential isolated singularity at $z = 0$.

By convention, a function $f(z)$ is said to possess a given type of singularity at ∞ according as $f(1/w)$ possesses that type of singularity at $w = 0$. Thus, $f(z) = 1 + z + z^2$ has no singularities in the finite complex plane but does have a pole of order 2 at ∞ since $f(1/w) = 1/w^2 + 1/w + 1$. Similarly, $f(z) = e^z$ possesses an essential isolated singularity at ∞.

We are now in a position to define what is meant by a <u>residue</u>. First, if $f(z)$ has an isolated singularity at $z = a$, then it possesses a unique Laurent expansion in a region $R = \{z: 0 < |z-a| < r\}$ about $z = a$. The coefficient a_{-1} in this Laurent expansion $f(z) = \sum_{-\infty}^{\infty} a_n(z-a)^n$ is termed the residue of $f(z)$ at $z = a$. Thus it follows that

$$a_{-1} = \frac{1}{2\pi i} \int_{\mathcal{C}} f(z)dz$$

where $\mathcal{C}$ is any s.c.c. enclosing the point $z = a$ and lying within R. The importance of the residue will become apparent shortly. First, we shall develop some various techniques for finding the residue of a function $f(z)$ at a point $z = a$.

<u>Method A</u> (Inspection of a_{-1} from the Laurent Series) It may be possible, by formal algebraic manipulations, to obtain the Laurent expansion of $f(z)$ about $z = a$ within R. In such a case, we obtain the residue of $f(z)$ at $z = a$ by simply reading off the coefficient a_{-1} of $1/(z-a)$ in this series.

<u>EXAMPLE 13.13</u> The residue of $f(z) = \dfrac{1}{z(1+z^2)}$ at $z = 0$ is 1 since within the region $R = \{z: 0 < |z| < 1\}$, $f(z) = 1/z - z + z^3 - \ldots$ (previously calculated), whence $a_{-1} = 1$.

<u>EXAMPLE 13.14</u> To find the residue of $f(z) = \frac{z-2}{z^2}\operatorname{Sin}(1/1-z)$ at $z = 1$ we first set $w = z-1$. We then investigate the residue of $g(w) = f(w+1)$ at $w = 0$. Now $g(w) = \frac{1-w}{(1+w)^2}\operatorname{Sin}(1/w)$ and for $0 < |w| < 1$:

$$\frac{1}{(1+w)^2} = \sum_0^\infty (-1)^n(n+1)w^n \quad \text{and}$$

$$\operatorname{Sin}(1/w) = \sum_0^\infty \frac{(-1)^n}{(2n+1)!}\left(\frac{1}{w}\right)^{2n+1} .$$

Thus,

$$g(w) = (1-w)\left\{\sum_0^\infty (-1)^n(n+1)w^n\right\}\left\{\sum_0^\infty \frac{(-1)^n}{(2n+1)!}\left(\frac{1}{w}\right)^{2n+1}\right\} .$$

Upon multiplying out and collecting coefficients of like powers of w we obtain the residue of $g(w)$ at $w = 0$ as the coefficient a_{-1} of $1/w$. A routine computation verifies that a_{-1} is:

$$a_{-1} = \sum_0^\infty \frac{(-1)^n(4n+1)}{(2n+1)!} = 2\sum_0^\infty \frac{(-1)^n}{2n!} - \sum_0^\infty \frac{(-1)^n}{(2n+1)!} = 2\operatorname{Cos}1 - \operatorname{Sin}1.$$

<u>EXERCISE 13.45</u> Using Method A, find the residues at $z = 0$ of the following functions: $f(z) = e^{1/z}$, $g(z) = z\operatorname{Cos}(1/z)$ and $h(z) = (1-e^{2z})/z^4$.

<u>Method B</u> (When $f(z)$ has a pole at $z = a$) If $f(z)$ has a pole of order k at $z = a$ and the residue of $f(z)$ at $z = a$ is required, the following method may be used. If $f(z)$ has a pole of order k at $z = a$ then the function $g(z) = (z-a)^k f(z)$ is regular in a neighborhood of $z = a$, possesses a Taylor Series expansion about $z = a$

in this neighborhood, and the residue a_{-1} of $f(z)$ at $z = a$ is the coefficient of $(z-a)^{k-1}$ in this expansion. Whence,

$$a_{-1} = g^{(k-1)}(a)/(k-1)!$$

<u>EXERCISE 13.46</u> Prove Method B is valid as stated.

<u>EXAMPLE 13.15</u> The residue of $f(z) = 1/(z-a)^n$ at $z = a$ is zero if $n \neq 1$ and one if $n = 1$.

<u>EXERCISE 13.47</u> Using Method B, evaluate the Residue of $f(z) = e^{2z}/(z-1)^2$ at $z = 1$ and the Residues of

$$f(z) = \frac{(3z^2+2)}{(z-1)(z^2+9)} \quad \text{at} \quad z = \pm 3i.$$

The next result is the fundamental Residue Theorem. In effect, it sets forth the 'Residue Method' for evaluating contour integrals of complex functions possessing a finite number of isolated singularities within the s.c.c. of integration. After establishing this result, we shall use it to evaluate certain complex and real integrals.

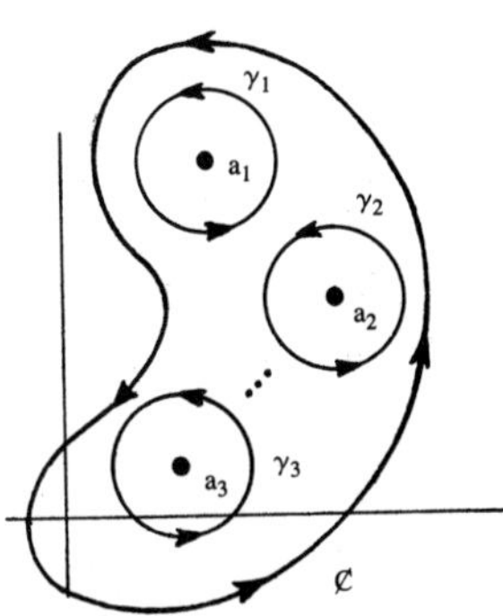

Figure 13-22

THEOREM 13.20 (Residue Theorem) Let $f(z)$ be Regular within and on a s.c.c. $¢$ except perhaps at a finite number $\{a_1, a_2, \ldots, a_n\}$ of isolated singularities lying inside $¢$. Then,

$$\int_{¢} f(z)dz = \sum_0^n 2\pi i\, R_j$$

where R_j is the residue of $f(z)$ at $z = a_j$, also written:
$R_j = \operatorname{Res}_{z=a_j} f(z)$, $j = 1,2,\ldots,n$.

PROOF. Let γ_j $(j=1,2,\ldots,n)$ be circular contours around $z = a_j$ $(j=1,2,\ldots,n)$ chosen so as to be disjoint and not intersecting $¢$. Under the hypotheses, this can always be done. From the Corollary (Theorem 13.8) to Cauchy's Theorem we may then write:

$$\int_{¢} f(z)dz = \sum_1^n \int_{\gamma_j} f(z)dz \ .$$

Now within each annular region about each a_j determined by γ_j the function $f(z)$ has a unique Laurent Expansion, say

$$f(z) = \sum_{-\infty}^{\infty} a_{nj}(z-a_j)^n \quad ,$$

where $a_{-1j} = R_j$ is the Residue of $f(z)$ at $z = a_j$

for $j = 1,2,\ldots,n$. However,

$$a_{-1j} = \frac{1}{2\pi i}\int_{\gamma_j} f(z)dz,$$

whence $2\pi i\, a_{-1j} = 2\pi i\, R_j = \int_{\gamma_j} f(z)dz$. Proof of the Theorem is completed upon summation over the index j.

The essence of the Residue Theorem is that, under the stated conditions, the process of contour integration can be accomplished by the process of evaluation of certain Residues of the integrand. The following Examples will illustrate this point.

EXAMPLE 13.16 To evaluate $I = \int_{\mathcal{C}} \frac{1}{z(1+z^2)}\,dz$ where $\mathcal{C} = \{z: |z| = 2\}$, note that the integrand has three isolated singularities (simple poles) within $\mathcal{C}$, namely $z = 0, i, -i$. A simple computation yields residues $1, -\frac{1}{2}, -\frac{1}{2}$ at these poles. Thus, $I = 2\pi i(1-\frac{1}{2}-\frac{1}{2}) = 0$. Now if $\mathcal{C}$ were replaced by $\mathcal{C}' = \{z: |z| = \frac{1}{2}\}$ then the only pole within $\mathcal{C}'$ would be $z = 0$, and therefore the value of the integral around $\mathcal{C}'$ would be $2\pi i(1) = 2\pi i$.

EXERCISE 13.48 Evaluate $I = \int_{\mathcal{C}} \frac{3z^2+2}{(z-1)(z^2+9)}\,dz$ where

(i) $\mathcal{C} = \{z: |z-2| = 2\}$, (ii) $\mathcal{C} = \{z: |z| = 4\}$.

EXERCISE 13.49 Evaluate $I = \int_{\mathcal{C}} \frac{1}{z^3(z+4)}\,dz$ with $\mathcal{C} = \{z: |z+2| = 3\}$.

EXAMPLE 13.17 Here we evaluate $I = \int_{\mathcal{C}} \frac{Csc(z^2)}{z^3} dz$ where $\mathcal{C}$ is the s.c.c. $\mathcal{C} = \{z: |z| = 1\}$. Now it is known that $Sinz = z - z^3/3! + z^5/5! - \ldots$ for all z. Since $Cscz = 1/Sinz$, a formal process of long division yields: $Cscz = \frac{1}{z} + \frac{1}{3!}z - [\frac{1}{5!} - (\frac{1}{3!})^2] z^3 + \ldots$ for $|z| > 0$. So the Laurent expansion of the integrand in the annular region $\{z: 0 < |z| < 1\}$ is: $Csc(z^2)/z^3 = \frac{1}{z^5} + \frac{1}{3!z} + \ldots$, whence by inspection $a_{-1} = 1/3!$. Since $z = 0$ is the only singularity of the integrand within $\mathcal{C}$ it follows that $I = 2\pi i(1/6) = \pi i/3$.

EXAMPLE 13.18 To evaluate $I = \int_{\mathcal{C}} e^{-z}/z^2 dz$ where $\mathcal{C} = \{z: |z| = 1\}$, it can easily be shown that the Laurent expansion of the integrand within $\{z: 0 < |z| < 1\}$ is given by $e^{-z}/z^2 = 1/z^2 - 1/z + 1/2! + \ldots$ whence $a_{-1} = -1$ and $I = -2\pi i$.

EXAMPLE 13.19 Now we evaluate $I = \int_{\mathcal{C}} \frac{1}{1+z^2} dz$, where $\mathcal{C}$ is the semi-circular contour illustrated in the accompanying diagram. Here the only singularity of the integrand within the s.c.c. determined by $\mathcal{C}$ is $z = i$, at which the residue is $\frac{1}{2i}$, whence, $I = 2\pi i (\frac{1}{2i}) = \pi$.

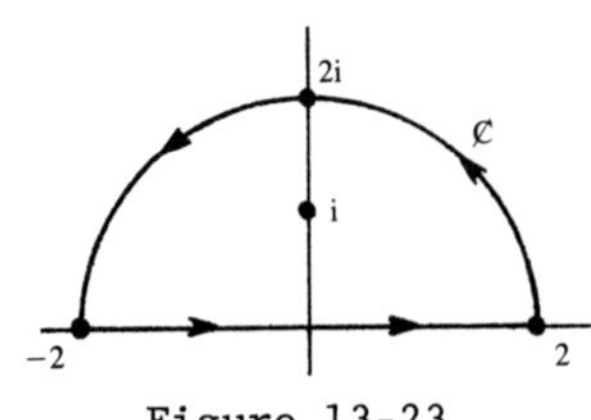

Figure 13-23

Finally, we use some simple Examples to illustrate a general method that enables us to evaluate a wide range of real Riemann integrals by considering the Integral of an appropriate complex function over a suitable contour in the complex plane.

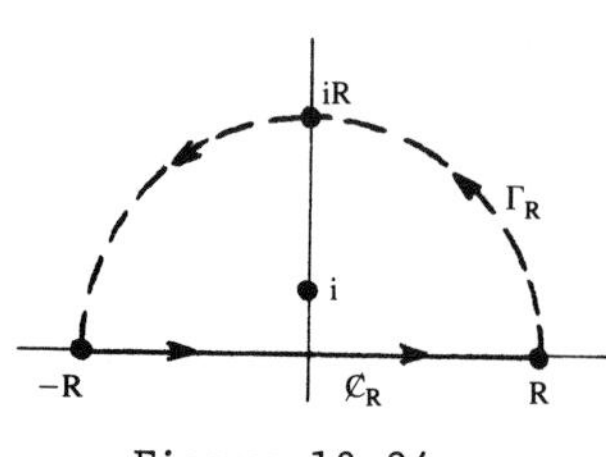

Figure 13-24

<u>EXAMPLE 13.20</u> To evaluate $\int_0^\infty \frac{1}{1+x^2}\,dx$ first consider $\int_{\mathcal{C}_R} \frac{1}{1+z^2}\,dz$, where $\mathcal{C}_R$ is the entire semi-circular (R>1) contour pictured here. By the Residue Theorem, this Integral has value π. Next, note that we may decompose the integral over $\mathcal{C}_R$ into two parts, namely the (real) integral from -R to R along the real axis plus the integral over the arc (-----) Γ_R. Thus,

$$\pi = \int_{\mathcal{C}_R} \frac{1}{1+z^2}\,dz = \int_{-R}^{R} \frac{1}{1+x^2}\,dx + \int_{\Gamma_R} \frac{1}{1+z^2}\,dz \quad .$$

Now since $|1+z^2| \geq |z|^2 - 1 \geq R^2 - 1 > 0$ on Γ_R.

$$\left| \int_{\Gamma_R} \frac{1}{1+z^2}\,dz \right| \leq \frac{\pi R}{R^2-1} \to 0 \quad \text{as} \quad R \to \infty \, .$$

Thus,

$$\pi = \lim_{R\to\infty} \int_{-R}^{R} \frac{1}{1+x^2}\,dx + \lim_{R\to\infty} \int_{\Gamma_R} \frac{1}{1+z^2}\,dz$$

whence, by symmetry,

$$\int_0^\infty \frac{1}{1+x^2}\,dx = \pi/2 \ .$$

It is important to realize that the limiting argument used in the preceding Example, namely $\lim_{R\to\infty} \int_{-R}^{R} f(x)dx$, yields the so-called Principal Value of the corresponding improper Riemann integral. This limit (Principal Value) may exist (be finite) even though the improper Riemann Integral $\int_{-\infty}^{\infty} f(x)dx$ does not. However, if the improper Riemann Integral exists, then so does the Principal Value, and the two are equal. In view of this, some caution should be exercised.

<u>EXERCISE 13.50</u> What is a simple sufficient condition on $f(x)$ which guarantees that the improper Riemann Integral $\int_{-\infty}^{\infty} f(x)dx$ exists? Give an example of a case where the Principal Value $\lim_{R\to\infty} \int_{-R}^{R} f(x)dx$ exists but the improper Riemann Integral $\int_{-\infty}^{\infty} f(x)dx$ fails to exist.

<u>EXAMPLE 13.21</u> To evaluate $\int_0^{\infty} \frac{x^2}{(x^2+9)(x^2+4)^2} dx$ we again use the contour $\mathcal{C}_R$ from the preceding Example, and begin by considering the contour integral of $f(z) = \frac{z^2}{(z^2+9)(z^2+4)^2}$ around $\mathcal{C}_R$. The singularities of $f(z)$ consist of simple poles at $z = \pm 3i$ and poles of order 2 at $z = \pm 2i$. However, our concern is only with those singularities in the upper half-plane. Straightforward computation yields residues $-(\frac{3}{50i})$ at $z = 3i$, $-(\frac{13}{200i})$ at $z = 2i$. Thus, by the Residue Theorem,

$$\int_{\mathcal{C}_R} f(z)dz = -2\pi i \left[\frac{3}{50i} + \frac{13}{200i}\right] = \frac{\pi}{100} \quad \text{for } R > 3.$$

As before, $\frac{\pi}{100} = \lim_{R\to\infty} \int_{-R}^{R} f(x)dx + \lim_{R\to\infty} \int_{\Gamma_R} f(z)dz$.

Now $\left|\int_{\Gamma_R} f(z)dz\right| \le \frac{R^3}{(R^2-9)(R^2-4)^2} \to 0$ as $R \to \infty$, whence, by symmetry,

$$\int_0^\infty \frac{x^2}{(x^2+9)(x^2+4)^2}\,dx = \frac{\pi}{200} .$$

EXERCISE 13.51 Evaluate $\displaystyle\int_{-\infty}^{\infty} \frac{x}{(x^2+1)(x^2+2x+2)}\,dx$.

EXERCISE 13.52 Evaluate $\displaystyle\int_0^\infty \frac{x^2}{(x^2+1)^2}\,dx$.

The above Examples were purposely simple, and avoided such complications as a pole on the real axis. However, they were intended only to illustrate the type of ideas underlying the general method. This technique generalizes to account for such complications (by suitable modification of contour), and extends to integrands involving Trigonometric functions as well. Frequently, the main problem is just choice of a suitable contour $\mathcal{C}_R$. (See References).

Hints and Answers to Exercises: Section 13

13.1 Hint: Use the multiplication rule with $z_1 = c + 0i$ and $z_2 = x + iy$.

13.2 Hint: The inequalities:

(i) $|z_1 + z_2| \le |z_1| + |z_2|$

(ii) $||z_1| - |z_2|| \le |z_1 - z_2|$

are equivalent to the geometric facts that (i) no side of a triangle can exceed in length the sum of the lengths of the two remaining sides or (ii) be less in length than the difference in lengths of the two remaining sides. Next observe that the inequality $|z_1| - |z_2| \le ||z_1| - |z_2||$ is immediately obvious. Therefore, the entire result of the Exercise follows upon considering inequality (ii) with z_2 replaced by $-z_2$.

13.3 Hint: Use the basic definitions of addition, multiplication, division and conjugation.

13.4 Hint: Use the definitions of modulus and conjugate.

13.5 Answer: Choosing $-\pi \le \theta < \pi$:

(i) $r = \sqrt{2}, \quad \theta = \pi/4$

(ii) $r = 1, \quad \theta = -\pi/2$

(iii) $r = 4, \quad \theta = -\pi$

(iv) $r = \sqrt{10}, \quad \theta = \mathrm{Tan}^{-1}(1/3)$.

13.6 Answer:

(i) $2\sqrt{2} + i2\sqrt{2}$

(ii) $5.796 + i1.554$

(iii) $\sqrt{3} - i$

13.7 Hint: $\mathrm{Sin}(\theta_1 + \theta_2) = \mathrm{Sin}\,\theta_1 \mathrm{Cos}\,\theta_2 + \mathrm{Cos}\,\theta_1 \mathrm{Sin}\,\theta_2$

$\mathrm{Cos}(\theta_1 + \theta_2) = \mathrm{Cos}\,\theta_1 \mathrm{Cos}\,\theta_2 - \mathrm{Sin}\,\theta_1 \mathrm{Sin}\,\theta_2$

13.8 Hint: Express the equation as:

$$r^n(\mathrm{Cos}\,\theta + i\mathrm{Sin}\,\theta)^n = q(\mathrm{Cos}\,\phi + i\mathrm{Sin}\,\phi) = r^n\,(\mathrm{Cos}\,n\theta + i\mathrm{Sin}\,n\theta),$$

hence $r^n = q$ and $n\theta = \phi + 2k\pi$ $(k = 0, \pm 1, \pm 2, \ldots)$ and accordingly, $r = q^{1/n}$ and $\theta = \phi/n + 2k\pi/n$ $(k = 0, \ldots, n-1)$ gives the distinct solutions.

13.9 Hint: For the first part expand $f = (x + iy)^2$, and for the second part note that $x = \frac{1}{2}(z + \bar{z})$ and $y = \frac{1}{2}(z - \bar{z})$.

13.10 Hint: Suppose $L_1 \neq L_2$ and set $d = |L_1 - L_2| > 0$. Now choose $\epsilon < d/2$ and establish a contradiction.

13.11 Hint: For $|z-a| < 1$ say, note that:

$$|z^3 - a^3| = |z-a||z^2 + az + a^2| \leq C|z-a|$$

for some fixed positive constant C.

13.12 Hint:

(i) $$|\alpha f(z) + \beta g(z) - \alpha f(a) - \beta g(a)| \leq \alpha|f(z) - f(a)| + \beta|g(z) - g(a)| .$$

(ii) $$|f(z)g(z) - f(a)g(a)| \leq |f(z)||g(z) - g(a)| + |g(z)||f(z) - f(a)| + |g(z) - g(a)||f(z) - f(a)| .$$

13.13 Hint: Let $\epsilon > 0$ be given. It must be shown that there exists a $\delta > 0$ sufficiently small so that if $|z-a| < \delta$ then $|g(f(z)) - g(f(a))| < \epsilon$. By continuity of g there exists a $\delta_1 > 0$ such that $|g(f(z)) - g(f(a))| < \epsilon$ whenever $|f(z) - f(a)| < \delta_1$ and by continuity of f there exists a $\delta_2 > 0$ such that $|f(z) - f(a)| < \delta_1$ whenever $|z-a| < \delta_2$. Now choose $\delta = \delta_2$.

13.14 Hint: For the first part, prove that given $\epsilon > 0$ there can always be found a $\delta > 0$ such that $|e^x - 1| < \epsilon$ whenever $\sqrt{x^2 + y^2} < \delta$. For the last part, note that:

$$\text{Cos}z = \tfrac{1}{2}(e^{iz} + e^{-iz})$$

and

$$\text{Sin}z = (e^{iz} - e^{-iz})/2i \quad .$$

13.15 Hint: Note that if $f'(a)$ exists then for $z \neq a$,

$$\frac{f(z)-f(a)}{z-a} = f'(a) + g(z)$$

where $g(z) \to 0$ as $z \to a$. Accordingly,

$$|f(z) - f(a)| \le |f'(a)||z-a| + |g(z)||z-a|$$

and the right-hand side clearly tends to zero as $z \to a$. By considering the difference quotient, it can be shown that the function $f(z) = \bar{z}$ is continuous but nowhere differentiable.

13.16 Hint: By considering the appropriate difference quotients, it can be shown that both $f(z) = \text{Re}(z)$ and $f(z) = |z|$ are nowhere differentiable (in the complex variable sense).

13.17 Hint: The points $z = x + iy$ on a straight line through the origin of the complex plane satisfy either $y = cx$ (c real) or $x = 0$. The limit does not exist, e.g., as z approaches the origin of the complex plane along the path determined by the complex points $z = x + iy$ satisfying: $y = x^3$ (others exist). The conclusion is that $f'(0)$ does not exist, since the limit must exist and be the same regardless of the path of approach.

13.18 Hint: Using the definition of a partial derivative as the limit of a suitable difference quotient, it can be shown that at $z = 0$, $u_x = v_y = 1$ and $u_y = -v_x = -1$ so that the Cauchy-Riemann Equations are satisfied. However, the limit of the difference quotient $f(z)/z$ is not independent of the path of approach to $z = 0$, hence $f'(0)$ does not exist. For example, the limit of $f(z)/z$ is $i/(1+i)$ along the line $y = x$ and it is $1 + i$ along the line $y = 0$.

13.20 Hint: For $f(z) = z^n$ either use induction or else the Polar form of the Cauchy-Riemann Equations found in Exercise 13.22. For the other two functions, note that: $e^z = e^x(\text{Cos}y + i\text{Sin}y)$ and $\text{Sin}z = \frac{1}{2}(e^y + e^{-y})\text{Sin}x + \frac{1}{2}(e^y - e^{-y})\text{Cos}x \cdot i$.

13.21 Hint: Show that the Cauchy-Riemann Equations are satisfied nowhere, whence the function is nowhere differentiable.

13.22 Hint: Use the transformation $x = r\text{Cos}\,\theta$, $y = r\text{Sin}\,\theta$ and then the chain rule for partial derivatives along with the standard form of the Cauchy-Riemann Equations.

13.24 Hint: Apply Theorem 13.2. For $f(z) = z^n$ care must be taken at $z = 0$ for $n = -1,-2,\ldots$.

13.25 Hint: Use the inequality contained in part (ii) of the Hint for Exercise 13.12.

13.26 Hint: The following series are used:

$$\sum_{0}^{\infty} a^n = 1/(1-a) \quad , \quad |a| < 1$$

$$\sum_{1}^{\infty} na^{n-1} = 1/(1-a)^2 \quad , \quad |a| < 1 \quad .$$

13.27 Hint: $\int_{\mathcal{C}} z^2 dz = \int_0^1 (t^2 + it)^2 (2t + i) dt$.

13.28 Hint: The first contour has parametric equations: $x = y = t$ $(0 \le t \le 1)$, whereas the second contour has for its first leg the parametric equations: $x = 0$, $y = t$ $(0 \le t \le 1)$, and for its second leg the parametric equations: $y = i$, $x = t$ $(0 \le t \le 1)$.

13.29 Hint: Compare approximating sums for the integral over "$\mathcal{C}$" and that over "$-\mathcal{C}$" .

13.30 Hint: An immediate (but perhaps not the best) bound M could be obtained by noting that $f(z) \le |y|+|x|+ 3|x|^2$, then maximizing the right-hand side over the appropriate contour.

13.31 Hint: Consider the integral (in the standard counter-clockwise fashion) of f over the s.c.c. determined by $\mathcal{C}_1$ and $\mathcal{C}_2$; then apply Theorem 13.6 and Exercise 13.29. (Assume, for simplicity, that $\mathcal{C}_1$ and $\mathcal{C}_2$ do not intersect. The result is true in general, however).

13.32 Hint: (Case $n = 2$: see accompanying diagram). Let $\mathcal{C}^*$ be the s.c.c. composed of: $\mathcal{C}$, L_1, $-\mathcal{C}_1$, $-L_1$, L_2, $-C_2$, $-L_2$. The interior of $\mathcal{C}^*$ is $D \backslash (D_1 \cup D_2)$ and f is assumed Regular within this domain as well as on its boundary $\mathcal{C}^*$. Now apply Theorem 13.6 and Exercise 13.29 along with the (almost) obvious additivity of the integral over the various parts forming the contour $\mathcal{C}^*$. The case $n > 2$ needs little extra argument.

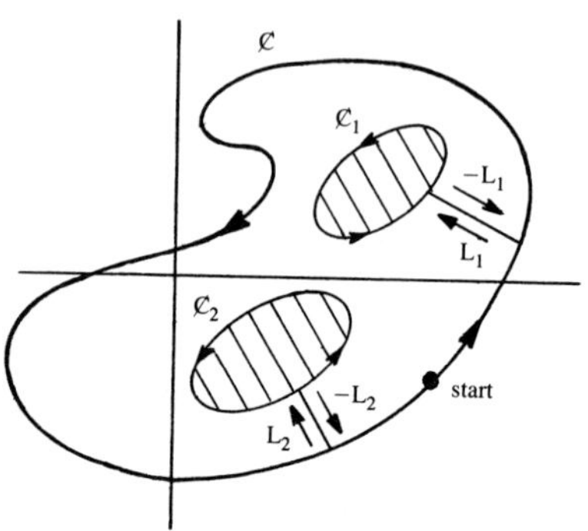

13.33 Answer: (i) $-4\pi i$ (ii) 0 .

13.34 Hint:

(i) $(4z^2 + 1) = 4(z - i/2)^2$

(ii) f is Regular within and on $\mathcal{C}$

(iii) $(z^2 + 2z + 2) = (z - i - 1)(z - i + 1)$

(iv) f is entire

13.35 Hint: For any point c within $\mathcal{C}$ we have:

$$f(c) = \int_{\mathcal{C}} \frac{f(c)}{z-c}\, dz = \int_{\mathcal{C}} \frac{g(c)}{z-c}\, dz = g(c) \quad .$$

13.36 Hint: It is sufficient to prove that:

(a) $S = \{(x,y): z = x + iy \in \mathcal{C}\}$ is a closed, bounded subset of E_2.

(b) $|f(z)| = (u^2 + v^2)^{\frac{1}{2}}$ is a continuous function of the two real variables (x,y) over E_2 hence $\mathcal{C}$.

(c) $|f(z)|$ is therefore bounded over S hence $\mathcal{C}$.

For the case $n = 1$, see the book 'Real Analysis' by H. L. Royden, The Macmillan Co., N.Y. (1963) page 36.

13.37 Hint:

(1a) Prove that if $\lim_{z \to p} f(z)$ exists then so does $\lim_{x \to p} f(x)$ and the two are equal.

(1b) $\lim_{x \to p} f(x)$ may exist when $\lim_{z \to p} f(z)$ does not.

For example, such is the case at every point on the real axis for the following function:

$$f(z) = \begin{cases} 0 & \text{if } \operatorname{Im}(z) = 0 \\ 1 & \text{if } \operatorname{Im}(z) \neq 0 \end{cases} \quad .$$

(2) The same comments and example applies for the case of 'continuity at p' and 'differentiability at p'.

(3) If f is continuous (in the complex sense) over an interval contour $\mathcal{C} = I$ on the real axis, and $\int_{\mathcal{C}} f(z)dz$ exists, then $\int_{I} f(x)dx$ (Riemann) exists and the two are equal. Consider approximating sums.

13.38 Hint: If necessary, see the Reference given in the Hint for Exercise 13.36.

13.39 Hint: Using the technique of Theorem 13.15, establish that in the Taylor Series representation $f(z) = \sum_{0}^{\infty} a_n z^n$ of f, $a_n = 0$ for $n > k$.

13.40 Hint: For the first part, establish more generally that if $h(z)$ and $f(z)$ are entire, then $g(z) = h(f(z))$ is entire. For the second part, note that $e^f = e^{u + iv}$ $e^u(\text{Cos}v + i\text{Sin}v)$.

13.42 Hint: By the straightforward generalization of Theorem 3.1 (Bolzano-Weierstrass) to the complex case, the set N possesses at least one limit point z_0. Then, for any positive integer n there exists an integer $n' > n$ and a point z' in N such that $1/(n'+1) \le |z'-z_0| < 1/n'$. Use this to construct the sequence $\{z_n\}$ converging to z_0. Since F is easily shown to be continuous at z_0, it then follows that $0 = \lim_{n\to\infty} F(z_n) = F(z_0)$. Being a limit point of zeros, z_0 cannot be isolated.

13.43 Hint: Assume f possesses two different (by at least one term) Laurent Expansions in the same annular region about $z = a$ and establish a contradiction.

13.44 Hint: (i) $1/(1+z^2) = 1 - z^2 + z^4 - z^6 + \ldots$ for $|z| < 1$ and $e^z = 1 + z + z^2/2! + z^3/3! + \ldots$ for all z. (ii) Formally divide $e^z - 1 = z + z^2/2! + z^3/3! + \ldots$ into 1.

13.45 Hint: (i) $e^w = 1 + w + w^2/2! + w^3/3! + \ldots$ for $|w| < \infty$. Now use the substitution $w = 1/z$ for $|z| > 0$.
(ii) $\cos w = 1 - w^2/2! + w^4/4! - w^6/6! + \ldots$ for $|w| < \infty$. Now use the substitution $w = 1/z$ for $|z| > 0$.
(iii) Formally divide $1 - e^{2z} = -2z - (2z)^2/2! - \ldots$ by z^4.

13.46 Hint: Use Theorem 13.13 (Taylor's Theorem).

13.47 Hint: (i) $g(z) = e^{2z}$ and $a_{-1} = g'(1)/1!$.
(ii) f has simple poles at $a = \pm 3i$. Therefore, $a_{-1} = g(a)$ where $g(z) = (z-a)f(z)$ for $a = \pm 3i$.

13.48 Hint: Use the results of Exercise 13.44. A diagram of the contour and poles is helpful.

13.49 Hint: $\mathcal{C}$ is a circular contour centered at $z = -2$ having radius 3. Accordingly, the triple pole at $z = 0$ and the simple pole at $z = -4$ are all included within the contour of integration. Use Method B to evaluate the Residues at these poles.

13.50 Hint: For instance, one sufficient condition is that: $f(x) = o(1/|x|^{1+d})$ as $x \to \pm\infty$ for some $d > 0$. The function $f(x)$ defined as follows suffices for part two:

$$f(x) = \begin{cases} 1/(1+x) & \text{for } x \geq 0 \\ 1/(x-1) & \text{for } x < 0 \end{cases} .$$

13.51 Answer: $-\pi/5$.

13.52 Answer: $\pi/4$.

References to Additional and Related Material: Section 13

1. Ash, R., "Complex Variables", Academic Press (1971).

2. Brand, L., "Advanced Calculus", John Wiley and Sons, Inc. (1958).

3. Churchill, R., "Complex Variables and Applications", 2nd Ed., McGraw-Hill, Inc. (1960).

4. Derrick, W., "Introductory Complex Analysis and Applications", Academic Press (1972).

5. Eves, H., "Functions of a Complex Variable", Prindle, Weber and Schmidt, Inc. (1966).

6. Forsyth, A., "Theory of Functions of a Complex Variable", Dover Publications (1965).

7. Kaplan, W., "Advanced Calculus", Addison-Wesley, Inc. (1952).

8. Kaplan, W., "Introduction to Analytic Functions", Addison-Wesley, Inc. (1966).

9. Leadbetter, M., Lecture Notes, University of North Carolina at Chapel Hill, Department of Statistics (1963).

10. Mitrinović, D., "Calculus of Residues", Noordhoff (1966).

11. Pennisi, L., "Elements of Complex Variables", Holt, Reinhart and Winston, Inc. (1963).

12. Pierpont, J., "Functions of a Complex Variable", Dover Publications, Inc.

13. Spiegel, M., "Schaum's Outline of Theory and Problems of Complex Variables", Schaum Publishing Co. (1964).

14. Titchmarsh, E., "The Theory of Functions", Second Edition, Oxford University Press (1960).

14. Matrices and Determinants

The notion of a matrix finds a wide variety of uses in Applied Mathematics. Here we shall examine some of the more important properties of matrices and determinants of complex numbers[1].

An $m \times n$ (read: m by n) matrix $A = (a_{ij})_{m\times n}$ is a rectangular array of complex numbers consisting of m rows and n columns:

$$A = \begin{bmatrix} a_{11} & a_{12} & \cdots & a_{1n} \\ a_{21} & a_{22} & \cdots & a_{2n} \\ \vdots & \vdots & & \vdots \\ a_{m1} & a_{m2} & \cdots & a_{mn} \end{bmatrix}$$

where a_{ij} denotes the element (entry) in the i-th row and j-th column $(i = 1,\ldots,m;\ j = 1,\ldots,n)$. The dimensions of A are m (rows) and n (columns) respectively. Two $m \times n$ matrices $A = (a_{ij})_{m\times n}$ and $B = (b_{ij})_{m\times n}$ are termed equal, written $A = B$, iff they are equal element-for-element, that is, $a_{ij} = b_{ij}$ $(i = 1,\ldots,m;\ j = 1,\ldots,n)$. The $m \times n$ matrix consisting of all zeros is denoted $0_{m\times n}$, or, where no confusion might arise, simply 0.

[1]Often we deal with matrices consisting of real numbers only. However, all results for matrices of complex numbers presented here are automatically valid - perhaps with minor modification - for this special case.

Two matrices $A = (a_{ij})_{mxn}$ and $B = (b_{ij})_{mxn}$ of the same dimensions may be added element-by-element to obtain their matrix sum defined by: $A + B = (a_{ij} + b_{ij})_{mxn}$. Clearly $A + B = B + A$. Furthermore, if k is any complex constant, we may define scalar multiplication of any matrix $A = (a_{ij})_{mxn}$ by k as follows: $kA = (ka_{ij})_{mxn}$. Clearly Ak is analogously defined and $kA = Ak$. Addition and scalar multiplication of $m \times n$ matrices obey the following algebraic properties: $(A + B) + C = A + (B + C)$, $A + (-1)A = A - A = 0_{mxn}$, $(k + r)A = kA + rA$, $k(A + B) = kA + kB$, $k(rA) = (kr)A$.

If A has dimensions $m \times n$ and B has dimensions $n \times p$ we say that A and B have equal contiguous dimension n. In this case we may define the matrix product $A \cdot B$ of A times B (here order <u>is</u> important) as follows: $A \cdot B = C$, where C is the $m \times p$ matrix with entries $c_{ij} = \sum_{k=1}^{n} a_{ik}b_{kj}$ $(i = 1,\ldots,m;\ j = 1,\ldots,p)$. Note that the matrix product $A \cdot B$ may be defined whereas the 'product' $B \cdot A$ may be meaningless. The process of performing matrix multiplication is often called 'row-by-column multiplication' for obvious reasons.

<u>EXERCISE 14.1</u> (Non-Commutativity of Matrix Product) Assuming that both $A \cdot B$ and $B \cdot A$ may be defined (when is this true?), show by example that in general the two are not equal.

EXERCISE 14.2 If both $A \cdot B$ and $B \cdot A$ are defined and equal, then the two matrices A and B are said to commute. Prove that A and B commute only if (but not if) they both are square (i.e. $m = n$) with the same dimensions. Construct a pair of 2x2 matrices that communte, and a pair that do not commute.

Subject to dimension requirements, the operations of matrix addition and matrix multiplication satisfy the following algebraic properties: $(A \cdot B) \cdot C = A \cdot (B \cdot C)$, $A \cdot (B + C) = A \cdot B + A \cdot C$, $(A + B) \cdot C = A \cdot C + B \cdot C$.

Next, the transpose A' of an $m \times n$ matrix A is defined to be the $n \times m$ matrix $A' = (a'_{ji})_{mxn}$ where $a'_{ji} = a_{ij}$ ($j = 1,\ldots,n$; $i = 1,\ldots,m$), that is, an interchange of rows with columns. The operation of transpose satisfies the following algebraic properties: $(A')' = A$, $(A + B)' = A' + B'$, $(A \cdot B)'$ $B' \cdot A'$, $(A \cdot B \cdot C)' = C' \cdot B' \cdot A'$, and so on.

Finally, the complex conjugate $\bar{A}$ of an $m \times n$ matrix A is defined as follows: $\bar{A} = (\bar{a}_{ij})_{mxn}$, where $\bar{a}_{ij}$ is the complex conjugate of the complex number a_{ij}. The operation of conjugation obeys the following algebraic properties: $\overline{(\bar{A})} = A$, $(\overline{A + B}) = \bar{A} + \bar{B}$, $(\overline{A \cdot B}) = \bar{A} \cdot \bar{B}$, $(\bar{A}') = (\bar{A})'$, $\overline{kA} = \bar{k}\bar{A}$.

EXERCISE 14.3 Prove that $A = \bar{A}$ iff the matrix A consists of real elements only.

We now specialize attention to square matrices, that is, matrices with an equal number of rows and columns (which will be taken as n). If $A = (a_{ij})_{nxn}$, a_{ii} ($i = 1,\ldots,n$) is defined as its <u>main diagonal.</u>

Associated with any square matrix A of complex numbers is a complex number $\det A$, termed the <u>determinant</u> of A, defined as:

$$\det A = \Sigma' \, (-1)^{f(j_1,\ldots,j_n)} \prod_{i=1}^{n} a_{ij_i} ,$$

where Σ' denotes summation over all $n!$ different permutations $(j_1,\ldots,j_n)$ of the integers $(1,\ldots,n)$, and $f(j_1,\ldots,j_n)$ denotes the number of transpositions required to change the latter into the former. A transposition consists of interchanging two numbers. Although $(1,\ldots,n)$ can be transposed into $(j_1,\ldots,j_n)$ in many different ways, it can be shown that the number of transpositions is always either even or odd, whence $(-1)^{f(j_1,\ldots,j_n)}$ is uniquely defined for each different permutation $(j_1,\ldots,j_n)$. This formal definition is usually not a practical way for evaluating $\det A$.

In order to develop a practical means for evaluating $\det A$, we introduce the following general notions as a preliminary. A <u>submatrix</u> of a matrix A is any rectangular array obtained from A by deleting certain rows and/or columns. A <u>minor</u> of A is the determinant of any square submatrix of A, and a <u>principal minor</u> is computed from a square submatrix of A whose main diagonal elements are main diagonal elements of A. Finally, the <u>minor of an element</u> a_{ij} of a square matrix A is the determinant of the square submatrix of A obtained by deleting its i-th row and j-th column; the <u>cofactor</u> A_{ij} of the element a_{ij} is defined to be $(-1)^{i+j}$ times the minor of a_{ij}. Using these notions, we can develop a practical rule for evaluating detA.

<u>EXAMPLE 14.1</u> (Illustrations of Preceding Definitions) Let

$$A = \begin{bmatrix} -1 & 1 & 0 & 2 \\ 1 & 1 & i & 0 \\ -i & 4 & 2 & i \\ 1 & 1 & 1 & 1 \end{bmatrix} .$$

Some submatrices of A are:

$$\begin{bmatrix} -1 & 0 \\ 1 & i \\ -i & 2 \end{bmatrix} \qquad \begin{bmatrix} 2 \\ 0 \\ 1 \end{bmatrix} \qquad \begin{bmatrix} -i & 4 & 2 & i \end{bmatrix}$$

Some minors of A are:

$$\det \begin{bmatrix} 1 & 0 \\ 1 & i \end{bmatrix} \qquad \det \begin{bmatrix} 1 & 1 & i \\ -i & 4 & 2 \\ 1 & 1 & 1 \end{bmatrix} \qquad \det \begin{bmatrix} i & 0 \\ 2 & i \end{bmatrix}$$

Some principal minors of A are:

$$\det \begin{bmatrix} -1 & 0 & 2 \\ -i & 2 & i \\ 1 & 1 & 1 \end{bmatrix} \qquad \det \begin{bmatrix} 2 & i \\ 1 & 1 \end{bmatrix} \qquad \det \begin{bmatrix} 1 & 0 \\ 1 & 1 \end{bmatrix}$$

Some minors of elements are:

$$\text{of } a_{11}: \quad \det \begin{bmatrix} 1 & i & 0 \\ 4 & 2 & i \\ 1 & 1 & 1 \end{bmatrix} \qquad \text{of } a_{23}: \quad \det \begin{bmatrix} -1 & 1 & 2 \\ -i & 4 & i \\ 1 & 1 & 1 \end{bmatrix} .$$

Some cofactors of elements are:

of a_{11}: $A_{11} = \det\begin{bmatrix} 1 & i & 0 \\ 4 & 2 & i \\ 1 & 1 & 1 \end{bmatrix}$ since $(-1)^{1+1} = 1$

of a_{23}: $A_{23} = -\det\begin{bmatrix} -1 & 1 & 2 \\ -i & 4 & i \\ 1 & 1 & 1 \end{bmatrix}$ since $(-1)^{2+3} = -1$.

EXERCISE 14.4 Display the cofactors of a_{34} and a_{22} for the matrix A of the preceding Example.

EXERCISE 14.5 Use the formal definition for detA to evaluate

$$\det\begin{bmatrix} a & b \\ c & d \end{bmatrix}.$$

The next Theorem provides a practical means for evaluating determinants whose orders exceed two; it is called the Row/Column expansion.

THEOREM 14.1 (Row/Column Expansion for Determinants)

$$\det A = \begin{cases} \sum_{j=1}^{n} a_{ij}A_{ij} & \text{(i-th row expansion)} \\ \sum_{i=1}^{n} a_{ij}A_{ij} & \text{(j-th column expansion)} \end{cases}$$

for any choice of row (i) or column (j).

PROOF. (See References).

If A is $n \times n$ then every cofactor in the above expansion is the determinant of an $(n-1)\times(n-1)$ matrix. Since the determinant of a 2×2 matrix is known already (see Exercise 14.5), the above method can be systematically employed to evaluate the determinants of 3×3, 4×4, etc. matrices. However, the time involved in performing the required computations increases rapidly with order.

EXERCISE 14.6 Evaluate $\det \begin{bmatrix} 1/2 & 0 & 1/2 \\ 1/3 & 1/3 & 1/3 \\ 1 & 0 & 0 \end{bmatrix}$ two ways; first expand by the 2nd row, then expand by the 3rd column. What other way would be quicker?

EXERCISE 14.7 Prove that if A and B are any 2×2 matrices then $\det(A \cdot B) = \det A \cdot \det B$. Is this result true for $n \times n$ matrices?

EXERCISE 14.8 Prove that for any square matrix A, $\det A = \det A'$ and $\overline{\det A} = \det \bar{A}$.

EXERCISE 14.9 Let A be any $n \times n$ matrix and A_k the matrix obtained from A by multiplying the elements of any one row or any one column by the complex number k. Prove that $\det(A_k) = k \cdot \det A$, whence prove that $\det(kA)$ $k^n \cdot \det A$.

EXERCISE 14.10 Prove that if two or more rows (columns) of the square matrix A are proportional, then $\det A = 0$.

<u>EXERCISE 14.11</u> Let $\hat{A}$ be the matrix obtained from the square matrix A by performing r row interchanges and c column interchanges. Prove that $\det(\hat{A}) = (-1)^{r+c} \det A$.

<u>EXERCISE 14.12</u> (Markov Matrices) An nxn matrix $M = (m_{ij})_{nxn}$ of non-negative real numbers is termed a Markov matrix iff the sum of elements in each row is one (i.e. $\sum_{j=1}^{n} m_{ij} = 1;\ i = 1,\ldots,n$). Prove that the product of two Markov matrices is again a Markov matrix.

An important special matrix is the <u>identity</u> matrix I. This nxn matrix possesses ones on its main diagonal and zeros elsewhere, formally: $I = (\delta_{ij})_{nxn}$, where $\delta_{ij} = 0\ (i \neq j)$ and $\delta_{ij} = 1\ (i = j)$. It is simple to verify that for any nxn matrix A, we have $AI = IA = A$. Furthermore, $\det I = 1$.

Now an nxn matrix A is termed <u>non-singular</u> iff $\det A \neq 0$ and termed <u>singular</u> otherwise. For any non-singular matrix A we define an nxn matrix A^{-1} as follows: $A^{-1} = (a^{ij})_{nxn}$, where $a^{ij} = A_{ji}/\det A\ (i,j = 1,\ldots,n)$. The matrix A^{-1} satisfies: $A \cdot A^{-1} = A^{-1} \cdot A = I$, and it is termed the <u>inverse</u> of A.

<u>EXERCISE 14.13</u> (Existence and Uniqueness of Inverse) For nxn matrices prove that there exists a matrix B such that $A \cdot B = I$ only if $\det A \neq 0$. Then prove that if $\det A \neq 0$ then A^{-1} as defined above is the only matrix satisfying this relationship. Thus an inverse exists iff $\det A \neq 0$; it is defined as above and is unique.

<u>EXERCISE 14.14</u> Find A^{-1} if $A = \begin{bmatrix} 1/2 & 0 & 1/2 \\ 1/3 & 1/3 & 1/3 \\ 1 & 0 & 0 \end{bmatrix}$.

<u>EXERCISE 14.15</u> (The Determinant of A^{-1}) If A is non-singular prove that $\det(A^{-1}) = 1/\det A$.

Subject to existence, inverse satisfies the following algebraic properties: $(A \cdot B)^{-1} = B^{-1} \cdot A^{-1}$, $(A')^{-1} = (A^{-1})'$, $(\bar{A})^{-1} = (\overline{A^{-1}})$, and clearly $I^{-1} = I$.

<u>EXERCISE 14.16</u> The matrix A of Exercise 14.14 is a Markov matrix. Need the inverse of a Markov matrix always exist?

<u>EXERCISE 14.17</u> (Markov Matrices: Continuation) For any Markov matrix M prove that the value of any element in the j-th column of $M^2 = M \cdot M$ lies between the largest and smallest elements in the j-th column of M.

<u>EXERCISE 14.18</u> (Doubly Stochastic Matrices) A Markov matrix is termed doubly stochastic iff its column sums are all also one (alternately, its transpose is a Markov matrix). Is the product of two doubly stochastic matrices again doubly stochastic? Need the inverse of a doubly stochastic matrix exist?

Some Special Types of Square Matrices.

Certain special types of square matrices tend to find frequent applications in the various areas of Applied Mathematics. Their importance stems from the special properties they satisfy. The following is a partial list of these various types of square nxn matrices along with their defining properties. Following this list we shall examine some properties and interrelationships that arise from the definitions.

Type of Matrix	Defining Property
Singular	$\det A = 0$
Non-singular	$\det A \neq 0$
Symmetric	$A = A'$
Hermetian	$A = \bar{A}'$
Skew-symmetric	$A = -A'$
Unitary	$A \cdot \bar{A}' = I$
Normal	$A \cdot \bar{A}' = \bar{A}' \cdot A$
Orthogonal	$A \cdot A' = I$
Idempotent	$A^2 = A$
Nilpotent	$A^2 = 0$
Diagonal	$a_{ij} = 0\ (i \neq j)$
Triangular	(upper) $a_{ij} = 0\ (i > j)$
	(lower) $a_{ij} = 0\ (i < j)$
Positive Definite	A real, A symmetric,
	$\underline{x}'A\,\underline{x} > 0$ for all real $\underline{x} \neq \underline{0}$ [1]
Positive Semi-definite	A real, A symmetric,
	$\underline{x}'A\,\underline{x} \geq 0$ for all real $\underline{x} \neq \underline{0}$

[1]This notation will be developed fully later. Briefly, here $\underline{x}'$ denotes the 1xn row matrix $\underline{x}' = (x_1,\ldots,x_n)$ of real numbers, and $\underline{x}$ denotes the nx1 column matrix which is its transpose; $\underline{0}$ denotes the nx1 column vector of all zeros. Thus, $\underline{x}'A\,\underline{x}$ is actually a single real number.

In what follows we shall examine some of the properties and interrelationships that follow from the above definitions. Orthogonal, Positive-Definite and Positive Semi-definite matrices, however, will be postponed until a later Section. We shall by no means exhaust the properties (this would require a considerable amount of material), but rather shall work with some of the more important properties, in an attempt to gain some experience in working with the various types of nxn matrices.

EXERCISE 14.19 Give a 2x2 example of each type of matrix.

Singular Matrices: $\det A = 0$.

Property 1. $A \cdot B$ singular iff A singular or B singular.

PROOF. $\det(A \cdot B) = \det A \cdot \det B$, whence $\det(A \cdot B) = 0$ iff $\det A = 0$ or $\det B = 0$.

Property 2. A singular iff A^n singular $(n = 1,2,\ldots)$.

PROOF. EXERCISE 14.20

Property 3. A singular iff $A\underline{x} = \underline{0}$ has a solution $\underline{x} \neq \underline{0}$.

PROOF. See Section 16.

Non-Singular Matrices: $\det A \neq 0$

Property 1. $A \cdot B$ non-singular iff A and B non-singular.

PROOF. EXERCISE 14.21

Property 2. A non-singular iff A^n non-singular $(n = 1,2,\ldots)$.

PROOF. EXERCISE 14.22

Property 3. A non-singular iff A^{-1} exists and is non-singular.

PROOF. Previously established.

Property 4. A non-singular iff $A\underline{x} = \underline{0}$ has only solution $\underline{x} = \underline{0}$.

PROOF. See Section 16.

Symmetric Matrices: $A = A'$

Property 1. A symmetric $\Rightarrow kA$ symmetric.

PROOF. $(kA)' = kA' = kA$.

Property 2. A symmetric, non-singular $\Rightarrow A^{-1}$ symmetric.

PROOF. $(A^{-1})' = (A')^{-1} = A^{-1}$.

Property 3. A and B symmetric $\Rightarrow A \pm B$ symmetric.

PROOF. $(A \pm B)' = A' \pm B' = A \pm B$.

Property 4. C_{mxn} arbitrary $\Rightarrow C'C$ symmetric nxn.

PROOF. <u>EXERCISE 14.23</u>

Property 5. A arbitrary square $\Rightarrow k(A + A')$ symmetric.

PROOF. <u>EXERCISE 14.24</u>

Property 6. D diagonal $\Rightarrow$ D symmetric.

PROOF. <u>EXERCISE 14.25</u>

Property 7. A symmetric $\Rightarrow$ all characteristic roots real.

PROOF. See Section 17.

Property 8. A real, symmetric $\Rightarrow$ $P \cdot A \cdot P' = D$ (diagonal) for some orthogonal matrix P.

PROOF. See Theorem 17.5 of Section 17.

Property 9. A symmetric $\Rightarrow$ A^n symmetric $(n = 1,2,\ldots)$.

PROOF. $(A^n)' = (A')^n = A^n$.

<u>Hermetian Matrices</u>: $A = \bar{A}'$

Property 1. A Hermetian $\Rightarrow$ $(A + \bar{A}')$ and cA (c real) Hermetian.

PROOF. First, $\overline{(A + \bar{A}')}' = (\bar{A} + A')' = (A + \bar{A}')$.
Next, $(\overline{cA})' = c\bar{A}' = cA$.

Property 2. A Hermetian $\Rightarrow$ A normal.

PROOF. $A \cdot \bar{A}' = A \cdot A = \bar{A}' \cdot A$.

Property 3. A arbitrary square $\Rightarrow$ $A \cdot \bar{A}'$ and $\bar{A}' \cdot A$ Hermetian.

PROOF. EXERCISE 14.26

Property 4. P_{mxn} arbitrary, A Hermetian $nxn \Rightarrow P \cdot A \cdot \bar{P}'$ Hermetian mxm.

PROOF. Dimension requirements are satisfied. Then, $(P \cdot A \cdot \bar{P}')' = (\bar{P} \cdot \bar{A} \cdot P')' = P \cdot \bar{A}' \cdot \bar{P}' = P \cdot A \cdot \bar{P}'$.

Property 5. A Hermetian $\Rightarrow$ all characteristic roots real.

PROOF. See Section 17.

Skew-symmetric Matrices: $A = -A'$

Property 1. A skew-symmetric $\Rightarrow$ $a_{ii} = 0$ $(i = 1,\ldots,n)$.

PROOF. EXERCISE 14.27

Property 2. A and B skew-symmetric $\Rightarrow$ $(A \pm B)$ skew-symmetric.

PROOF. $-(A \pm B)' = -A' \mp B' = (A \pm B)$.

Property 3. A arbitrary square $\Rightarrow$ $k(A - A')$ skew-symmetric.

PROOF. EXERCISE 14.28

Property 4. (Decomposition) A arbitrary square $\Rightarrow$ $A = B + C$ where B symmetric and C skew-symmetric.

PROOF. Take $B = \frac{1}{2}(A + A')$ and $C = \frac{1}{2}(A - A')$. Is this decomposition unique?

Property 5. A skew-symmetric $\Rightarrow$ A^2 symmetric.

PROOF. EXERCISE 14.29

Property 6. A skew-symmetric, k odd $\Rightarrow$ $\det A^k = 0$.

PROOF. EXERCISE 14.30

Property 7. A skew-symmetric $\Rightarrow$ real part of characteristic roots equal zero.

PROOF. See Section 17.

Unitary Matrices: $A \cdot \bar{A}' = I$

Property 1. A Unitary $\Rightarrow$ A non-singular, $A^{-1} = \bar{A}'$.

PROOF. EXERCISE 14.31

Property 2. A Unitary $\Rightarrow$ $|\det A| = 1$ [1]

PROOF. EXERCISE 14.32

Property 3. A Unitary $\Rightarrow$ A normal.

PROOF. $A \cdot A^{-1} = A \cdot \bar{A}' = A^{-1} \cdot A = \bar{A}' \cdot A$.

Property 4. A Unitary, real $\Rightarrow$ A orthogonal.

PROOF. $A \cdot A' = A \cdot \bar{A}' = I$ since $A = \bar{A}$.

Property 5. A Unitary $\Rightarrow$ characteristic roots have unit modulus.

PROOF. See Section 17.

[1] $|\ |$ denotes the modulus - in the sense of complex numbers - of detA. If detA is real, this reduces to ordinary absolute value.

<u>Normal Matrices</u>: $A \cdot \overline{A}' = \overline{A}' \cdot A$

Property 1. A normal, U unitary $\Rightarrow$ $\overline{U}' \cdot A \cdot U$ normal.

PROOF. <u>EXERCISE 14.33</u>

Property 2. A normal, all characteristic roots have unit modulus $\Rightarrow$ A Unitary.

PROOF. See Section 17.

<u>Orthogonal Matrices</u>: $A \cdot A' = I$

Property 1. A orthogonal $\Rightarrow A$ non-singular and $A^{-1} = A'$.

PROOF. <u>EXERCISE 14.34</u>

Property 2. A orthogonal $\Rightarrow$ $\det A = \pm 1$.

PROOF. <u>EXERCISE 14.35</u>

Property 3. A and B orthogonal $n \times n \Rightarrow A \cdot B$, A^{-1}, A^n $(n = 1,2,\ldots)$ orthogonal.

PROOF. <u>EXERCISE 14.36</u>

Property 4. A orthogonal, real $\Rightarrow$ A normal.

PROOF. <u>EXERCISE 14.37</u>

Property 5. A orthogonal $\Rightarrow$ characteristic roots have unit modulus.

PROOF. See Section 17.

Hints and Answers to Exercises: Section 14

14.2 Hint: For the products $A \cdot B$ and $B \cdot A$ both to be defined they must both be square with the same dimensions.

14.3 Hint: Note that $A = \bar{A}$ iff $a_{ij} = \bar{a}_{ij}$ for all i and j.

14.4 Answer: of a_{34}: $A_{34} = -\det\begin{bmatrix} -1 & 1 & 0 \\ 1 & 1 & i \\ 1 & 1 & 1 \end{bmatrix}$

of a_{22}: $A_{22} = \det\begin{bmatrix} -1 & 0 & 2 \\ -i & 2 & 0 \\ 1 & 1 & 1 \end{bmatrix}$.

14.5 Answer: ad-bc

14.6 Answer: -1/6

14.7 Hint: Use the result of Exercise 14.5. This is true in general.

14.8 Hint: Use the results of Theorem 14.1.

14.9 Hint: Use the formal definition of detA.

14.10 Hint: Use the results of Theorem 14.1.

14.11 Hint: It suffices to establish that a single row or column interchange alters the sign of the determinant of the square matrix involved.

14.13 Hint: For the first part note that $\det A \cdot \det B = 1$. For the second part observe that if $\det A \neq 0$ then $A \cdot A^{-1} = A \cdot B = I$; multiply by A^{-1} across the equality on the left.

14.15 Hint: $\det(A \cdot A^{-1}) = \det(A)\det(A^{-1}) = \det(I) = 1.$

14.16 Hint: No. For example, $\begin{bmatrix} \frac{1}{2} & \frac{1}{2} \\ \frac{1}{2} & \frac{1}{2} \end{bmatrix}$.

14.17 Hint: The elements of the j-th column of M^2 are: $n_{ij} = \sum_{k=1}^{n} m_{ik}\, m_{jk}$ $(i = 1,2,\ldots,n)$.

14.18 Hint: Use the transpose criterion for the first part, and use the 2 x 2 matrix of Exercise 14.16 for the second part.

14.20 Hint: $\det(A^n) = [\det A]^n$.

14.21 Hint: For two real numbers a and b note that $a \cdot b = 0$ iff $a = 0$ or $b = 0$.

14.22 Hint: Refer to the Hint for Exercise 14.20.

14.23 Hint: Clearly $(C'C)' = C'(C')' = C'C$.

14.27 Hint: Note that $x = -x$ iff $x = 0$.

14.29 Hint: $A^2 = (-A')^2 = (A^2)'$.

14.30 Hint: $\det(A) = \det(A')$.

14.31 Hint: Premultiply (i.e. multiply on the left) the equality $A \cdot \bar{A}' = I$ by A^{-1} .

14.32 Hint: If for a complex number z we have $z \cdot \bar{z} = 1$, then the modulus of z is one.

14.33 Hint: A straightforward application of the two definitions.

14.34 Hint: Premultiply the relation $A \cdot A' = I$ by A^{-1}.

14.35 Hint: $\det(A \cdot A') = \det(A)\det(A') = [\det(A)]^2 = 1$.

14.36 Hint: A straightforward application of the definition.

14.37 Hint: If A is real then $A = \bar{A}$.

References to Additional and Related Material: Section 14

1. Ayres, F., "Schaum's Outline of Theory and Problems of Matrices", Schaum Publishing Co. (1962).

2. Browne, E., "Introduction to the Theory of Determinants and Matrices", University of North Carolina Press (1958).

3. Cullen, C., "Matrices and Linear Transformations", Addison-Wesley, Inc. (1966).

4. Eves, H., "Elementary Matrix Theory", Allyn and Bacon, Inc. (1966).

5. Finkbeiner, D., "Introduction to Matrices and Linear Transformations", W. H. Freeman, Inc. (1960).

6. Gantmacher, F., "The Theory of Matrices", Chelsea Publishing Co. (1959).

7. Greybill, F., "Introduction to Matrices with Applications in Statistics", Wadsworth Publishing Co. (1969).

8. Hollingsworth, C., "Vectors, Matrices and Group Theory for Scientists and Engineers", McGraw-Hill, Inc. (1967).

9. Johnston, J., "Linear Equations and Matrices", Addison-Wesley, Inc. (1966).

10. Lancaster, P., "Theory of Matrices", Academic Press (1969).

11. Macduffee, C., "Vectors and Matrices", Carus Mathematical Monograph 7, Mathematical Association of America (1961).

12. Marcus, M., "A Survey of Matrix Theory and Matrix Inequalities", Allyn and Bacon, Inc. (1964).

13. Murdoch, D., "Linear Algebra for Undergraduates", John Wiley and Sons, Inc. (1957).

14. Pease, M., "Methods of Matrix Algebra", Academic Press (1965).

15. Pipes, L., "Matrix Methods for Engineers", Prentice-Hall, Inc. (1963).

16. Schkade, L., "Vectors and Matrices", C. E. Merrill Publishing Co. (1967).

17. Schwartz, J., "Introduction to Matrices and Vectors", McGraw-Hill, Inc. (1961).

18. Stoll, R., "Linear Algebra and Matrix Theory", McGraw-Hill, Inc. (1952).

19. Thrall, R., "Vector Spaces and Matrices", John Wiley and Sons, Inc. (1957).

20. Turnbull, H., "Theory of Determinants, Matrices, and Invariants", Dover Publications, Inc.

21. Wade, T., "The Algebra of Vectors and Matrices", John Wiley and Sons, Inc. (1951).

15. Vectors and Vector Spaces

Throughout what follows a <u>row vector</u> $\underline{a}' = (a_1, a_2, \ldots, a_n)$ is an ordered n-tuple of complex numbers.[1] Thus, a finite set $\underline{a}_1', \underline{a}_2', \ldots, \underline{a}_m'$ of such row vectors can conveniently be thought of as the rows of the $m \times n$ matrix of complex numbers:

$$A = \begin{bmatrix} \underline{a}_1' \\ \underline{a}_2' \\ \vdots \\ \underline{a}_m' \end{bmatrix} = \begin{bmatrix} a_{11} & a_{12} & \cdots & a_{1n} \\ a_{21} & a_{22} & \cdots & a_{2n} \\ \vdots & & & \\ a_{m1} & a_{m2} & \cdots & a_{mn} \end{bmatrix} .$$

Then the rows of A are termed <u>independent</u> iff no row vector of A can be expressed as a linear combination of the remaining row vectors of A. Otherwise, the rows (row vectors) of A are termed <u>dependent</u>. The following result is an alternate characterization of independence.

<u>THEOREM 15.1</u> (Characterization of Independence) The rows of A are independent (dependent) iff there does not (does) exist a non-null vector $\underline{c}' = (c_1, \ldots, c_m)$ of complex constants such that

$$\underline{c}'A = \underline{0}' .$$

[1] Most of the results of this Section are automatically valid in the case where we deal with real numbers and vectors of real numbers only. However, we shall consider the more general case.

PROOF. The matrix equation $\underline{c}'A = \underline{0}'$ can also be written as: $c_1\underline{a}_1' + c_2\underline{a}_2' + \ldots + c_m\underline{a}_m' = \underline{0}'$. Therefore, if $\underline{c}' \neq \underline{0}'$ there exists at least one index, say j, for which $c_j \neq 0$, and accordingly we may write: $\underline{a}_j' = -\sum_{i \neq j} (\frac{c_i}{c_j})\, \underline{a}_i'$, in which case the rows of A are dependent.

Conversely, if the rows of A are dependent, then for some index (row), say k, we have: $\underline{a}_k' = \sum_{i \neq k} c_i \underline{a}_i'$. It then follows that $\underline{c}'A = \underline{0}'$ with $\underline{c}' = (c_1, \ldots, c_{k-1}, -1, c_{k+1}, \ldots, c_m)$, which is clearly a non-null vector of complex constants. With this the proof is completed.

With the definition of addition and scalar multiplication of vectors (matrices) given in Section 14, the set C_n of all complex row vectors of length n forms a so-called <u>vector space</u>, meaning here that:

(i)	$\underline{0}' \in C_n$	C_n contains the null vector
(ii)	$\underline{x}', \underline{y}' \in C_n \Rightarrow \underline{x}' + \underline{y}' \in C_n$	C_n is closed under vector addition
(iii)	$\underline{x}' \in C_n$, c scalar $\Rightarrow c\underline{x}' \in C_n$	C_n is closed under complex scalar multiplication

With the above definition in mind, we term a subset T of C_n a <u>subspace</u> of C_n iff the vectors in T satisfy the three properties stated above.

<u>EXERCISE 15.1</u> Prove that the subset T_i of C_n, composed of all vectors for which the i-th component is zero, forms a subspace of C_n for each $i = 1,2,\ldots,n$.

We naturally say that the (row) vector $\underline{a}'$ <u>depends</u> upon (or is dependent upon) the (row) vectors $\underline{a}_1', \underline{a}_2', \ldots, \underline{a}_m'$ of A iff $\underline{a}'$ can be expressed as a linear combination of the (row) vectors forming A. This now provides us with an important example of a subspace of C_n.

<u>THEOREM 15.2</u> (Subspace Generated by the Rows of A) The set D(A) of all (row) vectors dependent upon the rows of A forms a Subspace of C_n.

PROOF. Any vector in D(A) can be expressed as $\underline{k}'A$ where $\underline{k}' = (k_1, k_2, \ldots, k_m)$ is some vector of m complex constants.

Clearly $\underline{0}' \in D(A)$ since $\underline{0}' = \underline{0}'A$. Next, if $\underline{x}_1' = \underline{k}_1'A$ and $\underline{x}_2' = \underline{k}_2'A$ belong to D(A) then so does their sum since $\underline{x}_1' + \underline{x}_2' = (\underline{k}_1 + \underline{k}_2)'A$. Finally, D(A) is closed under complex scalar multiplication since if $\underline{x}' = \underline{k}'A$ belongs to D(A) then clearly $c\underline{x}' = (c\underline{k})'A$ also belongs to D(A), for any complex constant c.

Thus, D(A) is a subspace of C_n and the proof is completed.

In the above situation, the rows of A are said to <u>generate</u> the subspace D(A) of C_n. Note, however, that the rows of A (termed the <u>generators</u> of the subspace) need not themselves be independent vectors. If, however, it so happens that the rows of A (i.e. the generators) <u>are</u> independent, then they are termed a <u>basis</u> of D(A).

<u>EXERCISE 15.2</u> Prove that a basis (hence also the generators) of a subspace is not unique by showing that both

(i) (1,0), (0,1)

and

(ii) (-i,0), (i,-i)

are bases (hence also generators) for all of C_2 .

As the preceding Exercise illustrates, there may be many different bases (hence generators) for a given vector space or subspace. In fact, given the generators (or a basis) of a specific subspace, the vectors generating the subspace may be altered in many different manners and still leave the generated subspace unchanged. The general result formalizing the above statements is contained in the following Theorem. Later on, our interest mainly will be with certain special cases of this result.

<u>THEOREM 15.3</u> (Alteration of Generators) As usual, let A be the m x n matrix consisting of row vectors $\underline{a}_1', \underline{a}_2', \ldots, \underline{a}_m'$ and let B be <u>any</u> non-singular m x m matrix of complex numbers. Then, the subspace generated by the rows of BA is identical with the subspace generated by the rows of A itself. That is, D(BA) = D(A).

PROOF. Suppose $\underline{x}' = \underline{k}'A$ belongs to D(A). Then $\underline{x}' = (B^{-1}{}'\underline{k})'BA$, which shows that $\underline{x}'$ belongs to D(BA). Conversely, suppose that $\underline{y}' = \underline{k}'BA$ belongs to D(BA). Then $\underline{y}' = (B'\underline{k})'A$, which shows that $\underline{y}'$ belongs to D(A) as well. This completes the proof.

Some special cases of the preceding Theorem with which we shall be concerned are contained in the following Theorem.

THEOREM 15.4 (Row Operations) The subspace D(A) generated by the row vectors of the matrix A, namely $\underline{a}_1', \underline{a}_2', \ldots, \underline{a}_m'$, remains unchanged if:

(i) $\underline{a}_i'$ is replaced by $c_i \underline{a}_i'$ $(c_i \neq 0)$

(ii) $\underline{a}_i'$ is replaced by $\underline{a}_i' + \underline{a}_j'$ $(i \neq j)$

(iii) $\underline{a}_i'$ is dropped if equal to $\underline{0}'$.

PROOF. Alteration (i) is accomplished by premultiplying A by B, where B is the non-singular diagonal matrix having $b_{ii} = c_i$ and $b_{jj} = 1$ $(i \neq j)$. Alteration (ii) is accomplished by premultiplying the matrix A by B, where B is the non-singular matrix obtained from the $m \times m$ identity matrix by adding its j-th row to its i-th row. Alteration (iii) requires no proof. Since all of these are special cases of the preceding Theorem, the proof is completed.

It is to be observed that the three 'row operations' indicated in the preceding Theorem may now be combined in various manners and performed any number of times leaving the generated subspace unchanged.

The following Example will illustrate use of the preceding Theorem.

EXAMPLE 15.1 Consider the subspace generated by the row vectors:

$$\underline{a}_1' = (2, -1, 1, 5)$$
$$\underline{a}_2' = (1, 2, -2, -3)$$
$$\underline{a}_3' = (7, 4, -4, 1) \quad .$$

If we replace $\underline{a}_1'$ by $\underline{b}_1' = \underline{a}_1' + \underline{a}_2'$ we obtain the new generators for the same subspace:

$$\underline{b}_1' = (3, 1, -1, 2)$$
$$\underline{a}_2' = (1, 2, -2, -3)$$
$$\underline{a}_3' = (7, 4, -4, 1) \quad .$$

If we next replace $\underline{a}_2'$ by $\underline{b}_2' = 2\underline{a}_2' + \underline{b}_1'$ we again obtain new generators for the same subspace:

$$\underline{b}_1' = (3, 1, -1, 2)$$
$$\underline{b}_2' = (5, 5, -5, -4)$$
$$\underline{a}_3' = (7, 4, -4, 1) \quad .$$

Finally, if we replace $\underline{a}_3'$ by $\underline{b}_3' = \underline{a}_3' - 3/2\underline{b}_1' - 1/2\underline{b}_2'$ (which is the null vector and can therefore be deleted) we obtain the following generators for the original subspace:

$$\underline{b}_1' = (3, 1, -1, 2)$$
$$\underline{b}_2' = (5, 5, -5, -4) \quad .$$

Later on, techniques will be developed that will enable us to actually prove that the final generators form a basis for the subspace. In fact, a systematic procedure based upon the above ideas will be given that will enable us to pass from a set of dependent generators to an independent basis for a given subspace.

Before we proceed further, we need the following useful and obvious result.

THEOREM 15.5 Any collection of $n+1$ or more (row) vectors in C_n is dependent.

PROOF. EXERCISE 15.3 (Note that the n row vectors $\underline{b}_1' = (1,0,\dots,0)$, $\underline{b}_2' = (0,1,0,\dots,0),\dots,\underline{b}_n' = (0,0,\dots,0,1)$ form a basis of C_n).

We are now in a position to prove the (perhaps obvious) result that every subspace T of C_n has a basis.

THEOREM 15.6 (Existence of a Basis) Every subspace T of C_n has a basis.

PROOF. To avoid the trivial case, assume T contains non-null vectors. Let $\underline{b}_1'$ be an arbitrary non-null vector in T, and let $D(\underline{b}_1')$ be the subspace of all vectors in T dependent upon $\underline{b}_1'$. If $T \equiv D(\underline{b}_1')$ we are done and $\underline{b}_1'$ is a basis of T. Otherwise, choose a non-null vector $\underline{b}_2'$ belonging to T but not in $D(\underline{b}_1')$ and let $D(\underline{b}_1',\underline{b}_2')$ denote the subspace of all vectors in T dependent upon $\underline{b}_1'$ and $\underline{b}_2'$ (which are clearly independent vectors). If $T \equiv D(\underline{b}_1',\underline{b}_2')$ then $\underline{b}_1'$ and $\underline{b}_2'$ form a basis of T and we are done. If not, continue the above process. However, in view of Theorem 15.5, the process must terminate in $r \le n$ steps, for otherwise we would have the impossible existence of $n+1$ or more indendent vectors in C_n. Thus we eventually arrive at a basis $\underline{b}_1', \ldots, \underline{b}_r'$ of r independent vectors for T. This completes the proof.

<u>EXERCISE 15.4</u> The rows of A generating $D(A)$ need not, of course, be independent. Prove, however, that it is always possible to select a basis for $D(A)$ from amongst the rows of A.

Although there may be many different bases for a given subspace T, one important fact is that, once a fixed basis is chosen, every vector in T can be expressed uniquely as a linear combination of the fixed basis vectors. This result is established in the following Theorem.

<u>THEOREM 15.7</u> (Unique Representation in Terms of Fixed Basis) Given a fixed basis of a subspace T of C_n, every vector in T can be expressed uniquely (i.e. in only one way) as a linear combination of these basis vectors.

PROOF. Let the fixed basis of T be $\underline{b}_1', \underline{b}_2', \ldots, \underline{b}_r'$ and let $\underline{b}'$ be an arbitrary vector in T. Since the basis vectors generate T, $\underline{b}'$ depends upon them whence: $\underline{b}' = \sum_1^r c_i \underline{b}_i'$ for certain values of the comples constants $c_1, c_2, \ldots, c_r$. Now suppose to the contrary that this representation is <u>not</u> unique, in which case there exists another representation, say $\underline{b}' = \sum_{1}^{r} d_i \underline{b}_i'$ where at least one of the differences $d_i - c_i$ is non-zero. However, if this were the case, then $\underline{0}' = \underline{b}' - \underline{b}' = \sum_{1}^{r} (d_i - c_i)\underline{b}_i'$, which would contradict the independence of the basis vectors. Thus, the representation must be unique, and the proof is completed.

It then follows immediately from the preceding Theorem that if the rows of the $m \times n$ matrix A are independent, they form a basis for $D(A)$ and every vector therein can be expressed uniquely as $\underline{c}'A$ for a suitable choice of $\underline{c}' = (c_1, c_2, \ldots, c_m)$.

Next we consider the question as to what happens to a basis of a subspace T when the vectors in the basis are altered by means of various row operations. The result is contained in the following Theorem.

<u>THEOREM 15.8</u> (Transforming a Basis Into a Basis) Let $\underline{b}_1', \underline{b}_2', \ldots, \underline{b}_r'$ denote a basis of a subspace T of C_n. Thus, $T \equiv D(B)$, where B is the $r \times n$ matrix having the basis vectors as rows. If C is <u>any</u> non-singular $r \times r$ matrix of complex numbers, then the rows of the $r \times n$ matrix CB also form a basis of T.

PROOF. It has already been proven in Theorem 15.3 that $D(B) \equiv D(CB)$. Thus we need only show that the rows of CB are independent. Suppose to the contrary that they are dependent. Then for some non-null vector $\underline{d}' = (d_1, d_2, \ldots, d_r)$ we would have $\underline{0}' = \underline{d}'CB$, whence $\underline{0}' = (C'\underline{d})'B$. But $(C'\underline{d})'$ is a non-null vector (why?) which contradicts independence of the rows of B. Thus we must conclude that the rows of CB form a basis of T. This completes the proof.

Therefore, of course, each of the 'row operations' described in Theorem 15.4 transforms a basis into a basis.

Now we have already seen that a basis for a subspace T of C_n is not unique. However, even though this is the case, it can be proved that the number of vectors forming the basis is unique. This unique number is termed the RANK of T, and the result is established in the following Theorem.

THEOREM 15.9 (Rank of a Subspace) For any non-null subspace T of C_n there exists a unique integer r ($0 < r \le n$), termed the rank of T, such that:

(i) any basis of T contains exactly r non-null vectors

and

(ii) any set of r independent vectors from T is a basis of T.

PROOF. We prove part (ii) first. Accordingly, let $\underline{b}_1', \underline{b}_2', \ldots, \underline{b}_r'$ be a basis of T (already proved to exist) and let $\underline{a}_1', \underline{a}_2', \ldots, \underline{a}_r'$ be any r independent vectors in T. By using only combinations of the three 'row' operations, we may pass from $\underline{b}_1', \underline{b}_2', \ldots, \underline{b}_r'$ to $\underline{a}_1', \underline{b}_2', \ldots, \underline{b}_r'$ then to $\underline{a}_1', \underline{a}_2', \underline{b}_3', \ldots, \underline{b}_r'$ and ultimately to $\underline{a}_1', \underline{a}_2', \ldots, \underline{a}_r'$. Since a basis is transformed into a basis under such operations, the second part of the Theorem is proved.

Finally, there can be no more than r independent vectors in T, for, if there were, any collection of r of them would form a basis of T and the others would depend upon them. This would contradict independence of the vectors in the original set, and, with this, the first part is established. This completes the proof.

If T_1 and T_2 are both subspaces of C_n, it may occur that, say, every vector of T_1 also belongs to T_2, that is, $T_1 \subseteq T_2$.[1] Naturally we call T_1 a subspace of T_2 in this case. (Both are subspaces of C_n of course).

There is a relationship between the ranks of T_1 and T_2 when T_1 is a subspace of T_2. This result is contained in the following Theorem.

THEOREM 15.10 (Relationship Between Ranks of Subspaces) If T_1 is a subspace of T_2 then $r_1 \leq r_2$, with equality iff $T_1 \equiv T_2$, where $r_1 = \text{Rank}(T_1)$ and $r_2 = \text{Rank}(T_2)$.

PROOF. Let $\underline{b}_1', \underline{b}_2', \ldots, \underline{b}_{r_1}'$ be a basis of T_1. Iff $T_1 \not\equiv T_2$ there exists a non-null vector, say $\underline{b}'$, belonging to T_2 but not to T_1. Since the vectors $\underline{b}_1', \underline{b}_2', \ldots, \underline{b}_{r_1}', \underline{b}'$ of T_2 are independent, it follows that $\text{Rank}(T_2) \geq r_1 + 1$. This completes the proof.

EXERCISE 15.5 Verify the statement in the preceding proof that the vectors $\underline{b}_1', \underline{b}_2', \ldots, \underline{b}_{r_1}', \underline{b}'$ are independent.

---

[1] Such need not be the case in general.

In view of the preceding Theorem, it is clear that if T_1 is a subspace of T_2 with basis $\underline{b}_1', \underline{b}_2', \ldots, \underline{b}_{r_1}'$, then this basis may be extended, by inclusion of vectors from the subspace T_2, so as to eventually form a basis for T_2. The number of new vectors that would be included must be, of course, $r_2 - r_1$.

We now consider a generalization of the notion of perpendicularity that is familiar from Euclidean Geometry. Of course, we are dealing here with vectors of complex numbers, so there is not the same Geometric interpretation as in the real case. However, the notions we consider should seem natural generalizations of this more familiar case.

If $\underline{a}' = (a_1, a_2, \ldots, a_n)$ and $\underline{b}' = (b_1, b_2, \ldots, b_n)$ are any two vectors in C_n, we define their inner product (also called 'dot' product) written $\underline{a}' \cdot \underline{b}'$ as follows:

$$\underline{a}' \cdot \underline{b}' = \sum_1^n a_i \bar{b}_i = \underline{a}' \underline{\bar{b}} ,$$

where $\bar{\ }$ denotes the operation of complex conjugation. In general, it can be seen that the inner product of two vectors in C_n is a complex number.

In terms of inner product, we now define what is meant by the length of a vector in C_n. If $\underline{a}' = (a_1, a_2, \ldots, a_n)$ is an arbitrary vector in C_n we define the length of $\underline{a}'$, written $||\underline{a}'||$, as follows:

$$||\underline{a}'|| = (\underline{a}' \cdot \underline{a}')^{\frac{1}{2}} .$$

EXERCISE 15.6 Prove that the length of a vector in C_n as defined above is a non-negative real number.

EXERCISE 15.7 (Properties of Length) Establish the following properties of length of vectors in C_n as defined above:

(i) $||\underline{a}'|| \geq 0$, with $||\underline{a}'|| = 0$ iff $\underline{a}' = \underline{0}'$.

(ii) $||c\underline{a}'|| = |c|\ ||\underline{a}'||$, c any complex constant

(iii) $||\underline{a}' + \underline{b}'|| \leq ||\underline{a}'|| + ||\underline{b}'||$

(iv) $||\bar{\underline{a}}'|| = ||\underline{a}'||$.

Now because:

$$|\underline{a}' \cdot \underline{b}'| = |\sum_1^n a_i\bar{b}_i| \leq \sum_1^n |a_i\bar{b}_i| = \sum_1^n |a_i||b_i| \leq ||\underline{a}'||\ ||\underline{b}'||,$$

it follows that the quantity:

$$\frac{\underline{a}' \cdot \underline{b}'}{||\underline{a}'||\ ||\underline{b}'||}$$

is a complex number with a modulus of at most one (provided $\underline{a}', \underline{b}' \neq \underline{0}'$) .

EXERCISE 15.8 Verify the above statements.

Hence, there exists a complex number θ with $0 \leq \text{Re}(\theta) < \pi$ such that:

$$\frac{\underline{a}' \cdot \underline{b}'}{||\underline{a}'||\ ||\underline{b}'||} = \text{Cos}\ \theta \quad .$$

In natural parallel with real case, we then define $\underline{a}'$ and $\underline{b}'$ to be orthogonal (perpendicular in an extended sense) iff $\cos\theta = 0$ or, equivalently, $\underline{a}' \cdot \underline{b}' = 0$. If either $\underline{a}' = \underline{0}'$ or $\underline{b}' = \underline{0}'$ then $\underline{a}' \cdot \underline{b}' = 0$ and $\underline{a}'$ and $\underline{b}'$ are automatically orthogonal.

EXERCISE 15.9 Is the complex number θ defined above unique?

We now begin to develop some useful results that follow from our definition of orthogonality. The first is contained in the following Theorem.

THEOREM 15.11 If the vector $\underline{b}'$ is orthogonal to each of the vectors forming the rows of A then $\underline{b}'$ is orthogonal to each vector in $D(A)$, in which case we say that $\underline{b}'$ is orthogonal to $D(A)$.

PROOF. For $\underline{b}'$ to be orthogonal to the rows of A it is necessary and sufficient that $A\underline{\bar{b}} = \underline{0}$. Since any vector in $D(A)$ can be expressed as $\underline{c}'A$ for suitable choice of $\underline{c}'$, it then follows that $\underline{c}'A\underline{\bar{b}} = 0$, whence any such vector is orthogonal to $\underline{b}'$. This completes the proof.

The preceding Theorem suggests that the set of all vectors orthogonal to a subspace T of C_n might itself possess certain important properties. This is indeed the case, and the result is contained in the following Theorem.

<u>THEOREM 15.12</u> (The Orthogonal Space to a Subspace) The set of all vectors orthogonal to a subspace T of C_n also forms a subspace T', termed the subspace orthogonal to T.

PROOF. Suppose the vectors forming the rows of A generate T. A necessary and sufficient condition that a vector $\underline{b}'$ be orthogonal to T is then that $A\bar{\underline{b}} = \underline{0}$. We must show that the set of all vectors $\underline{b}'$ satisfying this condition forms a subspace of C_n . We shall use the three defining properties of a (vector) subspace.

Certainly $\underline{0}'$ is orthogonal to T. Furthermore, if $\underline{a}'$ and $\underline{b}'$ are orthogonal to T then so is $\underline{a}' + \underline{b}'$ because $A(\bar{\underline{a}} + \bar{\underline{b}}) = \underline{0} + \underline{0} = \underline{0}$. Finally, if $\underline{b}'$ is orthogonal to T and c is any complex constant, then so is $c\underline{b}'$ because $A(\overline{c\underline{b}}) = \bar{c}A\bar{\underline{b}} = \underline{0}$. Thus, we complete the proof.

<u>EXERCISE 15.10</u> (Orthogonal Decomposition) Let T be a subspace of C_n and T' the subspace orthogonal to T. Prove that <u>any</u> vector $\underline{a}'$ of C_n can be expressed uniquely (decomposed) as: $\underline{a}' = \underline{b}' + \underline{c}'$, where $\underline{b}'$ belongs to T and $\underline{c}'$ belongs to T'.

<u>EXERCISE 15.11</u> (Ranks of Orthogonal Subspace) In general, two subspaces T_1 and T_2 are termed orthogonal iff every vector in T_1 is orthogonal to every vector in T_2, and conversely. Prove that $\text{Rank}(T_1) + \text{Rank}(T_2) \leq n$ with equality iff T_2 is the subspace orthogonal to T_1, i.e. $T_2 = T_1'$.

Finally, a vector $\underline{a}'$ in C_n is termed normal iff $||\underline{a}'|| = 1$, that is, iff $\underline{a}'$ has unit length. Note that every non-null vector $\underline{b}'$ in C_n may be normalized so as to become normal; this may be accomplished by multiplying $\underline{b}'$ by the scalar $1/||\underline{b}'||$.

We now conclude this section with a proof that any subspace T of C_n has an ortho-normal basis, that is, a basis consisting of vectors which are normal and orthogonal to one-another. This result is contained in the following Theorem.

THEOREM 15.13 (Gram-Schmidt) Every subspace T of C_n has an ortho-normal basis.

PROOF. (Constructive). It suffices to establish the existence of a basis consisting of mutually orthogonal vectors; an ortho-normal basis may then be obtained from it by normalizing each vector therein.

Begin with an arbitrary basis $\underline{b}_1', \underline{b}_2', \ldots, \underline{b}_r'$ (assume $r \geq 1$) of T. If $r = 1$ we are done. Therefore assume that $r \geq 2$. Recursively define vectors $\underline{e}_1', \underline{e}_2', \ldots, \underline{e}_r'$ as follows:

$$\underline{e}_1' = \underline{b}_1'$$

$$\underline{e}_2' = \underline{b}_2' - (\underline{b}_2' \cdot \underline{e}_1')\underline{e}_1'$$

$$\underline{e}_3' = \underline{b}_3' - (\underline{b}_3' \cdot \underline{e}_1')\underline{e}_1' - (\underline{b}_3' \cdot \underline{e}_2')\underline{e}_2'$$

$$\vdots$$

$$\underline{e}_r' = \underline{b}_r' - (\underline{b}_r' \cdot \underline{e}_1')\underline{e}_1' - \cdots\cdots - (\underline{b}_r' \cdot \underline{e}_{r-1}')\underline{e}_{r-1}' \quad .$$

It is not difficult to verify that this set of vectors is mutually orthogonal and a basis. This completes the proof.

<u>EXERCISE 15.12</u> Verify that the vectors $\underline{e}_1', \underline{e}_2', \ldots, \underline{e}_r'$ of the preceding proof form an orthogonal basis of T.

<u>EXERCISE 15.13</u> Use the Gram-Schmidt procedure to construct an ortho-normal basis for the subspace T of C_3 generated by the basis vectors: $\underline{b}_1' = (1, i, -1)$ and $\underline{b}_2' = (i, 0, 0)$. Is an ortho-normal basis unique? Prove, or disprove with a counterexample.

<u>EXERCISE 15.14</u> (Conjugate Subspace) Let T be a subspace of C_n with basis $\underline{b}_1', \underline{b}_2', \ldots, \underline{b}_r'$. Define $\overline{T}$ as the set of all vectors in C_n that are conjugates of vectors in T. Prove that $\overline{T}$ is indeed a subspace of C_n, termed the conjugate subspace of T, and that (i) Rank(T) = Rank($\overline{T}$), and in fact (ii) $\overline{\underline{b}}_1', \overline{\underline{b}}_2', \ldots, \overline{\underline{b}}_r'$ is a basis of $\overline{T}$.

<u>EXERCISE 15.15</u> (Continuation) If a basis consisting of vectors with real components only can be found for a subspace T, is it necessarily true that $T \equiv \overline{T}$? Either prove or provide a counterexample.

Hints and Answers for Exercises: Section 15

15.1 Hint: Verify that the three defining properties of a subspace are satisfied by T_i .

15.2 Hint: Show, for example, that for part (ii) the only solution of:

$$(c_1, c_2) \begin{bmatrix} -i & 0 \\ i & -i \end{bmatrix} = (0,0),$$

equivalently,

$$\begin{aligned} -ic_1 + ic_2 &= 0 \\ -ic_2 &= 0, \end{aligned}$$

is $c_1 = c_2 = 0$.

15.3 Hint: Use the comment accompanying the Exercise together with Theorem 15.4.

15.4 Hint: Use Theorem 15.6

15.5 Hint: Assume to the contrary that the vectors are dependent and establish a contradiction.

15.6 Hint: $||a||^2 = a_i\bar{a}_i$, and for any complex number z, $z\bar{z} = |z|^2$.

15.7 Hint: For part (iii) use the triangle inequality for complex numbers.

15.8 Hint: This follows from an extension of the triangle inequality, properties of the modulus of complex numbers, and the definition of length. See also Section 20.

15.9 Hint: Use the definition and properties of the Cosine of a complex number θ, namely $\text{Cos}\,\theta = \frac{1}{2}(e^{i\theta} + e^{-i\theta})$.

15.10 Hint: Assume the decomposition is not unique and establish a contradiction.

15.11 Hint: For the first part, establish that $T_1 \cap T_2$ contains only $\underline{0}'$. Then, for the second part, use, in addition, the definition of orthogonal subspace.

15.12 Hint: Use Theorem 15.8 and show that for a fixed but arbitrary k, $1 \le k \le n-1$, $\underline{a}_k'$ and $\underline{a}_{k+1}'$ are orthogonal.

15.13 Hint: An orthonormal basis is not unique.

15.14 Hint: Verify that the three defining properties of a subspace are satisfied by $\overline{T}$. Part (i) follows from part (ii).

References to Additional and Related Material: Section 15

1. Cullen, C., "Matrices and Linear Transformations", Addison-Wesley, Inc. (1966).

2. Hollingsworth, C., "Vectors, Matrices and Group Theory for Scientists and Engineers", McGraw-Hill, Inc. (1967).

3. Lipschutz, S., "Schaum's Outline of Theory and Problems of Linear Algebra", Schaum Publishing Co. (1968).

4. Macduffee, C., "Vectors and Matrices", Carus Mathematical Monograph 7, Mathematical Association of America (1961).

5. Murdoch, D., "Linear Algebra for Undergraduates", John Wiley and Sons, Inc. (1957).

6. Schkade, L., "Vectors and Matrices", C. E. Merrill Publishing Co. (1967).

7. Schwartz, J., "Introduction to Matrices and Vectors", McGraw-Hill, Inc. (1961).

8. Stoll, R., "Linear Algebra and Matrix Theory", McGraw-Hill Inc. (1952).

9. Thrall, R., "Vector Spaces and Matrices", John Wiley and Sons, Inc. (1957).

10. Wade, T., "The Algebra of Vectors and Matrices", Addison-Wesley, Inc. (1951).

16. Systems of Linear Equations and Generalized Inverse

A useful application of the results of the preceding Section will now be made. We will consider first the general solutions to systems of linear homogeneous equations, and then that of systems linear non-homogeneous equations. These results will then be used in finding a so-called generalized inverse of a matrix.

First consider the system of <u>m homogeneous linear equations in n (n≥ m) unknowns</u>:

$$\begin{aligned} a_{11}x_1 + a_{12}x_2 + \dots + a_{1n}x_n &= 0 \\ a_{21}x_1 + a_{22}x_2 + \dots + a_{2n}x_n &= 0 \\ \vdots \qquad \vdots \qquad\quad \vdots \quad & \\ a_{m1}x_1 + a_{m2}x_2 + \dots + a_{mn}x_n &= 0 \end{aligned}$$

or,

$$A\underline{x} = \underline{0}$$

where,

$$A = \begin{bmatrix} a_{11} & a_{12} & \cdots & a_{1n} \\ a_{21} & a_{22} & \cdots & a_{2n} \\ \vdots & & & \\ a_{m1} & a_{m2} & & a_{mn} \end{bmatrix}$$

and

$$\underline{x}' = (x_1, x_2, \dots, x_n), \quad \text{equivalently} \quad \underline{x} = \begin{bmatrix} x_1 \\ x_2 \\ \vdots \\ x_n \end{bmatrix} .$$

The coefficients a_{ij} are, in general, taken as complex numbers. We seek all (complex) vector solutions to the system $A\underline{x} = \underline{0}$. (Of course, there is always at least one solution, namely $\underline{x} = \underline{0}$, which is a solution to any homogeneous linear system). We now examine some properties of the solution vectors of the system.

First note that the two statements "$\underline{c}$ is a solution of $A\underline{x} = \underline{0}$" and "the conjugate $\overline{\underline{c}}'$ of the vector $\underline{c}'$ is orthogonal to $D(A)$" are equivalent. Both statements mean that $A\underline{c} = \underline{0}$. Therefore, the solutions of $A\underline{x} = \underline{0}$ consist of those column vectors obtained by transposing all row vectors in the subspace $\overline{D(A)^{\prime}}$, which is the subspace conjugate to the subspace orthogonal to $D(A)$. (See Section 15).

EXERCISE 16.1 Prove the statements contained in the preceding paragraph.

The preceding observations lead to the general solution of a homogeneous linear system of equations. The precise result is contained in the following Theorem.

THEOREM 16.1 (General Solution: Homogeneous Linear System) Let the homogeneous linear system of equations be $A\underline{x} = \underline{0}$. Suppose that Rank $D(A) = r$, and that $\underline{d}_1', \underline{d}_2', \ldots, \underline{d}_{n-r}'$ is a basis of the subspace orthogonal to $D(A)$ $(0 \le r \le n)$.

If $r < n$ then the general solution of the system is given by: $c_1\overline{\underline{d}}_1 + c_2\overline{\underline{d}}_2 + \ldots + c_{n-r}\overline{\underline{d}}_{n-r}$, where the c_i's are arbitrary complex constants. If $r = n$ then $\underline{x} = \underline{0}$ is the only solution.

PROOF. Since the solutions of $A\underline{x} = \underline{0}$ are the transposes of the row vectors of $\overline{D(A)^{\top}}$, which has as basis $\underline{d}_1', \underline{d}_2', \ldots, \underline{d}_{n-r}'$, the result follows immediately. In the special case $r = n$, $\overline{D(A)^{\top}}$ is the null space containing only $\underline{0}'$.

Of course, the preceding Theorem is not constructive (in contrast to the Gram-Schmidt Theorem 15.14). It does not actually provide an algorithm for obtaining the general solution referred to in the statement of the Theorem. Now, however, we shall consider a general algorithm which will provide the general solution to the homogeneous linear system $A\underline{x} = \underline{0}$, and, in the process, also provide us with a basis for D(A). The algorithm should be familiar from High School Algebra: one name for it is the "column sweep out method". It is, perhaps, best illustrated by an example.

EXAMPLE 16.1 Solve the homogeneous linear system of equations:

$$
\begin{aligned}
4x_2 + 8x_3 - 16ix_4 - 12ix_5 &= 0\\
3x_1 + 2x_2 - 5x_3 - 17ix_4 - 3ix_5 &= 0\\
4x_1 + 2x_2 - 8x_3 - 20ix_4 - 2ix_5 &= 0\\
x_1 - 3x_2 - 7x_3 + 13ix_4 + 8ix_5 &= 0
\end{aligned}
$$

Initial Stage: Equations and Coefficient Vectors as Given.

Equations	Coefficient Vectors	
$0x_1 + 4x_2 + 8x_3 - 16ix_4 - 12ix_5 = 0$	$(0,4,8,-16i,-12i) = \underline{a}_1'$	these vectors do generate D(A) but are dependent so are not a basis
$3x_1 + 2x_2 - 5x_3 - 17ix_4 - 3ix_5 = 0$	$(3,2,-5,-17i,-3i) = \underline{a}_2'$	
$4x_1 + 2x_2 - 8x_3 - 20ix_4 - 2ix_5 = 0$	$(4,2,-8,-20i,-2i) = \underline{a}_3'$	
$1x_1 + -3x_2 - 7x_3 + 13ix_4 + 8ix_5 = 0$	$(1,-3,-7,13i, 8i) = \underline{a}_4'$	

Stage 1: First Sweep Out Process (column two).

$0x_1 - 1x_2 - 2x_3 + 4ix_4 + 3ix_5 = 0$	$(0, -1, -2, 4i, 3i)$	these vectors also generate D(A) but are also dependent and so do not form a basis
$3x_1 + 0x_2 - 9x_3 - 9ix_4 + 3ix_5 = 0$	$(3, 0, -9, -9i, 3i)$	
$4x_1 + 0x_2 - 12x_3 - 12ix_4 + 4ix_5 = 0$	$(4, 0, -12, -12i, 4i)$	
$1x_1 + 0x_2 - 1x_3 + ix_4 - ix_5 = 0$	$(1, 0, -1, i, -i)$	

Stage 2: Second Sweep Out Process (column one).

$0x_1 - 1x_2 - 2x_3 + 4ix_4 + 3ix_5 = 0$	$(0, -1, -2, 4i, 3i)$	
$-1x_1 + 0x_2 + 3x_3 + 3ix_4 - ix_5 = 0$	$(-1, 0, 3, 3i, -i)$	
$0x_1 + 0x_2 + 0x_3 + 0x_4 + 0x_5 = 0$	$(0, 0, 0, 0, 0)$	may be deleted
$0x_1 + 0x_2 + 2x_3 + 4ix_4 - 2ix_5 = 0$	$(0, 0, 2, 4i, -2i)$	

Stage 3: Third Sweep Out Process (column three).

$0x_1 - 1x_2 + 0x_3 + 8ix_4 + ix_5 = 0$	$(0, -1, 0, 8i, i) = \underline{b}_1'$	these are independent vectors and generate D(A) so form a basis
$-1x_1 + 0x_2 + 0x_3 - 3ix_4 + 2ix_5 = 0$	$(-1, 0, 0, -3i, 2i) = \underline{b}_2'$	
$0x_1 + 0x_2 + 0x_3 + 0x_4 + 0x_5 = 0$		
$0x_1 + 0x_2 - 1x_3 - 2ix_4 + ix_5 = 0$	$(0, 0, -1, -2i, i) = \underline{b}_3'$	

The equations of Stage 3 are equivalent to:

$$x_1 = -3ix_4 + 2ix_5$$

$$x_2 = 8ix_4 + ix_5$$

$$x_3 = -2ix_4 + ix_5$$

where for arbitrary choices of x_4 and x_5, the vector of values $(x_1, x_2, x_3, x_4, x_5)$ satisfies the given system provided x_1, x_2, x_3 are chosen so as to satisfy the above three conditions. In purely algebraic terms, this amounts to a general solution of the system.

Now we relate the purely algebraic solution of the original system to the framework of vector (sub) spaces already established. First, since x_4 and x_5 are arbitrary, set

$$x_4 = c_1$$

$$x_5 = c_2$$

where c_1 and c_2 are arbitrary complex constants. Then the general solution of the original homogeneous system may be expressed as, say,

$$\begin{bmatrix} x_1 \\ x_2 \\ x_3 \\ x_4 \\ x_5 \end{bmatrix} = \begin{bmatrix} -3ic_1 + 2ic_2 \\ 8ic_1 + ic_2 \\ -2ic_1 + ic_2 \\ c_1 \\ c_2 \end{bmatrix} = c_1 \begin{bmatrix} -3i \\ 8i \\ -2i \\ 1 \\ 0 \end{bmatrix} + c_2 \begin{bmatrix} 2i \\ i \\ i \\ 0 \\ 1 \end{bmatrix} = c_1\underline{\bar{d}}_1 + c_2\underline{\bar{d}}_2 \ .$$

It is easily verified that the row vectors $\underline{\bar{d}}_1'$ and $\underline{\bar{d}}_2'$ are independent and form a basis for $\overline{D(A)'}$, the conjugate space of the subspace orthogonal to D(A). Furthermore, the row vectors $\underline{b}_1', \underline{b}_2'$ and $\underline{b}_3'$ remaining after the process of sweep out themselves form a basis of D(A), as previously noted. The respective ranks involved are: Rank D(A) = 3, Rank D(A)' = 2, Rank $\overline{D(A)'}$ = 2.

Therefore, the process described in this Example, although aimed at solution of a system of linear homogeneous equations, provides information concerning ranks, bases, etc. of the various types of subspaces with which we had previously been dealing in Section 15.

<u>EXERCISE 16.2</u> Obtain a basis for $D(A)'$.

The above Example illustrates the process of column sweep out, whether it be for the purpose of obtaining the general solution of a homogeneous linear system of equations, or for obtaining bases for various subspaces that underly the formal algebra of the situation.

It is now a bit easier to describe the general process of column sweep out. Suppose we begin with m equations in n unknowns $(m \leq n)$. Choosing columns as convenient in sequence for sweep out, the process will eventually terminate in, say, r stages, where there will be exactly:

r columns that have been swept out
and
$m - r$ identities (all zero coefficients) to be dropped.

The r variables in the swept out columns may then easily be expressed in terms of the remaining $n - r$ variables, whose values are arbitrary.

In terms of the subspaces involved, the preceding Example illustrates how we may use the formal algebraic solution described above to obtain:

(i) a basis of $D(A)$ which has rank $n - r$
(ii) a basis of $\overline{D(A)'}$ which has rank r
(iii) a basis of $D(A)'$ which has rank r .

<u>EXERCISE 16.3</u> Solve the following homogeneous linear system. Identify the various bases and ranks of subspaces associated with the algebraic solution.

$$
\begin{aligned}
2x_1 + x_2 + 5ix_3 &= 0 \\
3x_1 - 2x_2 + 2ix_3 &= 0 \\
5x_1 - 8x_2 - 4ix_3 &= 0 \quad .
\end{aligned}
$$

<u>EXERCISE 16.4</u> Solve the following homogeneous linear system. Identify the various subspaces, bases and ranks associated with the solution.

$$
\begin{aligned}
x_1 - x_2 + x_3 - x_4 + x_5 &= 0 \\
2x_1 - x_2 + 3x_3 \phantom{{}+x_4} + 4x_5 &= 0 \\
3x_1 - 2x_2 + 2x_3 + x_4 + x_5 &= 0 \\
x_1 \phantom{{}-x_2} + x_3 + 2x_4 + x_5 &= 0 \quad .
\end{aligned}
$$

Before proceeding with the next problem, namely obtaining the general solution to a system of non-homogeneous linear equations, we shall pause briefly to consider a topic that was postponed in Section 14 (Determinants and Matrices) but is appropriate to consider now. It deals with the so-called <u>Rank of an $m \times n$ matrix</u> of complex numbers, which is defined to be the order of the largest non-vanishing minor of the matrix. At first, there might seem to be little relation (and some confusion) with the notion of a <u>Rank of a subspace of C_n</u>, already defined in Section 15.

However, the two notions of 'Rank' are closely related, and the results we have just obtained about solution of homogeneous linear systems of equations enables us to explain the relationship. As it turns out, finding the rank of a subspace (using the preceding results) is quite straightforward, whereas 'Rank' as defined in matrix terminology seems difficult to work with. The following Theorem proves that we may choose either means to obtain 'Rank'.

THEOREM 16.2 (Rank of Matrix and Subspace) Let A be any $m \times n$ matrix with rows $\underline{a}_1', \underline{a}_2', \ldots, \underline{a}_m'$. The Rank of the matrix A equals the Rank of the subspace generated by the rows of A; that is, the two numbers are equal.

PROOF. Let A be any $m \times n$ matrix of complex numbers considered also as the coefficients in a set of m homogeneous linear equations in n unknowns. Let $\underline{A}$ be the $m \times n$ matrix of complex numbers representing the coefficients of the system (identity rows retained) upon completion of the process of column sweep out. Since only elementary row operations are used in the sweep out process, it follows from basic properties of determinants that the ranks of A and $\underline{A}$ are equal.

If the process of column sweep out requires exactly r stages, then A has exactly $m - r$ rows of zeros and exactly r columns that are swept out. Clearly, then, any minor of $\underline{A}$ of order exceeding r must vanish, whereas that minor consisting of the r swept out columns and the r non-zero rows has value ± 1. Thus the rank of the matrix A is r. But it has already been established that the rank of the subspace generated by the rows of A, that is D(A), is r. With this the proof is completed.

EXERCISE 16.5 Prove that the homogeneous linear system $A\underline{x} = \underline{0}$ has a non-trivial[1] solution iff Rank(A) < m. If Rank(A) = m then the rows of A are independent vectors, and the only solution is $\underline{x} = \underline{0}$.

Using whichever definition of Rank is appropriate, prove the following results:

EXERCISE 16.6 If A is any m x n matrix of complex numbers, then Rank(A) = Rank(A'), where A' is the transpose of the matrix A.

EXERCISE 16.7 If B is any non-singular n x n matrix of complex numbers, and A is any n x n matrix of complex numbers, then Rank(A) = Rank(AB) = Rank(BA).

EXERCISE 16.8 If A is any n x n matrix then Rank(A) = Rank(AA') = Rank(A'A).

EXERCISE 16.9 If $A_1, A_2, \ldots, A_k$ are n x n matrices of complex numbers having ranks $r_1, r_2, \ldots, r_k$ respectively, prove that $\text{Rank}(A_1 \pm A_2 \pm \ldots \pm A_k) \le r_1 + r_2 + \ldots + r_k$. Only the case k = 2 requires proof. Why?

EXERCISE 16.10 Under the conditions of the preceding Exercise, prove that $\text{Rank}(A_1 \cdot A_2 \cdots A_k) \le \min(r_1, r_2, \ldots, r_k)$. Only the case k = 2 needs proof. Why?

[1]That is, solutions other than $\underline{x} = \underline{0}$.

EXERCISE 16.11 Can it be proved that for any pair A,B of $n \times n$ matrices of complex numbers, $\mathrm{Rank}(A \cdot B) = \mathrm{Rank}(B \cdot A)$?

EXERCISE 16.12 For any $n \times n$ matrix A of complex numbers, prove that the following statements are equivalent:

(i) $\det A \neq 0$, that is, A is non-singular

(ii) the row vectors of A are independent

(iii) $\mathrm{Rank}(A) = n$

(iv) $A\underline{x} = \underline{0}$ has only the solution $\underline{x} = \underline{0}$.

Now we consider the problem of solving systems of non-homogeneous linear equations. A system of <u>m non-homogeneous linear equations in n $(n \geq m)$ unknowns</u> is a system of the form:

$$
\begin{array}{l}
a_{11}x_1 + a_{12}x_2 + \dots + a_{1n}x_n = d_1 \\
a_{21}x_1 + a_{22}x_2 + \dots + a_{2n}x_n = d_2 \\
\quad \vdots \qquad\quad \vdots \qquad\qquad\quad \vdots \qquad\; \vdots \\
a_{m1}x_1 + a_{m2}x_2 + \dots + a_{mn}x_n = d_m
\end{array}
$$

or,

$$A\underline{x} = \underline{d}$$

where,

$$
A = \begin{bmatrix}
a_{11} & a_{12} & \cdots & a_{1n} \\
a_{21} & a_{22} & \cdots & a_{2n} \\
\vdots & & & \\
a_{m1} & a_{m2} & \cdots & a_{mn}
\end{bmatrix}
$$

and

$$\underline{x}' = (x_1, x_2, \ldots, x_n), \text{ equivalently } \underline{x} = \begin{bmatrix} x_1 \\ x_2 \\ \cdot \\ \cdot \\ \cdot \\ x_n \end{bmatrix}$$

$$\underline{d}' = (d_1, d_2, \ldots, d_m), \text{ equivalently } \underline{d} = \begin{bmatrix} d_1 \\ d_2 \\ \cdot \\ \cdot \\ \cdot \\ d_m \end{bmatrix} \quad .$$

The constants a_{ij} and d_i are complex numbers; we seek any complex vector solutions $\underline{x}'$ (if any) to the system.

Unlike homogeneous systems, a non-homogeneous linear system of equations may have no solution. Such is the case, for example, with the system:

$$\begin{aligned} x_1 + x_2 + x_3 &= 0 \\ x_1 + x_2 + x_3 &= 1 \quad , \end{aligned}$$

in which case the system is termed inconsistent. On the other hand, the system may possess one or more solutions, in which case it is termed consistent. Naturally, then, we must seek conditions under which a non-homogeneous linear system is consistent (equivalently, inconsistent), and then find its general solution. The following Theorem is a first step.

THEOREM 16.3 (Consistency and Solution: Non-Homogeneous Systems) The non-homogeneous system of linear equations: $A\underline{x} = \underline{d}$ is consistent iff $\text{Rank}(A) = \text{Rank}(A,\underline{d})$, where $(A,\underline{d})$ is the $m \times (n+1)$ matrix obtained from the matrix A by adjoining to it the column $\underline{d}$ on its right. If the system proves consistent, then the general solution is obtained by adding to the general solution of the homogeneous system $A\underline{x} = \underline{0}$, any particular solution of the complete non-homogeneous system $A\underline{x} = \underline{d}$.

Before we proceed with a proof of the above Theorem, we shall illustrate a modified process of column sweep out that will (a) demonstrate existence of an inconsistency if such be the case, and (b) provide the general solution of a non-homogeneous linear system when it is consistent.

EXAMPLE 16.2 For convenience, we shall consider the following three systems of non-homogeneous linear equations simultaneously:

	(a)	(b)	(c)
$4x_2 - 8x_3 - 16x_4 - 12x_5 =$	20	20	d_1
$3x_1 + 2x_2 - 5x_3 - 17x_4 - 3x_5 =$	-5	-5	d_2
$4x_1 + 2x_2 - 8x_3 - 20x_4 - 2x_5 =$	-10	-7	d_3
$x_1 - 3x_2 - 7x_3 + 13x_4 + 8x_5 =$	-18	-18	d_4

where, in system (c), the constants d_1, d_2, d_3, d_4 are left unspecified for the time being. We shall demonstrate that:

(i) System (a) is consistent, and find its general solution,

(ii) System (b) is inconsistent, and no solutions exist,

(iii) System (c) may be used to establish necessary and sufficient conditions under which the above system is consistent, and obtain its general solution in all such cases.

The procedure amounts to performing the process of column sweep out on the matrix of constants $(A,\underline{d})$. We shall perform this process simultaneously for all three systems in this Example.

Initial Stage: Equations and Coefficient Vectors as Given

Equations	(a)	(b)	(c)	Coefficient vectors
$0x_1 + 4x_2 - 8x_3 - 16x_4 - 12x_5$	20	20	d_1	(0, 4, -8, -16, -12)
$3x_1 + 2x_2 - 5x_3 - 17x_4 - 3x_5$	-5	-5	d_2	(3, 2, -5, -17, -3)
$4x_1 + 2x_2 - 8x_3 - 20x_4 - 2x_5$	-10	-7	d_3	(4, 2, -8, -20, -2)
$1x_1 - 3x_2 - 7x_3 + 13x_4 + 8x_5$	-18	-18	d_4	(1,-3, -7, 13, 8)

Stage 1: First Sweep Out Process (column two)

	(a)	(b)	(c)	
$0x_1 - 1x_2 - 2x_3 + 4x_4 + 3x_5 =$	-5	-5	$-\frac{1}{4}d_1$	(0, -1, -2, 4, 3)
$3x_1 + 0x_2 - 9x_3 - 9x_4 + 3x_5 =$	-15	-15	$-\frac{1}{2}d_1+d_2$	(3, 0, -9, -9, 3)
$4x_1 + 0x_2 - 12x_3 - 12x_4 + 4x_5 =$	-20	-17	$\frac{1}{2}d_1+d_3$	(4, 0, -12, -12, 4)
$1x_1 + 0x_2 - 1x_3 + 1x_4 + 1x_5 =$	-3	-3	$\frac{3}{4}d_1+d_4$	(1, 0, -1, 1, -1)

Stage 2: Second Sweep Out Process (column one)

Equation	(a)	(b)	(c)	
$0x_1 - 1x_2 - 2x_3 + 4x_4 + 3x_5 =$	-5	-5	$-\frac{1}{4}d_1$	(0, -1, -2, 4, 3)
$-1x_1 + 0x_2 + 3x_3 + 3x_4 - 1x_5 =$	5	5	$\frac{1}{6}d_1-\frac{1}{3}d_2$	(-1, 0, 3, 3,-1)
$0x_1 + 0x_2 + 0x_3 + 0x_4 + 0x_5 =$	0	③	$\frac{1}{6}d_1-\frac{4}{3}d_2+d_3$	(0, 0, 0, 0, 0)
$0x_1 + 0x_2 + 2x_3 + 4x_4 - 2x_5 =$	2	2	$\frac{11}{12}d_1-\frac{1}{3}d_2+d_4$	(0, 0, 2, 4,-2)

Stage 3: Third Sweep Out Process (column three)

Equation	(a)	(b)	(c)	
$0x_1 - 1x_2 + 0x_3 + 8x_4 + 1x_5 =$	-3	-3	$\frac{2}{3}d_1-\frac{1}{3}d_2+d_4$	(0, -1, 0, 8, 1)
$-1x_1 + 0x_2 + 0x_3 - 3x_4 + 2x_5 =$	2	2	$-\frac{29}{24}d_1+\frac{1}{6}d_2-\frac{3}{2}d_4$	(-1, 0, 0,-3, 2)
$0x_1 + 0x_2 + 0x_3 + 0x_4 + 0x_5 =$	0	③	$\frac{1}{6}d_1-\frac{4}{3}d_2+d_3$	(0, 0, 0, 0, 0)
$0x_1 + 0x_2 - 1x_3 - 2x_4 + 1x_5 =$	-1	-1	$-\frac{11}{24}d_1+\frac{1}{6}d_2-\frac{1}{2}d_4$	(0, 0,-1,-2, 1)

In view of Stage 2, Row 3, Column (b), the system of non-homogeneous linear equations (b) is clearly inconsistent: any solution would have to satisfy the (impossible) condition:

$$0x_1 + 0x_2 + 0x_3 + 0x_4 + 0x_5 = 3 .$$

Such a situation is characteristic for an inconsistent system, at some stage of column sweep out process. This is a test for inconsistency.

Next, upon setting $x_4 = x_5 = 0$ in Stage 3 of system (a) and solving for $x_1, x_2,$ and $x_3,$ we obtain a particular solution

$$\underline{x}_0' = (-2, 3, 1, 0, 0)$$

to that system. In view of Theorem 16.3 and the knowledge we now have concerning the general solution of the associated homogeneous linear system $A\underline{x} = \underline{0}$, we may immediately write down the general solution of the complete non-homogeneous linear system (a) as follows:

$$\underline{x} = c_1 \begin{bmatrix} -3 \\ 8 \\ -2 \\ 1 \\ 0 \end{bmatrix} + c_2 \begin{bmatrix} 2 \\ 1 \\ 1 \\ 0 \\ 1 \end{bmatrix} + \begin{bmatrix} -2 \\ 3 \\ 1 \\ 0 \\ 0 \end{bmatrix} = c_1\overline{\underline{d}}_1 + c_2\overline{\underline{d}}_2 + \underline{x}_0 ,$$

say, where $\overline{\underline{d}}_1'$ and $\overline{\underline{d}}_2'$ form a basis of $\overline{D(A)^{\perp}}$, the conjugate subspace to the subspace orthogonal to $D(A)$, which, of course, is the subspace generated by the rows of A. Thus, the general solution to a consistent non-homogeneous linear system $A\underline{x} = \underline{d}$ consists of the sum of the general solution to the associated homogeneous linear system $A\underline{x} = \underline{0}$, plus any particular solution $\underline{x}_0$ of the complete non-homogeneous linear system $A\underline{x} = \underline{d}$.

In view of previous remarks, a necessary and sufficient condition that system (c) be consistent is that $\frac{1}{6}d_1 - \frac{4}{3}d_2 + d_3 = 0$. Subject to this condition the general solution of system (c) is obtained as above and is given by:

$$\underline{x} = c_1 \begin{bmatrix} -3 \\ 8 \\ -2 \\ 1 \\ 0 \end{bmatrix} + c_2 \begin{bmatrix} 2 \\ 1 \\ 1 \\ 0 \\ 1 \end{bmatrix} + \begin{bmatrix} 29/24d_1 - 1/6d_2 + 3/2d_4 \\ -2/3d_1 + 1/3d_2 - d_4 \\ 11/24d_1 - 1/6d_2 + 1/2d_4 \\ 0 \\ 0 \end{bmatrix} ,$$

where c_1 and c_2 are arbitrary complex constants. Now we prove Theorem 16.3

PROOF. First we prove that if the system $A\underline{x} = \underline{d}$ is indeed consistent, then the general solution has the form indicated in the statement of the Theorem. Accordingly, suppose that $\underline{x}_1'$ and $\underline{x}_2'$ are two different particular solutions. Then $A\underline{x}_1 = \underline{d}$ and $A\underline{x}_2 = \underline{d}$, whence $A(\underline{x}_1 - \underline{x}_2) = \underline{0}$, which shows that $\underline{x}_0 = \underline{x}_1 - \underline{x}_2$ is a solution of the homogeneous system. On the other hand, suppose $\underline{x}^*$ is the general solution of the homogeneous system and $\underline{x}_0$ is any particular solution of the complete non-homogeneous system. Then we have: $A(\underline{x}^* + \underline{x}_0) = \underline{0} + A\underline{x}_0 = \underline{d}$, showing that $\underline{x}^* + \underline{x}_0$ is a solution of $A\underline{x} = \underline{d}$. This completes the second part of the Theorem.

Consider now the first part of the Theorem, and suppose $\text{Rank}(A) = r$. Then $\text{Rank}(A,\underline{d}) = r$ or $r + 1$ according as the case may be. Now assume that the process of column sweep out is accomplished in r stages, and let A_r and $\underline{A}_r$ be the coefficient matrices corresponding to A and $(A,\underline{d})$ at this point. An inconsistency can arise iff a zero row vector of A_r corresponds to a non-zero row vector of $\underline{A}_r$. In such a case the elements in the i-th row (say) of A_r are all zeros, but the last element, namely $\underline{a}_{i,n+1}$, in the i-th row of $\underline{A}_r$ is <u>not</u> equal to zero. Accordingly, the $(r+1) \times (r+1)$ submatrix of $\underline{A}_r$ formed as follows:

(i) using as columns those columns that were swept out together with the (last) (n+1)-st column,

(ii) using as rows those rows that correspond to non-null rows in A_r together with the special i-th row,

has as its determinant $\pm\, \underline{a}_{i,n+1}$, so that in this case

Rank$(A,\underline{d}) = r+1$. Othersise, there is no inconsistency and Rank$(A,\underline{d}) = r$. This completes the proof.

EXERCISE 16.13 Keeping $\underline{d}$ general, solve the following system of non-homogeneous linear equations. What conditions on $\underline{d}$ insure consistency?

$$\begin{aligned} 2x_1 + x_2 + 5x_3 &= d_1 \\ 3x_1 - 2x_2 + 2x_3 &= d_2 \\ 5x_1 - 8x_2 - 4x_3 &= d_3 \quad . \end{aligned}$$

EXERCISE 16.14 Solve the following system. Is the system with $(d_1,d_2,d_3,d_4) = (1,2,1,0)$ consistent?

$$\begin{aligned} x_1 - x_2 + x_3 - x_4 + x_5 &= d_1 \\ 2x_1 - x_2 + 3x_3 + 4x_5 &= d_2 \\ 3x_1 - 2x_2 + 2x_3 + x_4 + x_5 &= d_3 \\ x_1 + x_3 + 2x_4 + x_5 &= d_4 \quad . \end{aligned}$$

EXERCISE 16.15 Solve the following non-homogeneous system, and give the conditions on $\underline{d}$ that insure consistency.

$$\begin{aligned} 2x_1 + x_2 + 5ix_3 &= d_1 \\ 3x_1 - 2x_2 + 2ix_3 &= d_2 \\ 5x_1 - 8x_2 + 4ix_3 &= d_3 \quad . \end{aligned}$$

Finally, we apply the above results to define a so-called generalized inverse of a matrix. The inverse A^{-1} of a non-singular square matrix A of complex numbers has already been defined and studied in Section 14. However, if A is either singular or not square, the inverse is not even defined. Here, a generalization of the notion of an inverse is introduced and studied. This notion finds use in many areas of Applied Mathematics.

For an arbitrary $m \times n$ matrix A of complex numbers, we define a generalized inverse A^* of A as any $n \times m$ matrix of complex numbers satisfying the relation: $AA^*A = A$.[1]

<u>EXERCISE 16.16</u> In general, a generalized inverse of a matrix is not unique. Prove, however, that if A is square and non-singular, then A^* is unique and, in fact, $A^* = A^{-1}$.

We now consider a result that provides a technique for finding a generalized inverse of an arbitrary $m \times n$ matrix A of complex numbers. The result is contained in the following Theorem.

<u>THEOREM 16.4</u> (Generalized Inverse of a Matrix) Consider the non-homogeneous linear system $A\underline{x} = \underline{d}$ where A is an arbitrary $m \times n$ matrix of complex numbers. If, for every $\underline{d}$ for which the system is consistent, $A^*\underline{d}$ is a solution, for some $n \times m$ matrix A^*, then A^* is a generalized inverse of A.

PROOF. Suppose $\underline{x} = A^*\underline{d}$ is a solution of $A\underline{x} = \underline{d}$ for every $\underline{d}$ for which the system is consistent, and let $\underline{y}$ be an arbitrary vector of n complex numbers. Upon setting $\underline{z} = A\underline{y}$ we see that the system $A\underline{x} = \underline{z}$ is consistent since

[1] This is also called a g-inverse or conditional inverse of A. Other types of generalized inverses have been studied and applied. Usually, they are defined by imposing added conditions on the generalized inverse A^* defined above. This is the most commonly occurring case. See References.

$\underline{y}$ is a solution. Thus, $AA^*\underline{z} = A\underline{x} = \underline{z}$ and accordingly, $AA^*A\underline{y} = A\underline{y}$. Since $\underline{y}$ is arbitrary, we conclude that $AA^*A = A$ and A^* is by definition a generalized inverse of A. This completes the proof.

<u>EXAMPLE 16.3</u> Find a generalized inverse of:

$$A = \begin{bmatrix} 0 & 4 & 8 & -16 & -12 \\ 3 & 2 & -5 & -17 & -3 \\ 4 & 2 & -8 & -20 & -2 \\ 1 & -3 & -7 & 13 & 8 \end{bmatrix}$$

The technique is to solve $A\underline{x} = \underline{d}$, keeping $\underline{d}$ general, and then express the solution in the form: $A^*\underline{d}$, for some matrix A^*. From Example 16.2, the system $A\underline{x} = \underline{d}$ has solution:

$$\underline{x} = \begin{bmatrix} x_1 \\ x_2 \\ x_3 \\ x_4 \\ x_5 \end{bmatrix} = \begin{bmatrix} -3c_1 - 2c_2 - 29/24d_1 - 1/6d_2 + 3/2d_4 \\ 8c_1 + c_2 \quad -2/3d_1 + 1/3d_2 \quad -d_4 \\ -2c_1 + c_2 + 11/24d_1 - 1/6d_2 + 1/2d_4 \\ c_1 \\ c_2 \end{bmatrix}$$

for arbitrary complex constants c_1 and c_2. Upon setting $c_1 = c_2 = 0$ we see that a solution may be expressed in the form: $A^*\underline{d}$ with

$$A^* = \begin{bmatrix} 29/24 & -1/6 & 0 & 3/2 \\ -2/3 & 1/3 & 0 & -1 \\ 11/24 & -1/6 & 0 & 1/2 \\ 0 & 0 & 0 & 0 \\ 0 & 0 & 0 & 0 \end{bmatrix}$$

Thus, this A^* is a generalized inverse of A.

EXERCISE 16.17 Verify that the A^* of the previous Example is actually a generalized inverse of the given A.

EXERCISE 16.18 Making use of the results of Exercise 16.15, find a generalized inverse of:

$$A = \begin{bmatrix} 2 & 1 & 5i \\ 3 & -2 & 2i \\ 5 & -8 & -4i \end{bmatrix} .$$

Verify that it is a generalized inverse of A.

EXERCISE 16.19 Using the results of Exercise 16.14, find a generalized inverse of:

$$A = \begin{bmatrix} 1 & -1 & 1 & -1 & 1 \\ 2 & -1 & 3 & 0 & 4 \\ 3 & -2 & 2 & 1 & 1 \\ 1 & 0 & 1 & 2 & 1 \end{bmatrix} .$$

Verify that it is a generalized inverse of A.

Hints and Answers to Exercises: Section 16

16.1 Hint: Use the definitions of orthogonality for vectors, and Orthogonal and Conjugate Subspaces.

16.2 Answer: $\underline{d}_1'$, $\underline{d}_2'$.

16.3 Hint: The process of 'column sweep out' applied to the first two columns yields the following system of equations:

$$\begin{aligned} 0x_1 - 1x_2 - (9/7)ix_3 &= 0 \\ -1x_1 + 0x_2 - (12/7)ix_3 &= 0 \end{aligned} .$$

16.4 Hint: The process of 'column sweep-out' applied to the first three columns yields the following system of equations:

$$\begin{aligned}
-1x_1 + 0x_2 + 0x_3 - 3x_4 + 1x_5 &= 0 \\
0x_1 - 1x_2 + 0x_3 - 3x_4 + 0x_5 &= 0 \\
0x_1 + 0x_2 - 1x_3 + 1x_4 - 2x_5 &= 0 \quad .
\end{aligned}$$

16.5 Hint: Use the results of Theorem 16.2 and Theorem 16.1.

16.6 Hint: Use the definition of Rank as the order of the largest non-vanishing minor.

16.7 Hint: Consider the rank of the Vector Spaces generated by the rows of A, AB and BA in view of Theorem 15.3.

16.8 Hint: One method of proof is to establish first that: (i) $D(A'A) \subseteq D(A)$, (ii) $D(AA') \subseteq D(A')$, (iii) $D(A') \subseteq D(AA')$ and (iv) $D(A) \subseteq D(A'A)$, then deduce the result from Theorem 15.11 and Exercise 16.6.

16.9 Hint: Consider $D(A_1 + A_2)$.

16.10 Hint: Since $\text{Rank}(A_1) = r_1$ there exists an $n \times (n-r_1)$ matrix H_1 of rank $n-r_1$ such that $H_1'A_1 = \underline{0}$, whence we have $H_1'A_1A_2 = \underline{0}$, so that $\text{Rank}(A_1A_2) \le r_1$. Similarly it can be established that $\text{Rank}(A_2A_1) = \text{Rank}(A_1A_2) \le r_2$ (see the next Exercise) from which the result follows.

16.11 Hint: Using different methods, this result can be deduced for two special cases from Exercises 17.13 and 17.14 of the next Section. However, prove that the result is true in general by considering D(AB) and D(BA).

16.12 Hint: This is a summary of results already established.

16.13 Answer:

$$\begin{bmatrix} x_1 \\ x_2 \\ x_3 \end{bmatrix} = c_1 \begin{bmatrix} -12/7 \\ -11/7 \\ 1 \end{bmatrix} + \begin{bmatrix} 2/7d_1 \quad 1/7d_2 \\ 3/7d_1 - 2/7d_2 \\ 0 \end{bmatrix},$$

where $2d_1 - 3d_2 + d_3 = 0$ is the consistency condition.

16.14 Answer:

$$\begin{bmatrix} x_1 \\ x_2 \\ x_3 \\ x_4 \\ x_5 \end{bmatrix} = c_1 \begin{bmatrix} -3 \\ -3 \\ 1 \\ 1 \\ 0 \end{bmatrix} + c_2 \begin{bmatrix} 1 \\ 0 \\ -2 \\ 0 \\ 1 \end{bmatrix} + \begin{bmatrix} d_1 - d_2 \\ 2d_1 - d_2 - d_3 + 3d_4 \\ 2d_1 \quad - d_3 + d_4 \\ 0 \\ 0 \end{bmatrix}$$

where $-3d_1 + d_2 + d_3 - 2d_4 = 0$ is the condition for consistency which is satisfied by (1,2,1,0).

16.15 Hint: Use the standard procedure. See also Exercise 16.2.

16.16 Hint: Pre and post-multiply the following relation on both sides by A^{-1}: $AA^*A = A$.

16.19 Answer:

$$A^* = \begin{bmatrix} 1 & -1 & 0 & 2 \\ 2 & -1 & -1 & 3 \\ 2 & 0 & -1 & 1 \\ 0 & 0 & 0 & 0 \\ 0 & 0 & 0 & 0 \end{bmatrix}.$$

References to Additional and Related Material: Section 16

1. Ben-Israel, A. and T. Greville, "Generalized Inverses: Theory and Applications", Wiley and Sons, Inc., New York (1974).

2. Bose, R., Unpublished Lecture Notes, University of North Carolina at Chapel Hill (1964).

3. Boullion, T. and P. Odell, "Generalized Inverse Matrices", Wiley-Interscience, (1971).

4. Cullen, C., "Matrices and Linear Transformations", Addison-Wesley, Inc. (1966).

5. Johnston, J., "Linear Equations and Matrices", Addison-Wesley, Inc. (1966).

6. Pease, M., "Methods of Matrix Algebra", Academic Press (1965).

7. Pipes, L., "Matrix Methods for Engineers", Prentice-Hall, Inc. (1963).

8. Paige, L. and O. Taussky (Editors), "Simultaneous Linear Equations and the Determination of Eigenvalues", National Bureau of Standards, Applied Mathematics Series No. 29, Washington, D. C. (1953).

9. Pringle, P. and A. Rayner, "Generalized Inverse Matrices with Applications to Statistics", Griffin Monograph 28 (1971).

10. Rao, C. and S. Mitra, "Conditional Inverse of Matrices and its Applications", John Wiley and Sons, Inc. (1971).

11. Wilkenson, J. and C. Reinsch, "Linear Algebra", Springer-Verlag, New York (1971).

17. Characteristic Roots and Related Topics

The solutions of many problems in Applied Mathematics involve the so-called characteristic roots of a related matrix. Here we consider some of the important properties and concepts related to characteristic roots.

Let A be an arbitrary $n \times n$ matrix of complex numbers. If for some complex scalar c and non-null complex vector $\underline{x}'$ the following relation is satisfied: $\underline{x}'A = c\underline{x}'$, then c is called a <u>characteristic root</u> (characteristic value, eigenvalue) of A and $\underline{x}'$ is called an associated <u>characteristic vector</u> (eigenvector).

Characteristic roots and vectors associated with a matrix possess many important properties which account for their frequent occurrence in solutions of problems in Applied Mathematics. For example, it will be demonstrated later that the characteristic roots of a matrix remain unchanged under many common transformations of the matrix. The following Exercise illustrates an immediate 'geometric' property of characteristic vectors.

<u>EXERCISE 17.1</u> (Geometry of Characteristic Vectors) For an $n \times n$ matrix A let c be a characteristic root. Prove that (upon inclusion of the null vector $\underline{0}'$) the set of characteristic vectors associated with c forms a subspace of rank r $(1 \leq r \leq n)$ of C_n. (Loosely speaking, vectors in this subspace are all 'stretched' by a factor of $|c|$, but not rotated, under the linear transformation

$\underline{y}' = \underline{x}'A$. Hence the name 'invariant' vectors is sometimes used in referring to characteristic vectors).

Note that the equation $\underline{x}'A = c\underline{x}'$, equivalently $\underline{x}'(A-cI) = \underline{0}'$, has a non-null vector solution iff c satisfies: $\det(A-cI) = 0$. (This follows from Theorem 16.12). Since $\det(A-cI)$ is actually a polynomial in c, termed the <u>characteristic polynomial</u> associated with A, it then follows from Theorem 13.16 (The Fundamental Theorem of Algebra) that we may express this polynomial as:

$$\det(A-cI) = (-1)^n c^n + b_1 c^{n-1} + \ldots + b_n = (-1)^n (c-c_1)(c-c_2)\ldots(c-c_n)$$

demonstrating the fact that the <u>characteristic equation</u> $\det(A-cI) = 0$ has exactly n roots (solutions) $c_1, c_2, \ldots, c_n$ which, by the way, are not necessarily all real and/or distinct. The following Exercise establishes that the solutions of the characteristic equation are indeed the characteristic roots associated with A.

<u>EXERCISE 17.2</u> (Characteristic Roots and Characteristic Equation) Prove that if c is a root of multiplicity r $(1 \le r \le n)$ of the characteristic equation associated with A, then c is a characteristic root of A, and conversely. Furthermore, show then that subspace of characteristic vectors associated with c has rank r.

<u>EXAMPLE 17.1</u> Find the characteristic roots and vectors associated with the matrix:

$$A = \begin{bmatrix} 1 & -2 \\ -2 & 1 \end{bmatrix} .$$

Consider first the characteristic equation:

$$\det(A-cI) = \det\begin{bmatrix} 1-c & -2 \\ -2 & 1-c \end{bmatrix} = c^2 - 2c-3 = (c-3)(c+1) = 0.$$

Thus the characteristic roots of A are: $c_1 = 3$ and $c_2 = -1$. The characteristic vectors associated with c_1 must satisfy $\underline{x}'A = 3\underline{x}'$ hence any such vector $\underline{x}' = (x_1, x_2)$ must be such that $x_1 + x_2 = 0$. Likewise any characteristic vector associated with c_2 must satisfy $\underline{x}'A = -\underline{x}'$ therefore we must have $x_1 - x_2 = 0$. Accordingly, we may characterize the subspace of rank 1 associated with c_1 as <u>all</u> scalar multiples of, say, the vector $(1,-1)$, and that associated with c_2 as <u>all</u> scalar multiples of, say, the vector $(1,1)$.

<u>EXERCISE 17.3</u> Establish that, in the expansion of $\det(A-cI)$, we have $b_n = \det(A)$ whence $\det(A) = \prod_1^n c_i$. This then establishes that A is non-singluar iff all its characteristic roots are non-zero.

<u>EXERCISE 17.4</u> (A Variance-Covariance Matrix) Prove that the characteristic roots of

$$\Sigma = \begin{bmatrix} 1 & \rho \\ \rho & 1 \end{bmatrix}$$

(where $-1 < \rho < 1$) are $1 + \rho$ and $1 - \rho$. Determine the associated subspaces of characteristic vectors. The special case $\rho = 0$ may have to be treated separately.

<u>EXERCISE 17.5</u> (Characteristic Roots of Markov Matrices) Find the characteristic roots and associated characteristic vectors for the following matrices:

$$M = \begin{bmatrix} 1-p & p \\ r & 1-r \end{bmatrix}$$

(where $0 \le p, r \le 1$) and

$$N = \begin{bmatrix} 0 & 0 & 0 & 1 \\ 0 & 0 & 0 & 1 \\ \frac{1}{2} & \frac{1}{2} & 0 & 0 \\ 0 & 0 & 1 & 0 \end{bmatrix} .$$

<u>EXERCISE 17.6</u> (Characteristic Roots of Transpose and Conjugate) Let A be an arbitrary n x n matrix of complex numbers. Establish the relationship between the characteristic roots and vectors of A and those of A' (transpose of A) and of $\bar{A}$ (conjugate of A).

As we mentioned earlier, the characteristic roots of a matrix remain unchanged (or changed in a known manner) under many common transformations of the matrix. As an illustration, we consider the notion of similarity. Two n x n matrices A and B of complex numbers are called <u>similar</u> iff there exists a non-singular matrix P such that $A = PBP^{-1}$. The following Theorem establishes that the characteristic roots of similar matrices are identical.

THEOREM 17.1 (Characteristic Roots of Similar Matrices) If A and B are similar then they have the same characteristic roots.

PROOF. First note that the characteristic polynomial associated with A can be written as:

$$\begin{aligned}\det(A-cI) &= \det(PBP^{-1}-cI)\\ &= \det(PBP^{-1}-PcIP^{-1})\\ &= \det(P)\cdot\det(B-cI)\cdot\det(P^{-1}).\end{aligned}$$

Because $\det(P)$ and $\det(P^{-1}) = 1/\det(P)$ are both non-zero, it follows that $\det(A-cI) = 0$ iff $\det(B-cI) = 0$. Therefore A and B have the same characteristic roots and the Theorem is proved.

EXERCISE 17.7 What can be said about the characteristic vectors of A and B associated with the same characteristic roots?

Because of the importance of the inverse of a non-singular matrix, it is natural to inquire about the relationship between the characteristic roots of a non-singular $n \times n$ matrix A of complex numbers and those of its inverse A^{-1}. The result is given below.

THEOREM 17.2 (Characteristic Roots of an Inverse) If A is non-singular then the characteristic roots of A^{-1} are the reciprocals of the characteristic roots of A.

PROOF. Because the characteristic roots of A are all non-zero. the result follows immediately from the identity: $\det(A^{-1}-cI) = (-c)^n \cdot \det(A^{-1}) \cdot \det(A-\frac{1}{c}I)$.

EXERCISE 17.8 Can a connection be established between the characteristic vectors of A and A^{-1}?

EXERCISE 17.9 (Scalar Multiples of a Matrix) Prove that for any complex scalar k the characteristic roots of kA are k times the characteristic roots of A.

EXERCISE 17.10 (Characteristic Roots of Triangular Matrices) Prove that the characteristic roots of a Triangular (hence also Diagonal) matrix are precisely the main diagonal elements.

In certain areas of Applied Mathematics we deal frequently with matrices which are real and symmetric (the so-called Variance-Covariance matrix of Exercise 17.4 is an example). The following Theorem establishes a special result about the characteristic roots of such matrices.

THEOREM 17.3 (Characteristic Roots of Real, Symmetric Matrices) The characteristic roots of a symmetric matrix of real numbers are all real.

PROOF. First, under the assumption, $A = A' = \bar{A}$. If c is a characteristic root of A and $\underline{x}'$ an associated characteristic vector, we have, of course, $\underline{x}'A = c\underline{x}'$, from which it follows that $\underline{x}'A\bar{\underline{x}} = c\underline{x}'\bar{\underline{x}}$. Obviously $\underline{x}'\bar{\underline{x}}$ is real, and so is $\underline{x}'A\bar{\underline{x}}$ because $\overline{\underline{x}'A\bar{\underline{x}}} = \bar{\underline{x}}'\bar{A}\underline{x} = \bar{\underline{x}}'A\underline{x} = (\underline{x}'A\bar{\underline{x}})' = \underline{x}'A\bar{\underline{x}}$. We must therefore conclude that c is real. This completes the proof.

EXERCISE 17.11 In the preceding proof, why is $(\underline{x}'A\bar{\underline{x}})' = \underline{x}'A\bar{\underline{x}}$?

In certain areas of Applied Mathematics, interest exists in the so-called trace of a matrix. Specifically, the trace $tr(A)$ of an $n \times n$ matrix A of complex numbers is defined to be the sum of its main diagonal elements, that is, $tr(A) = \sum_1^n a_{ii}$. The following Exercise establishes some elementary properties of trace.

EXERCISE 17.12 For $n \times n$ matrices A, B, and C and complex scalar k, prove that trace obeys the following properties: (i) $tr(A+B) = tr(A) + tr(B)$ (ii) $tr(kA) = k \cdot tr(A)$ (iii) $tr(AB) = tr(BA)$, $tr(ABC) = tr(BCA) = tr(CAB) = \ldots$ and so on.

The following Theorem shows that there is an important connection between the trace of a matrix and its characteristic roots.

THEOREM 17.4 (Trace and Characteristic Roots) If $c_1, c_2, \ldots, c_n$ are the characteristic roots of A then $tr(A) = \sum_1^n c_i$. Whence, if A and B are similar, $tr(A) = tr(B)$.

PROOF. Recall that the characteristic polynomial associated with A is given by: $\det(A-cI) = (-1)^n c^n + b_1 c^{n-1} + \ldots + b_n = (-1)^n (c-c_1)(c-c_2)\ldots(c-c_n)$. From inspection of $\det(A-cI)$ we conclude that the coefficient b_1 of c^{n-1}

must be $b_1 = (-1)^{n-1} \sum_1^n a_{ii}$, whereas from inspection of the factored polynomial the coefficient of c^{n-1} must be $(-1)^{n+1} \sum_1^n c_i$. Thus, $\sum_1^n a_{ii} = tr(A) = \sum_1^n c_i$, as asserted, and the proof is completed.

It can be shown that for any pair A,B of n x n matrices of complex numbers, both AB and BA have identical characteristic roots. The following Exercises establish this result in several special cases.

<u>EXERCISE 17.13</u> If at least one of the matrices A or B is non-singular, then AB and BA are similar and so have identical characteristic roots.

<u>EXERCISE 17.14</u> If both A and B are symmetric, then AB and BA have the same characteristic roots.

<u>EXERCISE 17.15</u> Using a Reference if necessary, establish the result that, in general, AB and BA have the same characteristic roots.

<u>EXERCISE 17.16</u> (Characteristic Roots of Markov Matrices) Prove that if c is a characteristic root of a Markov matrix then $|c| \le 1$. Furthermore, prove that $c = 1$ is always a characteristic root of a Markov matrix.

<u>EXERCISE 17.17</u> (Characteristic Roots of Powers of a Matrix) If c is a characteristic root of an n x n matrix A of complex numbers with an associated characteristic vector $\underline{x}'$, then for any positive integer m, c^m is a characteristic

root of A^m with associated characteristic vector $\underline{x}'$. What more may be concluded if A is a Markov matrix?

EXERCISE 17.18 (Ortho-Normal Characteristic Vectors) If c_1 and c_2 are distinct characteristic roots of a real, symmetric matrix A, show that we may always choose a characteristic vector $\underline{x}_1'$ associated with c_1 and a characteristic vector $\underline{x}_2'$ associated with c_2 in such a manner that $\underline{x}_1'$ and $\underline{x}_2'$ are ortho-normal, that is, $\underline{x}_1'\overline{\underline{x}}_2 = 0$ and $||\underline{x}_1'|| = 1$, $||\underline{x}_2'|| = 1$. Later we shall expand upon this result.

EXERCISE 17.19 (m-th Power of a Markov Matrix) Let M be an $n \times n$ Markov matrix with characteristic roots $c_1, c_2, \ldots, c_n$. Let $\underline{x}_i'$ be a non-null characteristic vector associated with c_i $(i = 1,2,\ldots,n)$, where for a root of multiplicity $r > 1$ the r associated characteristic vectors are chosen so as to be independent. Define two $n \times n$ matrices D and X as follows:

$$D = \begin{bmatrix} c_1 & & & \\ & c_2 & & 0 \\ & 0 & \ddots & \\ & & & c_n \end{bmatrix} \qquad X = \begin{bmatrix} \underline{x}_1' \\ \underline{x}_2' \\ \vdots \\ \underline{x}_n' \end{bmatrix} .$$

Establish the following results:

(i) X is non-singular, whence X^{-1} exists.

(ii) $XM = DX$, whence $M = X^{-1}DX$.

(iii) $M^m = X^{-1}D^mX$, $m = 1,2,\ldots$.

(Since D^m is trivial to compute, the process of finding M^m is apparently quite simple once X, D and X^{-1} are evaluated).

EXERCISE 17.20 Using the methods of the preceding Exercise, find M^n and N^n for the two Markov matrices of Exercise 17.5. Can $\lim_{n\to\infty} M^n$ and $\lim_{n\to\infty} N^n$ be defined in a natural manner? If so, what are they?

EXERCISE 17.21 (Characteristic Roots: Orthogonal and Unitary Matrices) Show by example(s) that the characteristic roots of neither an orthogonal nor unitary matrix need be real. Prove, however, that the modulus $|c|$ of any such characteristic root c is one, thereby establishing that the only possible real characteristic roots are ± 1.

EXERCISE 17.22 (Helmert's Transformation) Prove that the following linear transformation is orthogonal, i.e. is of the form $\underline{y}' = \underline{x}'C$ where C is an orthogonal matrix.

$$y_1 = (x_1 - x_2)/\sqrt{2}$$

$$y_2 = (x_1 + x_2 - 2x_3)/\sqrt{6}$$

$$\vdots$$

$$y_{n-1} = (x_1 + x_2 + \ldots + x_{n-1} - (n-1)x_n)/\sqrt{n(n-1)}$$

$$y_n = (x_1 + x_2 + \ldots + x_n)/\sqrt{n} \quad .$$

Display the orthogonal inverse transformation $\underline{x}' = \underline{y}'C^{-1}$.

EXERCISE 17.23 Prove that a Unitary transformation $\underline{y}' = \underline{x}'U$ is distance preserving in C_n.

As has been noted before, real, symmetric matrices occur frequently in certain areas of Applied Mathematics. The following result is a representation Theorem for such matrices that finds frequent use. The proof is based upon many of the results we have just developed concerning characteristic roots and associated characteristic vectors.

THEOREM 17.5 (Principal Axes Representation: Real, Symmetric) Let A be a real, symmetric $n \times n$ matrix. Then there exists a real, orthogonal matrix P such that $PAP^{-1} = PAP' = D$ (a diagonal matrix), the diagonal elements of D being the common characteristic roots of A and D.

PROOF. From Theorem 17.3 the characteristic roots of A are all real, though not necessarily all distinct. Let $\theta_1, \theta_2, \ldots, \theta_k$ $(1 \le k \le n)$ be the distinct characteristic roots, θ_i having multiplicity r_i so that $\sum_1^k r_i = n$. Finally, let $T_1, T_2, \ldots, T_k$ be the associated subspaces of corresponding characteristic vectors, T_i having rank r_i $(i = 1,2,\ldots,k)$ via Exercise 17.2.

In view of the special properties of the matrix A and its characteristic roots, in each subspace T_i we may choose r_i vectors with real components only, say $\underline{x}'_{ij}$ $(j = 1,2,\ldots,r_i)$, that form an ortho-normal basis for T_i $(i = 1,2,\ldots,k)$. From Exercise 17.18 it then follows that the entire set $\underline{x}'_{ij}$ $(j = 1,2,\ldots,r_i;$ $i = 1,2,\ldots,k)$ of n real vectors forms an ortho-normal basis of C_n. Thus, the matrix P defined by:

$$P = \begin{bmatrix} \underline{x}'_{11} \\ \underline{x}'_{12} \\ \cdot \\ \cdot \\ \cdot \\ \underline{x}'_{1r_1} \\ \cdot \\ \cdot \\ \cdot \\ \cdot \\ \cdot \\ \underline{x}'_{k1} \\ \underline{x}'_{k2} \\ \cdot \\ \cdot \\ \cdot \\ \underline{x}'_{kr_k} \end{bmatrix}$$

is real and orthogonal, and by the defining properties of the characteristic roots and ortho-normal real characteristic vectors involved, we have:

$$PAP' = D,$$

where the diagonal elements of the (diagonal) matrix D are precisely (in order):

$$\overset{\longleftarrow r_1 \longrightarrow}{\theta_1,\ldots,\theta_1};\ \overset{\longleftarrow r_2 \longrightarrow}{\theta_2,\ldots,\theta_2};\ \ldots\ldots\ldots;\ \overset{\longleftarrow r_k \longrightarrow}{\theta_k,\ldots,\theta_k},$$

the common characteristic roots of A and D. This completes the proof.

From the representation $PAP' = D$ of a real, symmetric matrix obtained above, it then follows that

$$A = P'DP$$

whence

$$A = \theta_1(\underline{x}_{11}\underline{x}_{11}' + \dots + \underline{x}_{1r_1}\underline{x}_{1r_1}') + \dots + \theta_k(\underline{x}_{k1}\underline{x}_{k1}' + \dots + \underline{x}_{kr_k}\underline{x}_{kr_k}')$$

where

$$I = P'P = (\underline{x}_{11}\underline{x}_{11}' + \dots + \underline{x}_{1r_1}\underline{x}_{1r_1}') + \dots + (\underline{x}_{k1}\underline{x}_{k1}' + \dots + \underline{x}_{kr_k}\underline{x}_{kr_k}') .$$

In view of the special ortho-normal properties of the real characteristic vectors above, it can be seen why the above expansion of A is termed an 'orthogonal decomposition' of the real, symmetric matrix. Further generalizations of results such as this, at a higher Mathematical level, also find applications. (See References).

EXERCISE 17.24 (Rank and Non-Zero Characteristic Roots) From the proof of the preceding Theorem, verify that $\text{Rank}(A) = \text{Rank}(D)$ so, since A and D have the same characteristic roots, $\text{Rank}(A)$ equals the number of its non-zero characteristic roots. Then prove that the same is true of any $n \times n$ matrix of complex numbers.

The next result is an immediate application of the Principal Axes Theorem.

THEOREM 17.6 If the real $m \times n$ matrix B has rank m then the real, symmetric matrix BB' has all positive characteristic roots (hence is non-singular).

PROOF. By the preceding Theorem there exists a real orthogonal matrix P such that: $P(BB')P' = D$, where D is a diagonal matrix with diagonal elements being the common characteristic roots of BB' and D.

Now define the matrix E by: $E = PB$ and observe that $EE' = D$. Since the i-th diagonal element of D is c_i (the i-th characteristic root of D and BB') and $c_i = \sum_1^n e_{ik}e'_{ki} = \sum_1^n e_{ik}^2$, it follows that $c_i > 0$, for otherwise, E and consequently B would have rank less than m. This completes the proof.

Closely connected with characteristic roots is what is termed a real quadratic form. Specifically, a real quadratic form is an expression of the form:

$$\underline{x}'A\underline{x}$$

where A is a real, symmetric $n \times n$ matrix and $\underline{x}' = (x_1, x_2, \ldots, x_n)$ is considered as a vector of real components.

EXERCISE 17.25 Prove that the assumption that A is symmetric actually represents no loss of generality since, for any real $n \times n$ matrix B there exists a real, symmetric $n \times n$ matrix A such that $\underline{x}'B\underline{x} \equiv \underline{x}'A\underline{x}$ for all real vectors $\underline{x}'$.

The concept of a real quadratic form arises naturally in various areas of Applied Mathematics. Furthermore, it is a preliminary for later defining positive (semi-) definite matrices. (Note that for any real vector $\underline{x}'$, $\underline{x}'A\underline{x}$ is a real number).

Real quadratic forms are categorized into two classes. The real quadratic form $\underline{x}'A\underline{x}$ (and the matrix A) are termed <u>positive semi-definite</u> iff $\underline{x}'A\underline{x} \geq 0$ for all real vectors $\underline{x}'$. The real quadratic form $\underline{x}'A\underline{x}$ (and the matrix A) are termed <u>positive definite</u> iff $\underline{x}'A\underline{x} > 0$ for all <u>non-null</u> real vectors $\underline{x}'$. (Obviously $\underline{x}'A\underline{x} = 0$ for $\underline{x}' = \underline{0}'$ in either case). Finally, the <u>rank</u> of the quadratic form is defined to be the rank of its associated matrix A.

In terms of the above concepts, it is possible to give a clear geometric interpretation to real quadratic forms. If we view the quadratic form $\underline{x}'A\underline{x}$ considered as a function of the n real variables $(x_1,x_2,\ldots,x_n)$, then the surface in n-dimensional Euclidean space defined by:

$$\underline{x}'A\underline{x} = \text{positive constant}$$

represents an n-dimensional <u>ellipsoid</u> if A is positive definite, and represents various types of <u>elliptical cylinders</u> if A is positive semi-definite (but not positive definite).

<u>EXERCISE 17.26</u> Express the ellipse $4x^2 + 9y^2 = 144$ in the form: $\underline{x}'A\underline{x}$ = positive constant, where A is a real, symmetric 2 x 2 matrix and $\underline{x}' = (x,y)$. Verify that A is positive definite.

<u>EXERCISE 17.27</u> Prove that the real quadratic form $\underline{x}'\underline{x} = \underline{x}'I\underline{x}$ is invariant under any real orthogonal transformation. Interpret this result geometrically in view of the fact that $(\underline{x}'\underline{x})^{\frac{1}{2}}$ represents length in Euclidean n-space.

Pursuing the geometry of real quadratic forms a bit further, we have the following application of Theorem 17.5 which suggests why the Theorem was originally termed the Principal Axes Representation.

<u>THEOREM 17.7</u> (Principal Axes Theorem) The n-dimensional surface $\underline{x}'A\underline{x}$ = positive constant, associated with the real quadratic form $\underline{x}'A\underline{x}$, is referred to its principal axes, i.e. the absence of cross-product terms, by means of the real orthogonal transformation: $\underline{y} = P\underline{x}$ where $PAP' = D$ (diagonal), P real, orthogonal.

PROOF. The proof follows immediately from the fact that $\underline{x}'A\underline{x} = \underline{y}'PAP'\underline{y} = \underline{y}'D\underline{y} = \sum_1^n c_i y_i^2$ under the stated transformation. The c_i's are, of course, the characteristic roots of A.

<u>EXERCISE 17.28</u> (Characteristic Roots of P.D. and P.S.D. Matrices) Deduce from the preceding Theorem that the characteristic roots of a positive definite (p.d.) matrix are all > 0 and that the characteristic roots of a positive semi-definite (p.s.d.) matrix are all ≥ 0.

In view of Theorem 17.7 and the preceding Exercise, we see that the semi-axes of the surface (ellipsoid, elliptical cylinder) $\underline{x}'A\underline{x} = \sum_1^n c_i y_i^2$ = positive constant, are indeed proportional to the reciprocals of the square roots of the characteristic roots of A. Furthermore, we can see that the rank of $\underline{x}'A\underline{x}$ (equivalently the rank of A, which is the number of its non-zero characteristic roots) is the smallest number of independent variables to which $\underline{x}'A\underline{x}$ may be brought by means of a real orthogonal transformation.

EXERCISE 17.29 (From Multivariate Analysis) Let $\underline{x}_1', \underline{x}_2', \ldots, \underline{x}_r'$ and $\underline{m}_1', \underline{m}_2', \ldots, \underline{m}_r'$ be row vectors with n real components. Define $\underline{x}'$ and $\underline{m}'$ as follows: $\underline{x}' = (1/r) \sum_1^r \underline{x}_i'$ and $\underline{m}' = (1/r) \sum_1^r \underline{m}_i'$. Finally, let V be any real, symmetric positive definite n x n matrix and suppose also that the following two n x n matrices are positive definite:

$$S^* = \sum_1^r (\underline{x}_i - \underline{m}_i)(\underline{x}_i - \underline{m}_i)'$$

$$S = \sum_1^r (\underline{x}_i - \underline{x})(\underline{x}_i - \underline{x})' \quad .$$

Establish the following results:

(i) $e = \sum_1^r (\underline{x}_i - \underline{m}_i)' V^{-1} (\underline{x}_i - \underline{m}_i)$ can be expressed

also as $e = \mathrm{tr}(V^{-1} S^*)$.

(ii) $\mathrm{tr}(V^{-1} S^*) = \mathrm{tr}(V^{-1} S) + n(\underline{x} - \underline{m})' V^{-1} (\underline{x} - \underline{m})$.

(iii) If $c_1, c_2, \ldots, c_n$ denote the characteristic roots of SS^{*-1}, prove that we then have: $\sum_1^n c_i = \mathrm{tr}(SS^{*-1})$ and $\prod_1^n c_i = \det(S)/\det(S^*)$.

The properties of p.d. and p.s.d. matrices are of such frequent use in certain areas of Applied Mathematics that we shall now develop some of their more important ones. This will be done in a series of Theorems and Exercises that follows.

THEOREM 17.8 If A is a p.d. (p.s.d.) n x n matrix and B is a real n x m matrix of rank m $(m \le n)$, then B'AB is a p.d. (p.s.d.) m x m matrix.

PROOF. Clearly $B'AB$ is real and symmetric. It must now be shown that for any real vector $\underline{y} \neq \underline{0}$, $\underline{y}'(B'AB)\underline{y} > 0$ (≥ 0). Accordingly, define $\underline{x} = B\underline{y}$ and note that since B has rank m, $\underline{y} \neq \underline{0}$ implies $\underline{x} \neq \underline{0}$. Then $\underline{y}'(B'AB)\underline{y} = (B\underline{y})'A(B\underline{y}) = \underline{x}'A\underline{x} > 0$ (≥ 0) since A is p.d. (p.s.d.) by hypothesis. This completes the proof.

EXERCISE 17.30 (Corollary of Theorem 17.8) Establish the following special case of the preceding Theorem. If A is a p.d. (p.s.d.) $n \times n$ matrix and B is any real, non-singular $n \times n$ matrix, then $B'AB$ is a p.d. (p.s.d.) $n \times n$ matrix.

The next result shows that the property of being p.d. is carried over from a matrix to its inverse.

THEOREM 17.9 If A is p.d. then A^{-1} is p.d.

PROOF. Clearly A^{-1} exists since A has all positive characteristic roots. Furthermore A^{-1} is symmetric because $AA^{-1} = I$ implies that $(A^{-1})'A' = (A^{-1})'A = I$, which shows that $A^{-1} = (A^{-1})'$. The proof then follows from Exercise 17.30 upon taking $B = A^{-1}$.

The next result deals with submatrices formed from p.d. (p.s.d.) matrices. As might be suspected, if the submatrix is formed in a suitable manner, the property of p.d. (p.s.d.) carries over to it from the original matrix.

<u>THEOREM 17.10</u> (Submatrices of P.D. or P.S.D. Matrices) Let C be the $m \times m$ submatrix obtained from a p.d. (p.s.d.) matrix A by deleting any $n-m$ rows and the <u>corresponding</u> $n-m$ columns $(1 \leq m < n)$. Then C is itself a p.d. (p.s.d.) matrix.

PROOF. Let I^* be the $n \times m$ matrix obtained from the $n \times n$ identity matrix I by deleting from it those columns corresponding to the ones deleted from the original matrix A in the formation of C. The proof is completed upon taking $B = I^*$ in Theorem 17.8.

The next Theorem summarizes results that are essentially established already concerning the characteristic roots of p.d. (p.s.d.) matrices.

<u>THEOREM 17.11</u> (Characteristic Roots of P.D. and P.S.D. Matrices) If the $n \times n$ matrix A is p.d. (p.s.d.) then all the characteristic roots of A are > 0 (≥ 0). The rank of A is r $(0 \leq r \leq n)$ iff exactly r characteristic roots are positive and the remaining $n-r$ characteristic roots are zero.

PROOF. <u>EXERCISE 17.31</u>.

Although the definition of a p.d. (p.s.d.) matrix is straightforward, the definition itself does not provide a practical method for establishing whether or not a given matrix is p.d. (or p.s.d.). The next Exercise provides a more practical test for such matrices.

EXERCISE 17.32 (Practical Test for P.D. (P.S.D.) Matrices) Prove that a real, symmetric $n \times n$ matrix A is p.d. (p.s.d.) provided the following sequence of its principal minors is positive (non-negative):

$$D_1 = a_{11},\ D_2 = \det\begin{bmatrix} a_{11} & a_{12} \\ a_{21} & a_{22} \end{bmatrix}, \ldots\ldots, \det\begin{bmatrix} a_{11} & a_{12} \cdots a_{1n} \\ a_{21} & a_{22} \cdots a_{2n} \\ \vdots & \vdots & \vdots \\ a_{n1} & a_{n2} & a_{nn} \end{bmatrix}.$$

The next few results deal with such things as existence of certain transformation matrices associated with p.d. matrices, as well as a characterization Theorem.

THEOREM 17.12 If A is a p.d. $n \times n$ matrix then there exists a non-singular matrix E such that $E'AE = I$.

PROOF. We may choose a real, orthogonal $n \times n$ matrix P such that $P'AP = D$ (diagonal) with the diagonal elements $c_1, c_2, \ldots, c_n$ of D being the (positive) characteristic roots of A. Define now the $n \times n$ diagonal matrix $D^{-\frac{1}{2}}$ having diagonal elements $c_1^{-\frac{1}{2}}, c_2^{-\frac{1}{2}}, \ldots, c_n^{-\frac{1}{2}}$. Upon setting $E = PD^{-\frac{1}{2}}$ it is easily shown that this non-singular matrix satisfies: $E'AE = I$. This completes the proof. (Note that the matrix E is not asserted to be orthogonal).

EXERCISE 17.33 (Generalization of Theorem 17.12) Prove, more generally, that if A is a p.s.d. $n \times n$ matrix of rank r $(0 \le r \le n)$ then there exists a non-singular $n \times n$

matrix E such that:

$$E'AE = \begin{bmatrix} I_r & O \\ O & O \end{bmatrix}$$

(where I_r is the $r \times r$ identity matrix).

The next result provides an interesting and frequently useful characterization Theorem for p.d. matrices.

<u>THEOREM 17.13</u> (Characterization: P.D. Matrices) A real, symmetric $n \times n$ matrix A is p.d. iff it can be expressed as: $A = FF'$, for some real, non-singular $n \times n$ matrix F.

PROOF. From Theorem 17.12 there exists a non-singular real matrix E such that $E'AE = I$. Choosing $F' = E^{-1}$, it follows that $FF' = (E^{-1})'E^{-1} = A$. On the other hand, if $A = FF'$, where F is real and non-singular, it follows from the fact that I is p.d. in Exercise 17.30 that $FIF' = FF' = A$ is p.d. This completes the proof.

<u>EXERCISE 17.34</u> (From Multivariate Analysis) Suppose S and T are both p.d. $n \times n$ matrices with $S = EE'$, E being real and non-singular. Define $T^* = E'TE$. Prove that: $tr(TS) = tr(T^*)$.

The next result may appear highly specialized, dealing with a transformation matrix associated with a p.d. matrix and an accompanying p.s.d. matrix. However, the result does find application in certain areas of Applied Mathematics.

<u>THEOREM 17.14</u> Let A be a p.d. and B a p.s.d. $n \times n$ matrix. Then, there exists a real, non-singular $n \times n$ matrix G such that <u>simultaneously</u> we have: $G'AG = I$ and $G'BG = D$ (diagonal).

PROOF. Since A is p.d. it follows from Theorem 17.12 that there exists a real, non-singular matrix E such that $E'AE = I$. Observe now that because $E'BE$ is real and symmetric, we may choose via Theorem 17.7 a real, orthogonal matrix P such that $P(E'BE)P' = D$ (diagonal). Upon selecting $G = EP'$, the Theorem is established.

<u>EXERCISE 17.35</u> Verify that the matrix G defined in the proof of the preceding Theorem has the required property.

<u>EXERCISE 17.36</u> (An Inequality) Prove that for any p.d. matrix A we have $\det(A) \leq \prod_{1}^{n} a_{ii}$. Is this true if A is p.s.d. but not p.d.?

The final topic related to characteristic roots that we shall be considering here is that of the so-called <u>largest characteristic root</u> of a matrix. For a real, symmetric matrix A, the largest amongst its characteristic roots is denoted $c_{max}(A)$. Interestingly enough, it finds application in certain areas of Applied Mathematics.

Many of the applications of largest characteristic root occur in the process of maximizing expressions involving quadratic forms. The following results will demonstrate some elementary results in this area. (See also References).

THEOREM 17.15 If A is a real, symmetric $n \times n$ matrix, then the maximum of the quadratic form $\underline{x}'A\underline{x}$ taken over all real vectors $\underline{x}'$ having unit length exists and is equal to $c_{max}(A)$. Alternately,

$$\max_{\underline{x}'\underline{x}=1} \underline{x}'A\underline{x} = c_{max}(A) .$$

PROOF. Certainly a maximum exists over the indicated set, viz. $\underline{x}'\underline{x} = 1$. We may choose a real, orthogonal matrix P such that $P'AP = D$ (diagonal). Set $\underline{x}' = \underline{y}'P$ and note that $\underline{x}'\underline{x} = \underline{y}'\underline{y}$, by orthogonality. Therefore,

$$\max_{\underline{x}'\underline{x}=1} \underline{x}'A\underline{x} = \max_{\underline{y}'\underline{y}=1} \underline{y}'PAP'\underline{y} = \max_{\underline{y}'\underline{y}=1} \underline{y}'D\underline{y} = c_{max}(D) = c_{max}(A) .$$

This completes the proof.

The next result deals with maximizing a ratio of quadratic forms. Again the notion of largest characteristic root enters into the solution.

THEOREM 17.16 If A is real and symmetric, and B is p.d. then

$$\max_{\underline{x}\neq\underline{0}} \frac{\underline{x}'A\underline{x}}{\underline{x}'B\underline{x}} = \max_{\underline{x}'\underline{x}=1} \frac{\underline{x}'A\underline{x}}{\underline{x}'B\underline{x}} = c_{max}(AB^{-1}) .$$

PROOF. The first part of the result comes from the fact that the ratio of quadratic forms given is homogeneous in the sense of Theory of Equations.

Since B is p.d. there exists, via Theorem 17.12, a real, non-singular matrix E such that $E'BE = I$.

Upon setting $\underline{x} = E\underline{y}$ it follows that:

$$\max_{\underline{x}\neq\underline{0}} \frac{\underline{x}'A\underline{x}}{\underline{x}'B\underline{x}} = \max_{\underline{y}\neq\underline{0}} \frac{\underline{y}'E'AE\underline{y}}{\underline{y}'E'BE\underline{y}}$$

$$= \max_{\underline{y}'\underline{y}=1} \frac{\underline{y}'E'AE\underline{y}}{\underline{y}'E'BE\underline{y}}$$

$$= \max_{\underline{y}'\underline{y}=1} \underline{y}'E'AE\underline{y}$$

$$= c_{max}(E'AE)$$

$$= c_{max}(AEE') = c_{max}(AB^{-1}),$$

since $E'AE$ and AEE' have the same characteristic roots. This completes the proof.

EXERCISE 17.37 (From Multivariate Testing) Under the conditions of the preceding Theorem, prove that:

$$c_{max}(AB^{-1}) \le c \text{ iff } \frac{\underline{x}'A\underline{x}}{\underline{x}'B\underline{x}} \le c \text{ for all } \underline{x} \neq \underline{0} .$$

EXERCISE 17.38 Prove that if A and B are at least p.s.d., then $c_{max}(AB) \le c_{max}(A) \cdot c_{max}(B)$.

Hints and Answers to Exercises: Section 17.

17.1 Hint: Verify that the set of vectors $\underline{x}'$ satisfying the relation $\underline{x}'A = c\underline{x}'$ (for some complex scalar c), satisfy the three defining properties of a vector (sub) space.

17.2 Hint: From its definition, it can be shown that c is a root of multiplicity r of the characteristic equation iff there exist r independent row vectors $\underline{x}_1', \ldots, \underline{x}_r'$ such that $\underline{x}_i'A = c\underline{x}_i'$ $(i = 1, \ldots\ldots, r)$.

17.3 Hint: One method is to set $c=0$ in the two different expansions for $\det(A-cI)$.

17.4 Hint: For example, the equation: $(\rho \neq 0)$

$$(x_1, x_2)\begin{bmatrix} 1 & \rho \\ \rho & 1 \end{bmatrix} = c(x_1, x_2)$$

equivalently:

$$(1-c)x_1 + \rho x_2 = 0$$

$$\rho x_2 + (1-c)x_2 = 0$$

has, for $c = 1-\rho$, solutions of the form $x_1 + x_2 = 0$, whence the subspace of rank 1 associated with the characteristic root $1-\rho$ has basis $(1,-1)$ and consists of all scalar multiples of this vector. Analogously for $c = 1+\rho$. In case $\rho=0$, the characteristic root $c=1$ has multiplicity 2 and must be treated separately.

17.5 Hint: Using the results of Section 16 on simultaneous linear equations, determine for what values of c the systems $\underline{x}'M = c\underline{x}'$ and $\underline{x}'N = c\underline{x}'$ have non-null vector solutions. Care must be taken in case of multiple roots.

17.6 Hint: First consider the characteristic equations associated with A' and $\bar{A}$ and compare them with the characteristic equation associated with A.

17.7 Hint: If c is a common characteristic root of A and B and for some non-null vector $\underline{x}'$ we have $\underline{x}'A = c\underline{x}'$, is it then true that $\underline{x}'B = c\underline{x}'$?

17.8 Hint: If c is a characteristic root of A and $\underline{x}'$ an associated non-null characteristic vector, is $\underline{x}'$ a characteristic vector associated with the characteristic root $1/c$ of A^{-1} ?

17.9 Hint: Consider the characteristic equation associated with kA.

17.10 Hint: Use the fact that the determinant of a triangular matrix is the product of its main diagonal elements.

17.11 Hint: The dimensions of the matrix $\underline{x}'A\bar{\underline{x}}$ are 1×1.

17.12 Hint: Properties (i) and (ii) are obvious. For property (iii) actually evaluate the sum of the main diagonal elements of both AB and BA and establish their equality.

17.13 Hint: Suppose B is non-singular. Then $AB = PBAP^{-1}$ for $P = B^{-1}$.

17.14 Hint: Consider the characteristic equations associated with both AB and BA, then use symmetry.

17.16 Hint: M' has the same characteristic roots as M. Accordingly, let c be a characteristic root of M' and $\underline{x}' = (x_1, x_2, \ldots, x_n)$ an associated non-null characteristic vector, whence $\underline{x}'M' = c\underline{x}'$. If $m = \max_i |x_i|$ prove that $|cx_i| \leq m$ $(i = 1, 2, \ldots, n)$.

17.17 Hint: For the first part, consider the characteristic equation associated with A^m. Clearly $|c^m| \leq 1$ for a Markov matrix, and $c = c^m = 1$ is one characteristic root.

17.18 Hint: Suppose $\underline{x}_1'A = c_1\underline{x}_1'$ and $\underline{x}_2'A = c_2\underline{x}_2'$. Accordingly, $\underline{x}_1'A\overline{\underline{x}}_2 = c_1\underline{x}_1'\overline{\underline{x}}_2$ and $\underline{x}_1'\overline{A'}\overline{\underline{x}}_2 = c_2\underline{x}_1'\overline{\underline{x}}_2$. Since c_1 and c_2 are distinct, and A is real and symmetric, the result follows upon algebraic manipulation of the two preceding relations. (Note: $A = \overline{A'}$).

17.19 Hint: That vectors $\underline{x}_1', \underline{x}_2', \ldots, \underline{x}_n'$ can be chosen with the given properties follows from Exercise 17.1, general properties of subspaces of a vector space and specific properties of the subspaces associated with distinct characteristic roots. Accordingly, independence of the entire set of characteristic vectors can be established guaranteeing that X^{-1} exists. The remaining parts follow easily.

17.21 Hint: The matrix $\begin{bmatrix} 3/5 & 4/5 \\ -4/5 & 3/5 \end{bmatrix}$ is both orthogonal and unitary and possesses non-real characteristic roots. Furthermore, if $\underline{x}'$ is a non-null characteristic vector associated with the characteristic root c of a unitary matrix A then it can be shown that $||\underline{x}'|| = ||\underline{x}'A||$.

17.22 Hint: First establish that $CC' = I$, and then for the inverse transformation note that $C^{-1} = C'$.

17.23 Hint: Refer to the Hint for Exercise 17.21.

17.24 Hint: The first part follows from Exercise 16.7. The last part can be deduced from Exercise 17.3 and Theorem 16.2.

17.25 Hint: Try $B = \frac{1}{2}(A + A')$.

17.26 Hint: The appropriate matrix is $A = \begin{bmatrix} 2 & 3 \\ 3 & 2 \end{bmatrix}$, and clearly $\underline{x}'A\underline{x} = 4x^2 + 9y^2 > 0$ for $(x,y) \neq (0,0)$.

17.27 Hint: If $\underline{y}' = \underline{x}'C$ where C is real and orthogonal, then $\underline{y}'\underline{y} = \underline{x}'CC'\underline{x} = \underline{x}'I\underline{x} = \underline{x}'\underline{x}$, hence such transformations correspond to 'rigid' motion in E_n.

17.28 Hint: It is easy to establish that $\sum_1^n c_i y_i^2 > 0 \quad (\geq 0)$ for all $\underline{y}' = (y_1, y_2, \ldots, y_n) \neq \underline{0}'$ if and only if $c_i > 0 \quad (\geq 0)$ for $i = 1,2,\ldots,n$.

17.29 Hint: (i) $\operatorname{tr}(\underline{x}_i - \underline{m}_i)'V^{-1}(\underline{x}_i - \underline{m}_i) = \operatorname{tr} V^{-1}(\underline{x}_i - \underline{m}_i)'(\underline{x}_i - \underline{m}_i)$.

(ii) Express S^* in terms of S using the definitions of $\underline{x}'$ and $\underline{m}'$.

(iii) This follows from Theorem 17.4 and Exercise 17.3.

17.31 Hint: This follows from Theorem 17.7 since A and D have the same rank and clearly $\operatorname{Rank}(D)$ equals the number of non-zero characteristic roots (diagonal elements) of D.

17.32 Hint: Try the case $n=2$ first. Show that the respective conditions imply that:

$$(x_1,x_2)\begin{bmatrix} a_{11} & a_{12} \\ a_{21} & a_{22} \end{bmatrix}\begin{pmatrix} x_1 \\ x_2 \end{pmatrix} > 0 \quad (\geq 0)$$

for all $(x_1,x_2) \neq (0,0)$. Observe that $a_{21} = a_{12}$.

17.33 Hint: Note that r characteristic roots are positive and $n-r$ are zero. The case $r=n$ corresponds to Theorem 17.12.

17.34 Hint: $\mathrm{tr}(ABC) = \mathrm{tr}(CAB)$.

17.36 Hint: First establish the inequality: $0 < \det(A) \leq a_{11}A_{11}$ where A_{11} is the cofactor of a_{11} .

17.37 Hint: There exist unit vectors $\underline{w}',\underline{y}'$ and $\underline{z}'$ such that: $c_{max}(AB) = \underline{w}'AB\underline{w}$, $c_{max}(A) = \underline{y}'A\underline{y}$ and $c_{max}(B) = \underline{z}'B\underline{z}$. Whence, $c_{max}(A) \cdot c_{max}(B) \geq \underline{w}'A\underline{ww}'B\underline{w}$. Since $A\underline{ww}'$ and $\underline{ww}'A$ have the same characteristic roots, we may replace $\underline{w}'A\underline{ww}'B\underline{w}$ by $\underline{w}'\underline{ww}'AB\underline{w} = \underline{w}'AB\underline{w} = c_{max}(AB)$.

References to Additional and Related Material: Section 17

1. Browne, E., "Introduction to the Theory of Determinants and Matrices", University of North Carolina Press (1958).

2. Finkbeiner, D., "Introduction to Matrices and Linear Transformations", W. H. Freeman and Co., Inc. (1960).

3. Gantmacher, F., "The Theory of Matrices", Chelsea Publishing Co. (1959).

4. Graybill, F., "Introduction to Matrices with Applications in Statistics", Wadsworth Publishing Co. (1969).

5. Hammarling, S., "Latent Roots and Latent Vectors", University of Toronto Press (1970).

6. Schwartz, J., "Introduction to Matrices and Vectors", McGraw-Hill, Inc. (1961).

7. Turnbull, H., "Theory of Determinants, Matrices and Invariants", Dover Publications, Inc.

18. Convex Sets and Convex Functions

Because of their useful properties, the notions of convex sets and convex functions find many uses in the various areas of Applied Mathematics. We begin with the basic definition of a convex set in n-dimensional Euclidean Space (E_n), where points are ordered n-tuples of real numbers such as $\underline{x}' = (x_1, x_2, \ldots, x_n)$ and $\underline{y}' = (y_1, y_2, \ldots, y_n)$.

<u>DEFINITION 18.1</u> (Convex Set) A subset S of E_n is termed convex iff for every pair of points $\underline{x}', \underline{y}'$ belonging to S, all points on the straight line joining $\underline{x}'$ and $\underline{y}'$ also belong to S. Equivalently, S is termed a convex subset of E_n iff for every pair $\underline{x}', \underline{y}'$ of points belonging to S, $p\underline{x}' + q\underline{y}'$ belongs to S, for all $p, q \geq 0$ with $p + q = 1$.

Convex subsets of E_2 are easily pictured. The following Diagram presents some convex subsets of E_2.

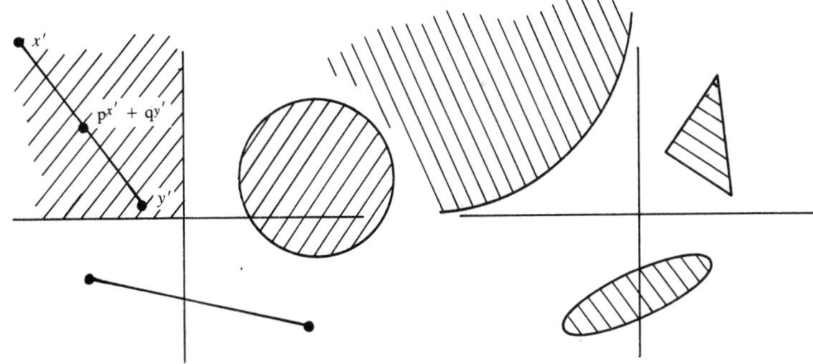

Figure 18-1

If $\underline{x}_1', \underline{x}_2', \ldots, \underline{x}_m'$ are points of E_n and $p_1, p_2, \ldots, p_m$ are non-negative real numbers with $\sum_1^n p_i = 1$, then it is customary to term the point $p_1\underline{x}_1' + p_2\underline{x}_2' + \ldots + p_m\underline{x}_m'$ of E_n a <u>convex linear combination</u> of the m given points.

Using the above notion, it can be proved that a subset S of E_n is convex iff for any integer $m \geq 2$, any convex linear combination of m points of S belongs to S. Equivalently, S is convex in E_n iff $\underline{x}_1', \underline{x}_2', \ldots, \underline{x}_m' \in S$ implies that $\sum_1^m p_i\underline{x}_i' \in S$ for all $0 \leq p_i \leq 1$ with $\sum_1^m p_i = 1$.[1]

<u>EXERCISE 18.1</u> Prove the assertion made in the preceding paragraph. Interpret the result geometrically and illustrate for the case $n=2$ and $m=3$.

<u>EXERCISE 18.2</u> (Combinations of Convex Sets) Prove that the intersection of <u>any</u> collection of convex subsets of E_n is again a convex subset of E_n. Provide an example to illustrate that this is not true for unions of convex subsets of E_n.

<u>EXERCISE 18.3</u> Demonstrate by examples that a convex subset of E_n need be neither open nor closed.

<u>EXERCISE 18.4</u> Describe the convex subsets of E_1 (the real line).

Of course, an arbitrary subset of E_n need not be convex. It is obvious, however, that any subset of E_n is <u>contained</u> in many different convex sets, which suggests that there may be a

[1]This result can be generalized. See References.

'smallest' (in some sense) convex set containing an arbitrary subset of E_n. This leads us to the following Definition.

DEFINITION 18.2 If S is an arbitrary subset of E_n, then the intersection of all convex sets containing S is denoted by S^*, and is termed the convex hull of S.

EXERCISE 18.5 In view of Exercise 18.2, S^* is convex. However, prove it is the 'smallest' convex set containing S, in the sense that for any other convex set S' containing S, $S^* \subseteq S'$.

Obviously, $S \equiv S^*$ if S is itself convex. The following Diagram illustrates the notion of a convex hull for the case of $n=2$ dimensional Euclidean Space.

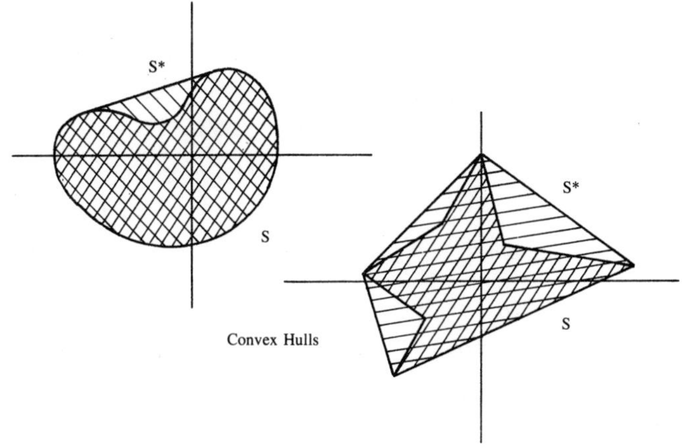

Figure 18-2

To describe adequately the unique properties possessed by convex sets, we need the following basic notions involving points, and sets of points, in E_n.

As is customary, the <u>inner product</u> $\underline{x}' \cdot \underline{y}'$ of two points $\underline{x}' = (x_1, x_2, \ldots, x_n)$ and $\underline{y}' = (y_1, y_2, \ldots, y_n)$ in E_n is defined by:

$$\underline{x}' \cdot \underline{y}' = \sum_1^n x_i y_i \quad .$$

Then, for any fixed point $\underline{a}' = (a_1, a_2, \ldots, a_n)$, and real scalar c, the set of points $\underline{x}'$ in E_n satisfying the condition: $\underline{a}' \cdot \underline{x}' = c$, is termed a <u>hyperplane</u>. For $n=3$ this is the familiar plane of 3-dimensional space.

<u>EXERCISE 18.6</u> Let $\underline{x}_0'$ be a fixed point in E_n. Prove that $\underline{a}' \cdot (\underline{x}' - \underline{x}_0') = 0$, or equivalently $\underline{a}' \cdot \underline{x}' = \underline{a}' \cdot \underline{x}_0'$ $(\underline{a}' \neq \underline{0}')$, is a hyperplane passing through the point $\underline{x}_0'$.

A hyperplane $\underline{a}' \cdot \underline{x}' = c$ is said to <u>separate</u> two subsets S and T of E_n iff $\underline{a}' \cdot \underline{x}' \leq c$ for all $\underline{x}'$ in S and $\underline{a}' \cdot \underline{x}' \geq c$ for all $\underline{x}'$ in T (or vice versa). Additionally, a hyperplane $\underline{a}' \cdot \underline{x}' = c$ is said to <u>support</u> a subset S of E_n at a boundary point[1] $\underline{x}_0'$ of the set, iff $\underline{a}' \cdot \underline{x}_0' = c$ (that is, the hyperplane passes through $\underline{x}_0'$) and $\underline{a}' \cdot \underline{x}_0' \geq c$ for all $\underline{x}'$ in S.[2]

In terms of the above notions, we may now state several of the important properties possessed by convex subsets of E_n.

[1] See page 1-15.

[2] See Figure 18-4.

These properties, in turn, make convex sets an important tool in many areas of Applied Mathematics. Although we shall not prove these properties here (some are a bit lengthy), only basic notions of points in E_n are required to establish each of them.

<u>THEOREM 18.1</u> (Separation) Any two disjoint convex subsets of E_n have at least one separating hyperplane.

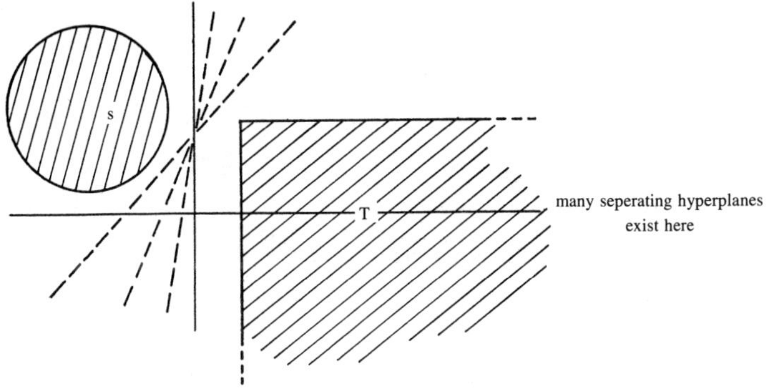

Figure 18-3

<u>EXERCISE 18.7</u> Provide an illustration where only one separating hyperplane exists for a pair of disjoint convex sets of E_2 .

<u>THEOREM 18.2</u> (Support). There is at least one supporting hyperplane at every boundary point of any convex subset of E_n .

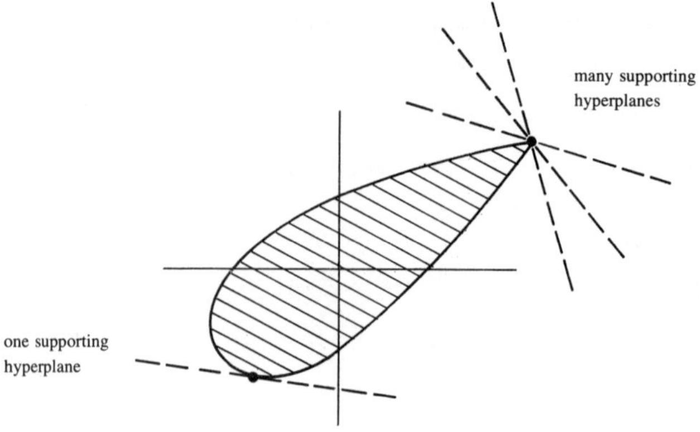

THEOREM 18.3 (Representation) Any point $\underline{s}^*$ in the convex hull S^* of a subset S of E_n can always be represented as a convex linear combination of at most $n + 1$ points of S. If $\underline{s}^*$ is a boundary point of S^* then this number can be reduced to at most n.

EXERCISE 18.8 Draw several figures to illustrate various cases of the preceding Theorem for $n=2$.

Proofs of the three preceding Theorems can be found in the book by Blackwell and Girshick, "Theory of Games and Statistical Decisions", John Wiley and Sons, (1954).

A related and equally important concept is that of a convex function. It is defined as follows:

DEFINITION 18.3 (Convex Funtion) A real-valued function $\underline{f}$ defined over a convex domain D of E_n is termed convex over D iff for every pair $\underline{x}', \underline{y}'$ of points in D, $\underline{f}(p\underline{x}' + q\underline{y}') \leq p\underline{f}(\underline{x}') + q\underline{f}(\underline{y}')$ for all $p,q \geq 0$ satisfying: $p + q = 1$.

EXERCISE 18.9 Prove that $\underline{f}$ is convex over the convex domain D of E_n iff for any positive integer $m \geq 2$, and any choice $\underline{x}_1', \underline{x}_2', \ldots, \underline{x}_m'$ of m points in D, $\underline{f}(\sum_1^m p_i \underline{x}_i') \leq \sum_1^m p_i \underline{f}(\underline{x}_i')$ for all $0 \leq p_i \leq 1$ satisfying: $\sum_1^m p_i = 1$.

Observe that an assumption of continuity for $\underline{f}$ was not made in the formal definition of a convex function. As a matter of fact, for $n = 1$ there do exist discontinuous convex functions. However, it can be proved[1] that if a function defined over an interval satisfies only the mild conditions that (i) it is bounded in an arbitrarily small subinterval, and (ii) satisfies $f(\frac{x+y}{2}) \leq \frac{1}{2}f(x) + \frac{1}{2}f(y)$ for points x,y in its domain, then the function is both continuous and convex. Furthermore, any convex function possesses a finite derivative at all but perhaps a countable number of points.

EXERCISE 18.10 Generalize the comments in the preceding paragraph to functions $\underline{f}$ defined over a convex subset of E_n $(n > 1)$.

Convexity of f for $n=1$ has the appealing geometric interpretation that the graph of f between any two points x and y in its interval domain always lies below the chord joining the points $(x,f(x))$ and $(y,f(y))$ on the graph. This is illustrated in the following Diagram.

[1]See Boas, R., "A Primer of Real Functions", Carus Mathematical Monograph No. 13, Wiley & Sons, New York (1961).

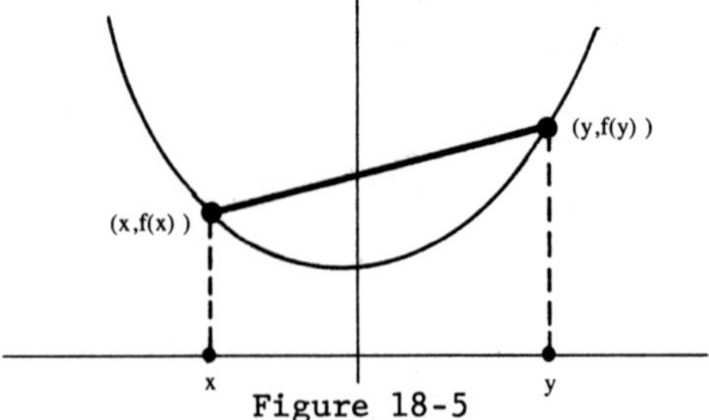

Figure 18-5

EXERCISE 18.11 Formulate the analogous geometric interpretation of convexity for a function of $n > 1$ variables. Illustrate for $n=2$.

For $n=1$, if we assume that a function f possesses a second derivative f'' at each point in its interval domain, then we have the following convenient characterization of convexity.

THEOREM 18.4 If f'' exists at each point of its interval domain D, then f is convex over D iff $f''(x) \geq 0$ for all $x \in D$.

PROOF. Suppose first that f is convex over D. Since f'' exists, it is known[1] that for each $x \in D$ we may write:

$$f''(x) = \lim_{\epsilon \downarrow 0} \frac{f(x+2\epsilon)-2f(x)+f(x-2\epsilon)}{(2\epsilon)^2} .$$

[1] Hobson, E., "Theory of Functions of a Real Variable", Vol. 1, Dover Publications, Inc.

By convexity, $f(x+2\epsilon) + f(x-2\epsilon) \geq 2f(x)$, whence we have established that $f''(x) \geq 0$.

Next suppose that $f''(x) \geq 0$ for all $x \in D$. From Taylor's Theorem it then follows that for any two points $z, w \in D$ we have $f(z) \geq f(w) + (z-w)f'(w)$, hence for all $x, y \in D$ and any non-negative numbers p, q such that $p + q = 1$, we have:

$$pf(x) \geq pf(px + qy) + p(x-px-qy)f'(px + qy)$$

and

$$qf(y) \geq qf(px + qy) + q(y-px-qy)f'(px + qy) .$$

Upon adding the two inequalities we see that $f(px + qy) \leq pf(x) + qf(y)$, hence f is convex over D. With this the proof is completed.

<u>EXERCISE 18.12</u> Over what intervals are the functions: $f(x) = a\text{Sin}bx$ and $f(x) = a\text{Tan}bx$, convex?

<u>EXERCISE 18.13</u> In Figure 18-5, the set of points in E_2 lying on or above the graph of f appears to be convex. Prove, in fact, that if f is continuous and convex over the interval D, then the set:

$$S = \{(x,y): \ x \in D \text{ and } y \geq f(x)\}$$

is indeed a convex subset of E_2. Generalize this result to the case of a function of $n \geq 2$ variables.

For a function f that is convex over the interval domain D of E_1 it follows from Theorem 18.2 (Support) and Exercise 18.13 that there exists at least one supporting hyperplane (in the case $n=2$ a straight line) at each point on the boundary of the set S defined above, namely, $S = \{(x,y): x \in D$ and $y \geq f(x)\}$. Accordingly, for each x_0 in D there exists a constant a_0 such that:

$$f(x) \geq f(x_0) + a_0(x-x_0)$$

for all x in D.

To generalize the above result, consider a continuous, convex function $\underline{f}$ defined over the convex domain D of E_n, $n > 1$. Then the set S defined by:

$$S = \{(\underline{x}',y): \underline{x}' \in D \text{ and } y \geq f(\underline{x}')\}$$

is a convex subset of E_{n+1}[1]. Furthermore, S has at least one supporting hyperplane at each of its boundary points. Therefore, for each $\underline{x}_0'$ in D there exists an n-tuple $\underline{a}_0'$ of constants such that:

$$f(\underline{x}') \geq f(\underline{x}_0') + \underline{a}_0' \cdot (\underline{x}'-\underline{x}_0')$$

for all $\underline{x}'$ in D.

<u>EXERCISE 18.14</u> Prove the preceding result in detail.

We next generalize the second-derivative criterion for convexity of a function of one variable to functions of n $(n > 1)$ variables. The result is contained in the following Theorem.

[1]See Exercise 18.13.

<u>THEOREM 18.5</u> Let $f(\underline{x}')$ be a real-valued function defined over a convex domain D of E_n $(n > 1)$. If all of the second-order partial derivatives: $f_{ij}(\underline{x}') = \frac{\partial^2}{\partial x_i \partial x_j} f(\underline{x}')$ $(i,j = 1,2,\ldots,n)$ exist (finite) throughout D and the $n \times n$ matrix $M(\underline{x}') = [f_{ij}(\underline{x}')]$ is p.s.d. throughout D, then $\underline{f}$ is convex over D.

PROOF. It follows from the multivariate version of Taylor's Theorem that, for any pair of points $\underline{z}',\underline{w}'$ in D, we have: $f(\underline{z}') = f(\underline{w}') + (\underline{z}'-\underline{w}')'f_*(\underline{w}') + (\underline{z}'-\underline{w}')'M(\underline{\theta}')(\underline{z}'-\underline{w}')$, where $f_*'(\underline{w}') = (f_1(\underline{w}'), f_2(\underline{w}'), \ldots, f_n(\underline{w}'))$ is the vector of n first-order partial derivatives of $\underline{f}$ evaluated at $\underline{w}'$, and $\underline{\theta}'$ is some point in D. Accordingly, since the matrix M is p.s.d. throughout D, it follows that $f(\underline{x}') \geq f(\underline{w}') + (\underline{z}'-\underline{w}')'f_*(\underline{w}')$.

Thus, for any pair of points $\underline{x}',\underline{y}'$ in D, and any $p,q \geq 0$ such that $p + q = 1$, we have:
$pf(\underline{x}') \geq pf(p\underline{x}'+q\underline{y}') + p(\underline{x}'-p\underline{x}'-q\underline{y}')'f_*(p\underline{x}'+q\underline{y}')$ and
$qf(\underline{y}') \geq qf(p\underline{x}' q\underline{y}') + q(\underline{y}'-p\underline{x}'+q\underline{y}')'f_*(p\underline{x}'+q\underline{y}')$.
Upon simple addition, the two inequalities yield:
$f(p\underline{x}'+q\underline{y}') \leq pf(\underline{x}') + qf(\underline{y}')$, whence $\underline{f}$ is convex over D, and the proof is completed.

<u>EXERCISE 18.15</u> Is the condition that the matrix $M(\underline{x}')$ be p.s.d. over D also necessary in order that $\underline{f}$ be convex over D? Consider the case n=2.

EXERCISE 18.16 Prove that the function $f(x,y) = 1/xy$ is convex over the domain $D = \{(x,y): x,y > 0\}$ of E_2.

We now generalize the familiar notion of a 'cross-section' of a set in order to expose some additional properties of convex sets and convex functions.

Let D be a convex subset of E_n, that is, a set of ordered n-tuples $\underline{x}' = (x_1, x_2, \ldots, x_n)$ of real numbers satisfying the convexity condition of Definition 18.1. For any choice of m indices $1 \leq i_1 < i_2 < \ldots < i_m \leq n$ $(m = 0,1,\ldots,n)$ and fixed values $x^o_{i_1}, x^o_{i_2}, \ldots, x^o_{i_m}$ of the corresponding coordinates, we define the section of D determined by these choices as follows:

$$D(x^o_{i_1}, x^o_{i_2}, \ldots, x^o_{i_m}) = \{\underline{x}' \in D: x_{i_1} = x^o_{i_1}, \ldots, x_{i_m} = x^o_{i_m}\} .$$

EXERCISE 18.17 Prove that every section of a convex set is again a convex set. Illustrate with several examples for $n=2$ and $n=3$.

Now let $\underline{f}$ be a convex function defined over the convex domain D of E_n. Then clearly $\underline{f}$ restricted to any section of D is also convex over that section. However, the converse is not true. That is, it is entirely possible that a function $\underline{f}$ be convex in any or all of the m $(< n)$ coordinates for any fixed values of the remaining $n-m$ coordinates, yet fail to be a convex function in E_n.

EXERCISE 18.18 Prove, then illustrate, the above statements for the case $n=2$.

We now consider a few applications of the preceding notions. The first is known as Jensen's Inequality, of which there are many versions and applications. We consider here only one case.

With respect to a given 1-dimensional c.d.f. F, we define the expected value $Ef(x)$ of a continuous function f as: $Ef(x) = \int_{-\infty}^{\infty} f(x)dF(x)$, provided this integral exists (finite). We may now state one form of Jensen's Inequality.

<u>THEOREM 18.6</u> (Jensen's Inequality) Suppose that g is continuous and convex, and that, with respect to a given c.d.f. F, both Ex and $Eg(x)$ exist. Then,

$$g(Ex) \le Eg(x) .$$

PROOF. Choosing $x_0 = Ex$, it follows that there exists a constant a_0 such that $g(x) \ge g(Ex) + a_0(x-Ex)$ for all x. Upon integration of both sides of the above inequality with respect to F, the desired result follows from the basic properties of the Riemann-Stieltjes Integral.

<u>EXERCISE 18.19</u> (Concave Functions) A function g is termed concave according as $-g$ is convex. Derive the following form of Jensen's Inequality for concave functions (under the appropriate conditions): $Eg(x) \le g(Ex)$.

<u>EXERCISE 18.20</u> (Arithmetic-Geometric-Harmonic Mean Inequality) For positive real numbers $x_1, x_2, \ldots, x_n$ we define their Arithmetic $\overline{X}$, Geometric $\overline{G}$ and Harmonic $\overline{H}$

means as follows:

$$\overline{X} = \frac{1}{n}(x_1+x_2+\ldots+x_n)$$

$$\overline{G} = (x_1 \cdot x_2 \cdot \ldots \cdot x_n)^{1/n}$$

$$\overline{H} = \frac{1}{\frac{1}{n}(\frac{1}{x_1}+\frac{1}{x_2}+\ldots+\frac{1}{x_n})} \quad .$$

By use of the preceding results concerning convex (concave) functions, prove that:

$$\overline{H} \le \overline{G} \le \overline{X} \quad .$$

For example, to prove that $\overline{G} \le \overline{X}$, take F as the 'step' c.d.f. with 'steps' of heights $1/n$ at the values x_i $(i = 1,2,\ldots,n)$, then use Jensen's Inequality with the concave function $g(x) = \log x$, $x > 0$. If the x_i's are not all distinct, F must be modified accordingly.

Hints and Answers to Exercises: Section 18

18.1 Hint: Use Mathematical Induction on the first part, and for the second part observe that for $n=2$ and $m=3$, all points of the form: $p_1\underline{x}_1' + p_2\underline{x}_2' + p_3\underline{x}_3'$, where the $\underline{x}_i'$'s are fixed, and the non-negative p_i's vary in such a manner that $p_1 + p_2 + p_3 = 1$, form a triangle.

18.2 Hint: Assume the intersection contains at least two points. Choose any pair of points therein, and prove that the line joining them also belongs to the intersection.

18.3 Answer: For $n=1$ the interval $(a,b]$ is convex but neither closed nor open.

18.4 Answer: Individual intervals of all types, including singleton points and, for completeness, the empty set.

18.5 Hint: Assume there exists a convex set S' that contains S and is properly contained in S^*,[1] and establish a contradiction.

18.6 Hint: Clearly the equation defines a hyperplane, and the point $\underline{x}_0'$ satisfies this equation.

18.7 Answer: The two half planes $S_1 = \{(x,y): x > 0\}$ and $S_2 = \{(x,y): x < 0\}$ are examples. The convex sets may be bounded or unbounded.

18.8 Hint: Use Mathematical Induction.

18.9 Hint: See, for example, the book titled "A Primer of Real Functions" by R. P. Boas, Jr., Carus Mathematical Monograph 13, Mathematical Association of America, (1961) pp. 142-144.

18.10 Hint: Use the notion of a hyperplane.

18.11 Hint: Sketch each function over one of its periods.

18.12 Hint: Use Definition 12.1. For generalizations use the notion of a hyperplane.

---

[1]That is, $S \subseteq S' \subseteq S^*$ but $S' \neq S^*$.

18.14 Hint: Use the definition of p.s.d. found in Exercise 17.32.

18.15 Hint: Use Theorem 12.5.

18.16 Hint: Use Definition 12.1.

References to Additional and Related Material: Section 18

1. Blackwell, D. and M. Girshick, "Theory of Games and Statistical Decisions", Wiley and Sons, Inc., New York (1954).

2. Bonnesen, T. and W. Fenchel, "Theorie der Konvexen Korper", Springer-Verlag, Berlin (1934) and Chelsea Publishing Co., New York (1948).

3. Eggleston, H., "Convexity", McGraw-Hill, Inc., New York (1964).

4. Eggleston, H., "Problems in Euclidean Space: Applications of Convexity", Pergamon Press, New York (1957).

5. Fan, K., "Convex Sets and Their Applications", Argonne National Laboratory, Applied Mathematics Summer Lectures (1959).

6. Fenchel, W., "Convex Cones, Sets and Functions", Princeton University Press (1953).

7. Garvin, W., "Introduction to Linear Programming", McGraw-Hill, Inc., New York (1960).

8. Hardy, G., Littlewood, J. and G. Pólya, "Inequalities", Cambridge University Press (1934).

9. Karlin, S., "Mathematical Methods and Theory in Games, Programming, and Economics", Addison-Wesley, Inc., Reading, Massachusetts (1959).

10. McKinsey, J., "Introduction to the Theory of Games", McGraw-Hill, Inc., New York (1952).

11. Mitrinović, D., "Analytic Inequalities", Springer-Verlag, Berlin (1970).

12. Neumann, J. von and O. Morgenstern, "Theory of Games and Economic Behavior", (Second Ed.), Princeton University Press (1947).

13. Roberts, A., "Convex Functions", Academic Press, New York (1973).

14. Stoer, J., "Convexity and Optimization in Finite Dimensions", Springer-Verlag, New York (1970).

15. Valentine, F., "Convex Sets", McGraw-Hill, Inc., New York (1964).

19. Max-Min Problems

A problem frequently encountered in Applied Mathematics is that of finding the maximum or minimum of a real valued function $f(x_1,...,x_n)$ of n real variables, where the variables $(x_1,...,x_n)$ are constrained to lie in some subset C of E_n.[1] Depending upon the nature of f and the manner in which the subset C is specified, there are various techniques for solving the above problem. First we will be concerned with two important cases:

<u>Case A</u> (Unconstrained max-min) Find the maximum or minimum of a function $f(x_1,...,x_n)$ as the variables range over all of E_n .

<u>Case B</u> (Constrained max-min) Find the maximum or minimum of a function $f(x_1,...,x_n)$ where the variables are constrained so as to satisfy m $(< n)$ equations called 'side conditions'. These side conditions then specify the subset C of E_n .

Mention will be made later of various other cases of this problem and of methods applicable in these cases.

Before proceeding, it is important to recall from Elementary Calculus the difference between a local (or relative) max or min and a global (or absolute) max or min. The techniques we shall consider here provide direct means for locating local

[1] A maximum or minimum need not exist; for our purposes, we assume this is not the case.

(or relative) max or min's of a function; some additional investigation may then be necessary to locate a global (or absolute) max or min of the function.

Case A (Unconstrained max-min)

Functions of one variable. Let $f(x)$ be a regular[1] real function of the real variable x. The critical points of f are those points x satisfying the condition $f'(x) = 0$. A necessary and sufficient condition that a critical point $x=a$ be a relative extrema[2] is that there exist an even integer n such that $f'(a) = f^{(2)}(a) = \dots = f^{(n-1)}(a) = 0$, and $f^{(n)}(a) \neq 0$. The critical point is a relative minimum if $f^{(n)}(a) > 0$, and a relative maximum of $f^{(n)}(a) < 0$. If no such even integer exists, then the point is an inflection point.

EXAMPLE 19.1 The regular function $f(x) = x^3$ has $f'(0) = f^{(2)}(0) = 0$, and $f^{(3)}(0) \neq 0$. Thus $x=0$ is an inflection point. On the other hand $f(x) = x^4$ has $f'(0) = f^{(2)}(0) = f^{(3)}(0) = 0$, and $f^{(4)}(0) > 0$. Thus $x=0$ is a relative minimum. The function $f(x) = e^{-1/x^2}$ does have a minimum at $x=0$, but the above method fails to apply in locating it because this function is not regular at $x=0$.

[1] To say that $f(x)$ is regular at a point $x=a$ means that the function possesses a convergent Taylor Series representation about $x=a$. See also Section 13. This condition can be weakened to assume only existence of $f''(x)$. However, in this case, we would have only a sufficient condition for existence of an extrema.

[2] That is, max or min.

EXERCISE 19.1 (Maximum Likelihood Estimator) Prove that the maximum value of the function:

$$f(\mu) = \prod_{1}^{n} \frac{1}{\sqrt{2\pi}} e^{-\frac{1}{2}(x_i - \mu)^2}$$

(the x_i's being fixed) occurs at $\mu = \frac{1}{n} \Sigma x_i$.

EXERCISE 19.2 (Mode of the Beta Distribution) Find the value of x within the interval $[0,1]$ maximizing:

$$f(x) = \frac{\Gamma(a+b)}{\Gamma(a)\Gamma(b)} x^{a-1}(1-x)^{b-1} ,$$

where a and b are positive constants.

Functions of $n > 1$ variables. Let $f(x_1,\ldots,x_n)$ be a function of the n variables $(x_1,\ldots,x_n)$ possessing all first and second order partial derivatives: $f_i = \frac{\partial f}{\partial f_i}$ and $f_{ij} = \frac{\partial^2 f}{\partial x_i \partial x_j}$. The critical points of f are those satisfying the n equations: $f_1 = f_2 = \ldots = f_n = 0$. Now let $(a_1,a_2,\ldots,a_n)$ be a critical point and $D_1,D_2,\ldots,D_n$ the following sequence of determinants of partial derivatives all evaluated at the point $(a_1,a_2,\ldots,a_n)$:

$$D_1 = f_{11}, \quad D_2 = \begin{vmatrix} f_{11} & f_{12} \\ f_{21} & f_{22} \end{vmatrix}, \quad D_3 = \begin{vmatrix} f_{11} & f_{12} & f_{13} \\ f_{21} & f_{22} & f_{23} \\ f_{31} & f_{32} & f_{33} \end{vmatrix},$$

$$\ldots\ldots\ldots, \quad D_n = \begin{vmatrix} f_{11} & f_{12} & \cdots & f_{1n} \\ f_{21} & f_{22} & \cdots & f_{2n} \\ \vdots & \vdots & & \vdots \\ f_{n1} & f_{n2} & \cdots & f_{nn} \end{vmatrix}.$$

A sufficient condition that $(a_1,a_2,\ldots,a_n)$ be a relative minimum is that $D_1 > 0$, $D_2 > 0,\ldots,D_n > 0$. A sufficient condition that $(a_1,a_2,\ldots,a_n)$ be a relative maximum is that $D_1 < 0$, $D_2 > 0,\ldots,(-1)^n D_n > 0$. Otherwise the situation is indeterminate.[1]

EXAMPLE 19.2 Classify, if possible, the critical point $(x,y,z) = (1,1,1)$ of the function $f(x,y,z) = x^4 + y^4 + z^4 - 4xyz$. The necessary conditions for a critical point are equivalent to the three equations $x^3 - yz = y^3 - xz = z^3 - xy = 0$, which are clearly satisfied by $(1,1,1)$. Evaluation of the second-order partial derivatives at this point yields the following sequence of determinates:

$$D_1 = 12, \quad D_2 = \begin{vmatrix} 12 & -4 \\ -4 & 12 \end{vmatrix}, \quad D_3 = \begin{vmatrix} 12 & -4 & -4 \\ -4 & 12 & -4 \\ -4 & -4 & 12 \end{vmatrix}.$$

Since $D_1 = 12 > 0$, $D_2 = 128 > 0$, $D_3 = 1024 > 0$, it follows that $(1,1,1)$ is a relative minimum of f.

[1] See the Reference by Gillespie for a proof.

<u>EXAMPLE 19.3</u> The function $f(x,y,z) = xyz(1-x-y-z)$ has a relative maximum at $(x,y,z) = (1/4,1/4,1/4)$. First note that the necessary conditions for a critical point, namely $f_x = f_y = f_z = 0$, are equivalent to the three equations: $yz-2xyz-y^2z-yz^2 = xz-x^2z-2xyz-xz^2 = xy-x^2y-xy^2-2xyz = 0$. These are clearly satisfied by $(1/4,1/4,1/4)$. Evaluation of the second-order partial derivatives at this point yields the following sequence of determinates:

$$D_1 = -1/8, \quad D_2 = \begin{vmatrix} -1/8 & -1/16 \\ -1/16 & -1/8 \end{vmatrix}, \quad D_3 = \begin{vmatrix} -1/8 & -1/16 & -1/16 \\ -1/16 & -1/8 & -1/16 \\ -1/16 & -1/16 & -1/8 \end{vmatrix}.$$

Since $D_1 = -1/8 < 0$, $D_2 = 3/256 > 0$, $D_3 = -1/1024 < 0$, it follows that the given point is a relative maximum of f.

<u>EXERCISE 19.3</u> (Joint Maximum Likelihood Estimators) Find the value of (μ, θ) maximizing the function:

$$f(\mu, \theta) = \prod_1^n \frac{1}{\sqrt{2\pi\theta}} e^{-\frac{1}{2}\left(\frac{x_i - \mu}{\theta}\right)^2}$$

(the x_i's being fixed). Prove it is the maximum.

<u>EXERCISE 19.4</u> (Least Squares Estimators) Find the value of (m,b) minimizing the function $g(m,b) = \sum_1^n (y_i-mx_i-b)^2$, where the x_i's and y_i's are fixed. Prove it is the minimum.

Case B (Constrained max-min)

A problem frequently encountered in Applied Mathematics is that of maximizing or minimizing a function of several variables, where the variables involved must satisfy one or more equations called 'side conditions'.

EXAMPLE 19.4 (Maximum Entropy) Find the values of $(p_1,\ldots,p_n)$ maximizing $H(p_1,\ldots,p_n) = -\Sigma\, p_i \log p_i$ subject to the condition $p_1+\ldots+p_n = 1$. (Assume $p_i > 0$).

Although for any specific problem it may be possible to eliminate some variables by suitable substitution, and thereby obtain an ordinary, unconstrained, max-min problem, we shall consider an alternate, general approach using Lagrange Multipliers; it is useful and convenient in a variety of situations.

First, we shall investigate how and why the method of Lagrange Multipliers works by considering the three-variable case in detail. Note that maximizing or minimizing the function $f(x,y,z) = f$ subject to the two side conditions $g(x,y,z) = 0$ and $h(x,y,z) = 0$ say, is geometrically equivalent to maximizing or minimizing the value of $f(x,y,z)$ as the coordinates (x,y,z) vary along the curve in 3-space produced by the (assumed) intersection of the two surfaces $g=0$ and $h=0$.

From properties of a relative extrema, it is known that, at such a point, the directional derivative of f along the tangent line to the curve at the point must be zero. Since this directional derivative is the component of $\Delta f = \frac{\partial f}{\partial x}i + \frac{\partial f}{\partial y}j + \frac{\partial f}{\partial z}k$ along the tangent line, we see that Δf is normal[1] to the tangent line at the point. Also, since the gradient vectors $g = \frac{\partial g}{\partial x}i + \frac{\partial g}{\partial y}j + \frac{\partial g}{\partial z}k$ and $h = \frac{\partial h}{\partial x}i + \frac{\partial h}{\partial y}j + \frac{\partial h}{\partial z}k$ always satisfies this property, we see that the three vectors f, g, h must be coplanar, and hence dependent[2], at the point of extrema. Thus there exist constants c_1 and c_2 such that $f + c_1\ g + c_2\ h = \underline{0}$ (zero vector). This vector equation is equivalent to the three scalar equations:

$$\frac{\partial f}{\partial x} + c_1 \frac{\partial g}{\partial x} + c_2 \frac{\partial h}{\partial x} = 0$$

$$\frac{\partial f}{\partial y} + c_1 \frac{\partial g}{\partial y} + c_2 \frac{\partial h}{\partial y} = 0$$

$$\frac{\partial f}{\partial z} + c_1 \frac{\partial g}{\partial z} + c_2 \frac{\partial h}{\partial z} = 0 \quad .$$

These equations can be arrived at by simply setting the successive partial derivatives of the auxiliary Lagrange Function, namely $L = f + c_1 g + c_2 h$, equal to zero. The constants c_1 and c_2 are usually called Lagrange Multipliers.

Together with the two side conditions, there are five equations in the five unknowns (x, y, z, c_1, c_2), which can then presumably be solved simultaneously to locate the co-ordinates of potential relative maxima and minima. These points must then be tested further to determine their nature.

---

[1] That is, perpendicular

[2] This follows from the results of Section 15 (Vectors and Vector Spaces).

It is not difficult to see that the method described above will generalize naturally to more than three variables and differing numbers of side conditions.

Accordingly, assume the needed regularity conditions are fulfilled[1], and let $f(x_1,\ldots,x_n)$ be the function to be maximized or minimized subject to the side conditions $g^k(x_1,\ldots,x_n) = 0$ $(k = 1,2,\ldots,m)$, where $m < n$. First form the auxiliary Lagrange Function $L = f + c_1 g^1 +\ldots+ c_m g^m$. From this we obtain the n Lagrange Equations: $f_r + c_1 g_r^1 +\ldots+ c_m g_r^m = 0$ $(r = 1,2,\ldots,n)$, where $f_r = \frac{\partial f}{\partial x_r}$ and $g_r^k = \frac{\partial g^k}{\partial x_r}$. Together with the m side conditions we then have $m+n$ equations in the $m+n$ unknowns $(x_1,\ldots,x_n,c_1,\ldots,c_m)$. Solutions to these equations then provide potential relative maxima and minima. We now consider a method for classifying these critical points.

Let $(a_1,\ldots,a_n,c_1',\ldots,c_m')$ be a solution of the $m+n$ equations, so that $(a_1,\ldots,a_n)$ is a potential relative maximum or minimum of f subject to the given side conditions. Let Q denote the function $Q = f + c_1'g^1+\ldots+ c_m'g^m$ and $Q_{rs} = \frac{\partial^2 Q}{\partial x_r \partial x_s}$. Form the following sequence $E_1,\ldots,E_{n-m}$ of determinants of partial derivatives evaluated at the point $(a_1,\ldots,a_n)$:

[1]Relating to existence of mixed partial derivatives.

$$E_1 = \begin{vmatrix} Q_{11} & \cdots & Q_{1n} & g_1^1 & \cdots & g_1^m \\ \vdots & & \vdots & \vdots & & \vdots \\ Q_{n1} & \cdots & Q_{nn} & g_n^1 & \cdots & g_n^m \\ g_1^1 & \cdots & g_n^1 & 0 & \cdots & 0 \\ \vdots & & \vdots & \vdots & & \vdots \\ g_1^m & \cdots & g_n^m & 0 & \cdots & 0 \end{vmatrix}$$

$$E_2 = \begin{vmatrix} Q_{22} & \cdots & Q_{2n} & g_2^1 & \cdots & g_2^m \\ \vdots & & \vdots & \vdots & & \vdots \\ Q_{n2} & \cdots & Q_{nn} & g_n^1 & \cdots & g_n^m \\ g_2^1 & \cdots & g_n^1 & 0 & \cdots & 0 \\ \vdots & & \vdots & \vdots & & \vdots \\ g_2^m & \cdots & g_n^m & 0 & \cdots & 0 \end{vmatrix}$$

$$\vdots$$

$$E_{n-m} = \begin{vmatrix} Q_{n-m,n-m} & \cdots & Q_{n-m,n} & g_{n-m}^1 & \cdots & g_{n-m}^m \\ \vdots & & \vdots & \vdots & & \vdots \\ Q_{n,n-m} & \cdots & Q_{n,n} & g_n^1 & \cdots & g_n^m \\ g_{n-m}^1 & \cdots & g_n^1 & 0 & \cdots & 0 \\ \vdots & & \vdots & \vdots & & \vdots \\ g_{n-m}^m & \cdots & g_n^m & 0 & \cdots & 0 \end{vmatrix} .$$

A sufficient condition that the point $(a_1,\ldots,a_n)$ be a **relative** minimum is that:

either (i) m be even and $E_1 > 0,\ E_2 > 0,\ldots,\ E_{n-m} > 0$
or (ii) m be odd and $E_1 < 0,\ E_2 < 0,\ldots,\ E_{n-m} < 0$.

A sufficient condition that the point $(a_1,\ldots,a_n)$ be a **relative** maximum is that:

either (i) n be even and $E_1 > 0,\ E_2 < 0,\ldots,(-1)^{n-m}E_{n-m} < 0$
or (ii) n be odd and $E_1 < 0,\ E_2 > 0,\ldots,(-1)^{n-m}E_{n-m} > 0$.
Otherwise the situation is indeterminate.[1]

EXAMPLE 19.5 Find, and classify if possible, the critical values of the function $f = yz + zx + xy$ subject to the side condition $x^2 + y^2 + z^2 = 3$. Here the Lagrange function is $L = yz + zx + xy + c(x^2 + y^2 + z^2 - 3)$ and the Lagrange equations $L_x = L_y = L_z = 0$ together with the side condition are equivalent to: $y + z + 2cx = z + x + 2cy = y + x + 2cz = 0$, $x^2 + y^2 + z^2 = 3$. The solutions to these four equations in four unknowns are: $(x,y,z,c) = (1,1,1,-1),\ (-1,-1,-1,-1)$. Note that $n=3$, $m=1$.

For the first solution, $Q = yz + zx + xy - (x^2 + y^2 + z^2 - 3)$ and, upon evaluation of its partial derivatives at $(1,1,1)$, yields the following sequence of $n-m=2$ determinants:

---

[1]See the Reference by Gillespie for a proof.

$$E_1 = \begin{vmatrix} -2 & 1 & 1 & 1 \\ 1 & -2 & 1 & 2 \\ 1 & 1 & -2 & 2 \\ 2 & 2 & 2 & 0 \end{vmatrix}, \quad E_2 = \begin{vmatrix} -2 & 1 & 2 \\ 1 & -2 & 2 \\ 2 & 2 & 0 \end{vmatrix}.$$

Now since n is odd and $E_1 = -108 < 0$, $E_2 = 24 > 0$, it follows that the point $(1,1,1)$ is a relative maximum of f so constrained.

EXERCISE 19.5 Classify the critical point $(-1,-1,-1)$ in the preceding Example.

EXERCISE 19.6 (Maximum Variance in Poisson-Bernoulli Sums) Prove that the maximum of the function: $V(p_1,\ldots,p_n) = \sum_1^n p_i(1-p_i)$ subject to the side condition $\sum_1^n p_i = np$ occurs at $(p_1,\ldots,p_n) = (p,\ldots,p)$. (Assume $0 < p_i < 1$). What is the maximum?

EXERCISE 19.7 (Maximum Entropy) Prove that the maximum value of the function $H(p_1,\ldots,p_n) = -\Sigma\, p_i \log p_i$, subject to the side condition $p_1 +\ldots+ p_n = 1$, is $\log n$, and occurs at $(p_1,\ldots,p_n) = (1/n,\ldots,1/n)$. What about a minimum? (Assume $p_i > 0$).

EXAMPLE 19.6 Consider the critical points of the function $f(x,y,z) = x^2 + y^2 + z^2$ subject to the two side conditions $z(x+y) = -2$ and $xy = 1$. Here the auxiliary Lagrange function L is given by: $L = x^2 + y^2 + z^2 + c_1(z(x+y)+ 2) + c_2(xy-1)$. The Lagrange equations together with the two side conditions are equivalent to the five equations: $2x + c_1z + c_2y = 2y + c_1z + c_2x = 2z + c_1(x+y) = 0$, and $z(x+y) = -2$, $xy = 1$. The following values of the five unknowns (x, y, z, c_1, c_2) are solutions of this system: $(1,1,-1,1,-1)$, $(-1,-1,1,1,-1)$.

For the first solution, $Q = x^2 + y^2 + z^2 + (z(x+y) + 2) - (xy-1)$, and, upon evaluating the appropriate partial derivatives at $(x,y,z) = (1,1,-1)$, we obtain the following determinant $(n=3,\ m=2,$ hence $n-m=1)$:

$$E_1 = \begin{vmatrix} 2 & -1 & 1 & -1 & 1 \\ -1 & 2 & 1 & -1 & 1 \\ 1 & 1 & 2 & 2 & 0 \\ -1 & -1 & 2 & 0 & 0 \\ 1 & 1 & 0 & 0 & 0 \end{vmatrix} .$$

Now since m is even and $E_1 = 24 > 0$, it follows that $(1,1,-1)$ is a relative minimum of f so constrained.

EXERCISE 19.8 Classify the remaining critical point in the preceding Example.

EXERCISE 19.9 (Limitations of the Method) Find the six critical points of $f(x,y,z) = x^2 + y^2 + z^2$ subject to: $x^2 + 2y^2 + 3z^2 = 6$, and show that only two of them can be classified by the present method. For the other 4, the situation is indeterminate. (In such cases a common sense alternative is to explore the geometry of the situation.)

EXERCISE 19.10 Prove that the function $f(x,y,z) = \log x^2y^2 + 4$ subject to the condition $x^2 + y^2 = 2$ takes on its global maximum at $(x,y) = (\pm 1, \pm 1)$. Sketch the function to see why a minimum does not exist.

There are methods for treating other general types of max-min problems. For example, the subject of Linear Programming treats the problem of maximizing or minimizing a <u>linear</u> function $f(x_1,\ldots,x_n) = d + e_1x_1 +\ldots+ e_nx_n$ of the n variables $(x_1,\ldots,x_n)$ subject to <u>linear</u> constraints of the form:

$$b_j + a_{1j}x_1 +\ldots+ a_{nj}x_n \le 0 \quad (j = 1,\ldots,m_1)$$

$$c_j + d_{1j}x_1 +\ldots+ d_{nj}x_n = 0 \quad (j = 1,\ldots,m_2) .$$

where $m_1 + m_2 < n$ and certain of the variables might also be constrained to be non-negative.

Permitting the function to be maximized or minimized, or some of the constraints themselves, to be non-linear leads to the subject of Non-linear Programming. These specialized subjects can be found treated in many of the References given at the end of the present section. Some techniques employ the properties of convex sets presented in Section 18.

Finally, when faced with a non-standard situation, a common sense approach is to explore the geometry of the specific problem at hand. For instance, if a univariate function is not regular at certain points, then these are logical points to investigate for relative extrema; if a multivariate function fails to possess mixed partial derivatives at certain points, then similar reasoning applies.

Hints and Answers to Exercises: Section 19

19.1 Hint: Solve the equivalent problem of maximizing $\log f(\mu)$.

19.2 Answer:

(i) $x = 0$ if $b < 1 < a$

(ii) $x = 1$ if $a < 1 < b$

(iii) $x = \frac{1}{2}$ if $a = b$

(iv) $x = (a-1)/(b-a)$ if $1 < a < b$ or $b < a < 1$.

19.3 Hint: Solve the equivalent problem of maximizing $\log f(\mu, \theta)$. Answer:

$$\mu = (1/n)\ \Sigma\ x_i = \bar{x}$$

$$\theta^2 = (1/n)\ \Sigma\ (x_i - \bar{x})^2$$

19.4 Answer: (Assuming the x_i's are not all equal)

$$m = \frac{\Sigma(x_i-\bar{x})(y_i-\bar{y})}{\Sigma\ (x_i-\bar{x})^2}$$

$$b = \bar{y} - m\bar{x}$$

where $\bar{x} = (1/n)\ \Sigma\ x_i$ and $\bar{y} = (1/n)\ \Sigma\ y_i$.

19.5 Answer: $E_1 = -108 < 0$ and $E_2 = 24 > 0$: a relative maximum.

19.6 Hint: Rewrite V as $np - \Sigma\ p_i^2$.

19.8 Answer: $E_1 = 24 > 0$: a relative minimum.

19.9 Answer: Indeterminate cases are: $x = \pm\sqrt{6}$, $y = z = 0$, $c_1 = -1$ where $E_1 = -192 < 0$ and $E_2 = 0$; $y = \pm\sqrt{5}$, $x = z = 0$, $c_1 = -\frac{1}{2}$ where $E_1 = E_2 = 48 > 0$. For $z = \pm\sqrt{2}$, $x = y = 0$, $c_1 = -(1/3)$ we have $E_1 = -(16/9)$ and $E_2 = -(4/3)$ whence a relative maxima.

References to Additional and Related Material: Section 19

1. Bazaraa, M. and C. Shetty, "Foundations of Optimization", Springer-Verlag, New York (1976).

2. Berman, Al, "Cones, Matrices and Mathematical Programming", Springer-Verlag, New York (1973).

3. Brent, R., "Algorithms for Minimization Without Derivatives", Prentice-Hall, Inc. (1973).

4. Collantz, L. and W. Wetterling, "Optimization Problems", Springer-Verlag, New York (1975).

5. Cooper, L. and D. Steinberg, "Methods of Optimization", W. B. Saunders Co., Philadelphia (1970).

6. Danø, S., "Linear Programming in Industry", Springer-Verlag, New York (1974).

7. Danø, S., "Non-Linear and Dynamic Programming", Springer-Verlag, New York (1975).

8. Dantzig, G., "Linear Programming and Extensions", Princeton University Press (1963).

9. Demyanov, V. and R. Rubinov, "Approximate Methods in Optimization Problems", American Elsevier Publishing Co. (1970).

10. El-Hodiri, M., "Constrained Extrema: Introduction to the Differential Case with Economic Applications", Springer-Verlag, New York (1971).

11. Gillespie, R., "Partial Differentiation", Oliver and Boyd, Ltd. (1951).

12. Hancock, H., "Theory of Maxima and Minima", Dover Publications, Inc. (1974).

13. Hestenes, M., "Optimization Theory: the Finite Dimensional Case", Dover Publications, Inc. (1975).

14. Kaplan, W., "Advanced Calculus", Addison-Wesley, Inc. (1952).

15. Koo, D., "Elements of Optimization", Springer-Verlag, New York (1978)

16. Kowalik, J., "Methods for Uncontrained Optimization Problems", American Elsevier Publishing Co. (1968).

17. Reinfeld, N. and W. Vogel, "Mathematical Programming", Prentice-Hall, Inc. (1958).

18. Spiegel, M., "Schaum's Outline of Theory and Problems of Advanced Calculus", Schaum Publishing Co. (1963).

19. Spivey, W. and R. Thrall, "Linear Optimization", Holt, Reinhart and Winston, Inc. (1970).

20. Zukhovitskiy, S. and L. Avdeyeva, "Linear and Convex Programming", W. B. Saunders, Inc. (1966).

20. Some Basic Inequalities

For purposes of problem-solving in Applied Mathematics, an appropriate inequality may be as valuable as, or a workable substitute for, an exact relationship. There are, of course, many different types of inequalities, some very general, others highly specialized. Some inequalities involve sequences or series of real or complex numbers, some involve integrals, others involve determinants of matrices, and so on.

In this Section we shall introduce and prove several general Classical inequalities that prove of constant use in many different fields of application. Further results concerning inequalities may be found in the References.

(i) The Cauchy Inequality. There are many different versions of this Classical inequality. We consider first that version dealing with finite series of real numbers.

(a) Cauchy Inequality: Finite Series Version. Let $\sum_1^n a_i$ and $\sum_1^n b_i$ be two finite series of real numbers.[1]

Then,

$$\left[\sum_1^n (a_i b_i)\right]^2 \leq \left[\sum_1^n a_i^2\right] \left[\sum_1^n b_i^2\right] .$$

PROOF. View the terms of the two series as row-vectors $\underline{a}' = (a_1, \ldots, a_n)$ and $\underline{b}' = (b_1, \ldots, b_n)$

[1] Assuming that the two series have the same number of terms is no loss of generality. Should this not be the case, 'dummy' zero terms may always be added to the 'shorter' series.

with n real components each. Accordingly, in the notation of Section 15 (Vectors and Vector Spaces), we have,

$$[\sum_1^n (a_i b_i)]^2 = |\underline{a}' \cdot \underline{b}'|^2 ,$$

$$\sum_1^n a_i^2 = \|\underline{a}'\|^2 ,$$

and

$$\sum_1^n b_i^2 = \|\underline{b}'\|^2 .$$

It has already been proved in Section 15 that,

$$|\underline{a}' \cdot \underline{b}'| \le \|\underline{a}'\| \cdot \|\underline{b}'\| .$$

The result then follows upon squarring.

<u>EXAMPLE 20.1</u> Since inequality (a) is valid for any real numbers, it remains valid if the terms of each finite series are replaced by their absolute values. Accordingly, we obtain the following inequality:

$$[\sum_1^n |a_i b_i|]^2 \le [\sum_1^n a_i^2]\,[\sum_1^n b_i^2] .$$

Should some of the terms of the original series be negative, this inequality might be more useful than the form found in (a). Why?

<u>EXERCISE 20.1</u> (Application: Correlation) Let (x_1, y_1), $(x_2, y_2), \ldots, (x_n, y_n)$ be n pairs of real numbers, with $\bar{x} = (1/n) \Sigma x_i$ and $\bar{y} = (1/n) \Sigma y_i$. The sample correlation r between the first and second components of the pairs is defined as follows:

$$r = \frac{\Sigma (x_i-\bar{x})(y_i-\bar{y})}{\sqrt{\Sigma (x_i-\bar{x})^2 \cdot \Sigma (y_i-\bar{y})^2}}\ .^{1}$$

Prove that $|r| \leq 1$.

EXAMPLE 20.2 (Attainment of Equality in Cauchy's Inequality) Upon specializing the results in Section 15 to the case of vectors with real components only, note that:

$$\text{Cos } \theta = \frac{\Sigma a_i b_i}{\sqrt{\Sigma a_i^2 \cdot \Sigma b_i^2}} \quad {}^{2}$$

actually represents the (real) Cosine of the angle θ between the vectors $\underline{a}' = (a_1,\ldots,a_n)$ and $\underline{b}' = (b_1,\ldots,b_n)$. Accordingly, $\text{Cos}^2\theta = 1$ iff $\theta = 0$ or π, in which case one of the vectors $\underline{a}'$, $\underline{b}'$ is a real scalar multiple of the other, that is, iff $\underline{a}' = \lambda\underline{b}'$ or $\underline{b}' = \delta\underline{a}'$ for some real number λ or δ. Therefore, equality is attained in (a) iff either $a_i = \lambda b_i$ or $b_i = \delta a_i$ for $i = 1,\ldots,n$.

EXERCISE 20.2 (Special Values of the Correlation) Prove that $r^2 = 1$ iff $x_i = ay_i + b$ or $y_i = cx_i + d$ $(i = 1,\ldots,n)$ for some pair of real numbers a,b or c,d. Further, prove that $r = 0$ iff $\underline{x}' = (x_1,\ldots,x_n)$ and $\underline{y}' = (y_1,\ldots,y_n)$ are orthogonal vectors.

---

[1]Assume that at least two x_i's and two y_i's differ. Why?

[2]Assume that neither $\underline{a}'$ nor $\underline{b}'$ is the zero vector. Otherwise the inequality in (a) is degenerate.

<u>EXERCISE 20.3</u> The Cauchy Inequality is sometimes referred to as the Cauchy-Schwarz Inequality. Deduce the following sequence of inequalities from results established so far:

$$\sum_1^n a_i b_i \le \sum_1^n |a_i b_i| \le [\sum_1^n a_i^2]^{\frac{1}{2}} \cdot [\sum_1^n b_i^2]^{\frac{1}{2}} .$$

(b) <u>Cauchy Inequality: Infinite Series Version</u>. Let $\sum_1^\infty a_n$ and $\sum_1^\infty b_n$ be two infinite series of real numbers (not necessarily convergent). If the two series $\sum_1^\infty a_n^2$ and $\sum_1^\infty b_n^2$ both converge, then the series $\sum_1^\infty (a_n b_n)$ converges and,

$$[\sum_1^\infty (a_n b_n)]^2 \le [\sum_1^\infty a_n^2] \cdot [\sum_1^\infty b_n^2] .$$

PROOF. The result follows from a simple limiting process applied to the appropriate partial sums, using the Finite Series Version of Cauchy's Inequality.

<u>EXERCISE 20.4</u> (Extension) Prove that in the Infinite Series Version of Cauchy's Inequality, the statement "$\sum_1^\infty (a_n b_n)$ converges" may be replaced by the statement "$\sum_1^\infty (a_n b_n)$ converges absolutely", in which case the inequality may be written as,

$$[\sum_1^\infty |a_n b_n|]^2 \le [\sum_1^\infty a_n^2] \cdot [\sum_1^\infty b_n^2] .$$

As noted before, this form of the inequality might prove more useful should either series contain negative terms. Why?

(c) Cauchy Inequality: Complex Number Version. If $\{a_n\}$ and $\{b_n\}$ are two sequences of complex numbers for which the series of squared moduli: $\Sigma |a_n|^2$ and $\Sigma |b_n|^2$ both converge, then the series of moduli: $\Sigma |a_n b_n|$ converges and,

$$[\Sigma |a_n b_n|]^2 \le [\Sigma |a_n|^2] \cdot [\Sigma |b_n|^2] .$$

PROOF. EXERCISE 20.5.

(d) Cauchy Inequality: Riemann-Stieltjes Integral Version. Let f and g be two continuous functions, and F a 1-dimensional c.d.f.[1] If the Riemann-Stieltjes Integrals of both f^2 and g^2 exist with respect to F over I, then so does the Riemann-Stieltjes Integral of $|fg|$ exist with respect to F over I and,

$$[\int_I |fg| dF]^2 \le [\int_I f^2 dF] \cdot [\int_I g^2 dF] .$$

Before proceeding with a proof, observe the following. It is not assumed that either the Riemann-Stieltjes Integral of f or g exists with respect to F over I. This might be the case if these functions became unbounded over I. In some such cases, however, the integrals of the squared functions do exist, in which case the inequality applies. However, if both f and g are bounded over I, all integrals above exist with respect to any c.d.f. F.

[1]This result can be generalized to include Bounded Variation Functions.

PROOF. First note that $|fg|$ is continuous over I. Assume, for simplicity, that this interval is of the basic bounded type $[a,b)$.[1] Next, let $\{P_n\}$ be any sequence of refinements partitioning $[a,b)$ for which $\|P_n\| \to 0$ as $n \to \infty$. By applying the Finite Series Version of the Cauchy Inequality, we then obtain the following inequality, which is valid for all approximating sums corresponding to the integral for $|fg|$:

$$\{\sum_1^n |f(x_{i,n}^*)g(x_{i,n}^*)|[F(x_{i,n}^*)-F(x_{i-1,n}^*)]\}^2 \equiv$$

$$\{\sum_1^n |f(x_{i,n}^*)|\,|F(x_{i,n}^*)-F(x_{i-i,n}^*)|^{\frac{1}{2}}|g(x_{i,n}^*)|\,|F(x_{i,n}^*)-F(x_{i-1,n}^*)|^{\frac{1}{2}}\}^2$$

$$\leq \sum_1^n f^2(x_{i,n}^*)[F(x_{i,n}^*)-F(x_{i-1,n}^*)]+\sum_1^n g^2(x_{i,n})[F(x_{i,n}^*)-F(x_{i-1,n}^*)] .$$

As $n \to \infty$, the last two approximating sums on the right-hand side tend, respectively, to the integrals (assumed to exist) of:

$$\int_I f^2 dF \quad \text{and} \quad \int_I g^2 dF \quad .$$

Accordingly, the sequence of approximating sums corresponding to the integral for $|fg|$ are all bounded, whence, according to Exercise 9.2, the integral of $|fg|$ with respect to F exists over I. The required result then follows immediately upon allowing $n \to \infty$ in the above inequality.

[1]The result is valid, however, for the most general case of the Riemann-Stieltjes Integral considered in Section 9.

Note that the inequality in (d) remains valid if the integrand $|fg|$ is replaced by fg. The former form is, however, conceivably more useful should f and/or g fail to be non-negative.

EXERCISE 20.6 Give an example of a situation where the integrals of the continuous functions f and g fail to exist over an interval I with respect to some c.d.f. F, yet the integrals of f^2 and g^2 both exist with respect to F over I, and the inequality in (d) therefore applies to the (existent) integral of $|fg|$ with respect to F over I.

Alternate and/or more advanced versions of the Cauchy Inequality exist; many of these are found in the References.

EXERCISE 20.7 Give a sufficient condition that equality be attained in the Riemann-Stieltjes Version of the Cauchy Inequality (more general than $f \equiv g$).

The Cauchy Inequality is actually a special case of another Classical inequality known as Hölder's Inequality. To prove this result we begin first with the following Lemma.

LEMMA 20.1 (The p-q Inequality) Let p and q be any fixed real numbers greater than 1 such that:

$$1/p + 1/q = 1 .$$

Then, for any real numbers $a,b \geq 0$, we have:

$$a^{1/p}b^{1/q} \leq a/p + b/q .$$

PROOF. We shall establish the inequality using the notions of convexity for functions of two real variables developed in Section 18. (There are many other ways.)

Consider the function $f(a,b)$, of the two real variables (a,b) defined over the convex domain $D = \{(a,b): a,b \geq 0\}$ of E_2, in the following manner:

$$f(a,b) = a/p + b/q - a^{1/p}b^{1/q} .$$

Since $f(a,0) = a/p$ and $f(0,b) = b/q$ are clearly both non-negative for all $a,b \geq 0$, it will be sufficient to establish that $f(a,b)$ is convex over D. (Why?). The 2×2 matrix of second-order partial derivatives of f referred to in Theorem 18.5 can be shown to be p.s.d. over D by using the criterion for p.s.d. matrices established in Exercise 17.32, thereby establishing convexity of f over D.

<u>EXERCISE 20.8</u> Establish the p-q Inequality using a different technique.

(ii) <u>The Hölder Inequality</u>. As with the Cauchy Inequality, there are many different versions of the Hölder Inequality. We shall first state and prove that version dealing with finite series of real numbers. Other versions will be left as Exercises. The results and proofs of (i) should prove useful with these Exercises.

(a) <u>Hölder Inequality: Finite Series Version.</u> Let $\sum_1^n a_i$ and $\sum_1^n b_i$ be two finite series of real numbers, and p,q any fixed real numbers greater than 1 such that: $1/p + 1/q = 1$. Then,

$$\sum_1^n |a_i b_i| \le [\sum_1^n |a_i|^p]^{1/p} \cdot [\sum_1^n |b_i|^q]^{1/q} .$$

PROOF. First assume that $\underline{a}' = (a_1,\ldots,a_n)$ and $\underline{b}' = (b_1,\ldots,b_n)$ are not zero vectors, for otherwise the inequality is obvious. Next, for simplicity of notation, set:

$$A_p = [\sum_1^n |a_i|^p]^{1/p} \quad \text{and} \quad B_q = [\sum_1^n |b_i|^q]^{1/q} .$$

Next, for $i = 1,\ldots,n$, define real numbers x_i and y_i as follows:

$$x_i = \frac{|a_i|^p}{\sum_1^n |a_i|^p} , \quad y_i = \frac{|b_i|^q}{\sum_1^n |b_i|^q} .$$

Applying the p-q Inequality with $a = x_i$ and $b = y_i$, we then have:

$$\frac{|a_i b_i|}{A_p \cdot B_q} \le \frac{|a_i|^p}{pA_p^p} + \frac{|b_i|^q}{qB_q^q} .$$

Upon summation over the index $i = 1,\ldots,n$ we obtain:

$$\frac{\sum_{1}^{n} |a_i b_i|}{A_p \cdot B_q} \le \frac{1}{p} + \frac{1}{q} = 1 ,$$

thereby establishing the required result.

It is clear that the Cauchy Inequality: Finite Series Version is a special case of the preceding result with $p = q = 2$. Furthermore, the use of absolute values is necessary, in general, in Hölder's Inequality because certain non-integral powers of negative numbers, e.g. $(-2)^{\pi}$, are not meaningful within the framework of real numbers.

(b) <u>Hölder Inequality: Infinite Series Version</u>. Let $\sum_{1}^{\infty} a_n$ and $\sum_{1}^{\infty} b_n$ be two infinite series of real numbers (not necessarily convergent), and p,q any fixed real numbers greater than 1 such that $1/p + 1/q = 1$. If the two series $\sum_{1}^{\infty} |a_n|^p$ and $\sum_{1}^{\infty} |b_n|^q$ both converge, then the series $\sum_{1}^{\infty} (a_n b_n)$ converges absolutely and,

$$\sum_{1}^{\infty} |a_n b_n| \le [\sum_{1}^{\infty} |a_n|^p]^{1/p} \cdot [\sum_{1}^{\infty} |b_n|^q]^{1/q} .$$

PROOF. <u>EXERCISE 20.9</u>.

<u>EXERCISE 20.10</u> Give an example of a pair of non-absolutely convergent infinite series, and a pair p,q of real numbers greater than 1 with $1/p + 1/q = 1$, for which the series $\sum_{1}^{\infty} |a_n|^p$ and $\sum_{1}^{\infty} |b_n|^q$ both converge, and apply the above inequality.

EXERCISE 20.11 Formulate a variation of the preceding Infinite Series Version of Hölder's Inequality in which one of the series is finite and the other infinite. Illustrate the result with an example.

(c) Hölder Inequality: Complex Number Version.

EXERCISE 20.12 Formulate and prove a Complex Number Version of Hölder's Inequality.

(d) Holder Inequality: Riemann-Stieltjes Integral Version. Let f and g be two continuous functions and F a 1-dimensional c.d.f.[1] Also, let p,q be any fixed real numbers greater than 1 for which $1/p + 1/q = 1$. If the Riemann-Stieltjes Integrals of both $|f|^p$ and $|g|^q$ exist with respect to F over I, then the Riemann-Stieltjes Integral of $|fg|$ exists with respect to F over I, and:

$$\int_I |fg|dF \le [\int_I |f|^p dF]^{1/p} \cdot [\int_I |g|^q dF]^{1/q} .$$

EXERCISE 20.13 Employ Hölder's Inequality to prove (d).

EXERCISE 20.14 Review the definition of the n-dimensional Riemann-Stieltjes Integral of a continuous function $\underline{g}$ with respect to an n-c.d.f.[1] $\underline{F}$ over a rectangle R. Then, establish that a result analogous to (d) holds in this case. State it.

[1]The results generalize to the case of Bounded Variation Functions.

The preceding inequalities have dealt with products; either sums or integrals thereof. There are several basic inequalities dealing with sums themselves. The first of these results has previously been introduced in an elementary form (see Exercise 13.2).

(iii) The Triangle Inequality. In various settings, this is one of the more useful and frequently applied inequalities in Applied Mathematics. In essence, it provides an often useful bound on the modulus of a sum of real or complex numbers.

(a) Triangle Inequality: Finite Sum Version. Let $a_1 + a_2 + \ldots + a_n$ be a (finite) sum of real or complex numbers. Then,

$$|a_1 + a_2 + \ldots + a_n| \le |a_1| + |a_2| + \ldots + |a_n| \quad .$$

PROOF. This follows immediately, by use of Mathematical induction, from the results in Exercise 13.2.

It is clear that this inequality remains valid whether the terms are functions (real or complex) defined over a common domain, integrals (real or complex), and so on.

(b) Triangle Inequality: Infinite Sum Version. Let $\sum_1^\infty a_n$ be a convergent series of real or complex numbers. Then,

$$\left|\sum_1^\infty a_n\right| \le \sum_1^\infty |a_n| \quad .$$

PROOF. This follows immediately from the Finite Sum Version by an obvious limiting argument, and is useful only if the series on the right converges.

The remark following (a) is also appropriate for the Infinite Sum Case.

(iv) <u>The Minkowski Inequality</u>. This Classical generalization of the Triangle Inequality also has several versions. We first state and prove that version dealing with finite sums.

(a) <u>Minkowski Inequality: Finite Sum Version</u>. Let $\sum_1^n a_i$ and $\sum_1^n b_i$ be two finite sums of real or complex numbers, and p any real number greater than 1. Then,

$$\left[\sum_1^n |a_i + b_i|^p\right]^{1/p} \leq \left[\sum_1^n |a_i|^p\right]^{1/p} + \left[\sum_1^n |b_i|^p\right]^{1/p} .$$

PROOF. Using a previously established notation, the two terms on the right-hand side of the inequality are denoted by A_p and B_p respectively. Denote the term on the left-hand side by $[A + B]_p$. Now

$$\begin{aligned} [A + B]_p^p &= \sum_1^n |a_i + b_i|^p \\ &= \sum_1^n |a_i + b_i|^{p-1} |a_i + b_i| \\ &\leq \sum_1^n |a_i| \cdot |a_i + b_i|^{p-1} + \sum_1^n |b_i| \cdot |a_i + b_i|^{p-1} . \end{aligned}$$

Applying Hölder's Inequality: Finite Series Version to each of the sums in the last expression (with q defined by the relationship: $1/p + 1/q = 1$), we have,

$$\sum_1^n |a_i|\cdot|a_i + b_i|^{p-1} \le A_p \cdot [A + B]_p^{p/q}$$

and

$$\sum_1^n |b_i|\cdot|a_i + b_i|^{p-1} \le B_p \cdot [A + B]_p^{p/q} \quad .$$

Accordingly,

$$[A + B]_p^p \le (A_p + B_p)\ [A + B]_p^{p/q} \ .$$

The required inequality then follows upon simple division[1] and use of the relationship between p and q.

(b) <u>Minkowski Inequality: Infinite Sum Version</u>. Let $\sum_1^\infty a_n$ and $\sum_1^\infty b_n$ be two (not necessarily convergent) infinite series of real or complex numbers, and p any real number greater than 1. If the two series $\sum_1^\infty |a_n|^p$ and $\sum_1^\infty |b_n|^p$ both converge, then so also does the series $\sum_1^\infty |a_n + b_n|^p$, and:

$$[\sum_1^\infty |a_n + b_n|^p]^{1/p} \le [\sum_1^\infty |a_n|^p]^{1/p} + [\sum_1^\infty |b_n|^p]^{1/p} \ .$$

PROOF. This follows by the standard limiting argument from the Finite Sum Version.

---

[1]Assume $[A + B]_p \neq 0$. Otherwise the inequality is degenerate.

As with the Triangle Inequality, the terms involved in Minkowski's Inequality could also be functions (real or complex) defined over a common domain, Integrals (real or complex), and so on.

EXERCISE 20.15 Illustrate the Minkowski Inequality with a pair of non-convergent infinite series of real numbers and a suitable real number $p > 1$.

EXERCISE 20.16 Formulate the obvious generalization of Minkowski's Inequality to the case where $k > 2$ series are involved, that is, to the case where the general term of the series on the left is: $|a_{1n} + \ldots + a_{kn}|^p$.

EXERCISE 20.17 Show that the Triangle Inequality actually corresponds to the case $p=1$ in the generalization developed in Exercise 20.16.

(c) Minkowski Inequality: Riemann-Stieltjes Integral Version. Let F be a 1-dimensional c.d.f.[1] and f and g two continuous (though not necessarily integrable) functions over an interval I. If for some real number $p > 1$ the Riemann-Stieltjes Integrals of both $|f|^p$ and $|g|^p$ exist with respect to F over I, then so does that of $|f + g|^p$, and:

$$[\int_I |f + g|^p dF]^{1/p} \leq [\int_I |f|^p dF]^{1/p} + [\int_I |g|^p dF]^{1/p}$$

PROOF. EXERCISE 20.18.

[1] The result generalizes to the case of Bounded Variation functions and $n > 1$ dimensions.

<u>EXERCISE 20.19</u> State the obvious generalization of (c) for the case where the integrand on the left consists of $k > 2$ functions.

<u>EXERCISE 20.20</u> Verify that the case $p=1$ of the preceding generalization (Exercise 20.19) could be deduced from the Triangle Inequality.

<u>EXERCISE 20.21</u> Give sufficient conditions that exact equality hold in (a), (b) and (c). Do these differ from those for the Cauchy Inequality? Why?

(v) <u>The c_r Inequality</u>. This final basic inequality finds use in applications similar to that of the Triangle Inequality. The reason for this should be apparent from its statement.

(a) <u>c_r Inequality: Basic Version</u>. Let a and b be any pair of real or complex numbers. Then for $r > 0$:

$$|a + b|^r \leq c_r|a|^r + c_r|b|^r \quad ,$$

where

$$c_r = \begin{cases} 1 & \text{for } 0 < r < 1 \\ 2^{r-1} & \text{for } 1 \leq r < \infty . \end{cases}$$

PROOF. <u>EXERCISE 20.22</u>.

<u>EXERCISE 20.23</u> How can the c_r Inequality be extended in the case where $k > 2$ real or complex numbers occur on the left-hand side?

Finally, note that the c_r Inequality can be used in the case where its terms are functions (real or complex) defined over a common domain, or integrals (real or complex), and can be used to construct an integral inequality somewhat of the form of the Minkowski Inequality: Riemann-Stieltjes Integral Version.

<u>EXERCISE 20.24</u> Verify the last statement.

<u>Hints and Answers to Exercises: Section 20</u>

20.1 Hint: $r^2 = \dfrac{[\Sigma(a_i b_i)]^2}{[\Sigma\ a_i^2][\Sigma\ b_i^2]}$,

where $a_i = (x_i - \overline{x})$ and $b_i = (y_i - \overline{y})$ for $i = 1,2,\ldots,n$.

20.2 Hint: The condition $\underline{a}' = \lambda\underline{b}'$ is equivalent to the n conditions: $a_i = \lambda\ b_i$ $(i = 1,2,\ldots,n)$. Since we assume that $\underline{a}' \neq \underline{0}'$ and $\underline{b}' \neq \underline{0}'$, it follows that the conditions $\underline{a}' = \lambda\underline{b}'$ and $\underline{b}' = \delta\underline{a}'$ $(\lambda, \delta$ real numbers) are equivalent. Why?

20.3 Hint: This is appropriate for real numbers only. Note that $x \leq |x|$ for any real number x. See also Example 18.1.

20.4 Hint: See Example 20.1. If either series contains negative terms, the inequality:

$$[\Sigma\ |a_i b_i|]^2 \leq [\Sigma\ a_i^2]\ [\Sigma\ b_i^2]$$

is conceivably more informative than:

$$[\Sigma\ (a_i b_i)]^2 \le [\Sigma\ a_i^2]\ [\Sigma\ b_i^2]$$

because in general:

$$[\Sigma\ (a_i b_i)]^2 \le [\Sigma\ |a_i b_i|]^2 \quad .$$

Recall that absolute convergence implies convergence, but not conversely.

20.5 Hint: This follows immediately from Exercise 20.4, since $|a_n|$, $|b_n|$ and $|a_n b_n|$ are ordinary real numbers.

20.6 Hint: As an example, consider the following:

$$F(x) = \begin{cases} 0 & \text{for } x \le 0 \\ x & \text{for } 0 < x \le 1 \\ 1 & \text{for } x > 1 \end{cases} ,$$

$$f(x) = \begin{cases} 1/(1-x) & \text{for } x < 1 \\ 0 & \text{for } x \ge 1 \end{cases} ,$$

$$g(x) = \begin{cases} x/(x^2-1) & \text{for } x < 1 \\ 0 & \text{for } x \ge 1 \end{cases} .$$

We then have the following results (integrals on the left are Riemann-Stieltjes and on the right are Improper Riemann).

$$\int_{0-}^{1-} f\,dF = \int_{0}^{1} 1/(1-x)\,dx \quad \text{does not exist.}$$

$$\int_{0-}^{1-} g\,dF = \int_{0}^{1} x/(x^2-1)\,dx \quad \text{does not exist.}$$

$$\int_{0-}^{1-} f^2\,dF = \int_{0}^{1} 1/(1-x)^2\,dx \quad \text{exists.}$$

$$\int_{0-}^{1-} g^2\,dF = \int_{0}^{1} x^2/(x^2-1)^2\,dx \quad \text{exists.}$$

Therefore,

$$\int_{0-}^{1-} |fg|\,dF = \frac{1}{2}\int_{0}^{1} [1/(1-x)^2 + 1/(1-x)(1+x)]\,dx \quad \text{exists,}$$

and:

$$\int_{0-}^{1-} |fg|\,dF \le \left[\int_{0-}^{1-} f^2\,dF\right]^{\frac{1}{2}} \left[\int_{0-}^{1-} g^2\,dF\right]^{\frac{1}{2}} .$$

Verify the above statements. Construct a different example.

20.7 Hint: A simple sufficient condition is that $f(x) \equiv \lambda g(x)$ (λ a real constant) for all x in I. More generally it is sufficient to require that $f(x) \equiv \lambda g(x)$ for all subintervals J of I for which $\int_J dF > 0$. For, if $\int_J dF = 0$, then $\int_J f\,dF = \int_J g\,dF = 0$, so that f and g may have arbitrary, not necessarily proportional, values over J, consistent with the assumed continuity of the functions.

20.8 Hint: Using Max-Min techniques, an alternate approach would be to prove that the function $f(x,y)$ of two real variables defined as follows:

$$f(x,y) = x/p + y/q - x^{1/p}y^{1/q}\ , \quad x,y \geq 0$$

is non-negative, and assumes its (absolute) minimum over $x,y \geq 0$ at $x = y = 0$. Accordingly,

$$x/p + y/q - x^{1/p}y^{1/q} \geq 0 \quad \text{for all} \quad x,y \geq 0,$$

which is equivalent to the required result.

20.9 Hint: This follows immediately from the Finite Series Version upon using a suitable limiting process applied to the sequences of partial sums corresponding to the three series involved.

20.10 Hint: One example is: $a_n = 1/n^{3/5}$, $b_n = 1/n^{4/5}$, $p = q = 2$.

20.11 Hint: Assume, for example, $a_n = 0$ for $n > N$. If, now, we set $a_1 = a_2 = \ldots = a_N = 1$, then Hölder's Inequality yields:

$$\sum_{1}^{N} |b_n| \leq N^{1/p} [\Sigma |b_n|^q]^{1/q} .$$

Note: $[\Sigma |b_n|^q]^{1/q}$ is decreasing in q for $q > 1$.

20.12 Hint: Hölder's Inequality remains valid for complex numbers since only moduli (which are real) are involved.

20.13 Hint: Use Hölder's Inequality on the approximating sums corresponding to $\int_I |fg|dF$. Refer also to the proof of the Cauchy Inequality: Riemann-Stieltjes Version.

20.14 Hint: Use the approximating sum approach. There is little real difference from the case $n=1$ considered in Exercise 20.13.

20.15 Hint: An example is: $a_n = 1/n^a$, $b_n = 1/n^b$ $(a,b > 0)$ with $p > \max(1/a, 1/b)$.

20.16 Hint: Use Mathematical Induction.

20.17 Hint: If $a_{jn} = 0$ $(j = 1,2,\ldots,k;\ n > 1)$ and $p=1$, we obtain the Triangle Inequality, namely,

$$|a_{11} + a_{21} + \ldots + a_{k1}| \le |a_{11}| + |a_{21}| + \ldots + |a_{k1}| .$$

Note, however, that the proof given for Minkowski's Inequality in this Section is <u>not</u> valid for $p=1$.

20.18 Hint: Apply Minkowski's Inequality: Finite Series Version to the partial sums corresponding to the integral:

$$\left[\int_I |f + g|^p dF\right]^{1/p} .$$

20.20 Hint: Apply the Triangle Inequality to the partial sums corresponding to the integral:

$$\int_I |f + g| dF .$$

20.21 Hint: (a) $a_i \equiv b_i$ $(i = 1,2,\ldots,n)$

(b) $a_i \equiv b_i$ $(i = 1,2,\ldots\ldots)$

(c) $f(x) \equiv g(x)$ for all x in I, or more generally, $f(x) \equiv g(x)$ for all subintervals J of I for which $\int_J dF > 0$. See also the Hint for Exercise 20.7.

20.22 Hint: First prove the result for non-negative real numbers as follows: if $x,y \geq 0$ and $r > 1$ then:

$$(x + y)^r \leq 2^r \max(x^r,y^r) \leq 2^r \tfrac{1}{2}(x^r + y^r) = 2^{r-1}x^r + 2^{r-1}y^r.$$

On the other hand, if $x,y \geq 0$ and $0 < r \leq 1$, it is not difficult to prove that the function:

$$f(x,y) = x^r + y^r - (x + y)^r$$

is non-negative (using Max-Min techniques). This, then, establishes the c_r Inequality for x,y real and non-negative. Now use these results to establish the c_r Inequality for the general case.

20.23 Hint: For example, $|a + b + c|^r \leq c_r|a + b|^r + c_r|c|^r \leq c_r^2|a|^r + c_r^2|b|^r + c_r|c|^r$.

20.24 Hint: Under suitable conditions:

$$\int_I |f + g|^r dF \leq c_r \int_I |f|^r dF + c_r \int_I |g|^r \, dF .$$

The result generalizes to the case of Bounded Variation Functions and $n > 1$ dimensions.

References to Additional and Related Material: Section 20

1. Beckenbach, E., "An Introduction to Inequalities", Random House, New York (1961).

2. Beckenbach, E. and R. Bellman, "Inequalities", Springer-Verlag, N.Y. (1971).

3. Hardy, G. and J. Littlewood, "Inequalities", Cambridge University Press, (1934, 1952).

4. Hille, E., "Analytic Function Theory", Vols. I, II, Ginn and Co., Boston, New York (1959, 1962).

5. Kazarinoff, N., "Analytic Inequalities", Holt, Reinhart and Winston, Inc., New York (1961).

6. Marcus, M. and H. Minc, "A Survey of Matrix Theory and Matrix Inequalities", Allyn-Bacon, Inc., Boston (1964).

7. Mitrinović, D., "Analytic Inequalities", Springer-Verlag, Berlin, (1970).

8. Pólya, G. and G. Szegő, "Isoperimetric Inequalities in Mathematical Physics", Princeton University Press (1951).

9. Timan, A., "Theory of Approximation of Functions of a Real Variable", Oxford University Press (1963).

10. Uspensky, J., "Theory of Equations", McGraw-Hill, Inc. New York, (1948).

11. Wilf, H., "Finite Sections of Some Classical Inequalities", Springer-Verlag, New York (1970).

12. Whittaker, E. and G. Watson, "A Course in Modern Analysis", Cambridge University Press (1952).

Index